Sensations

Sensations

FRENCH ARCHAEOLOGY BETWEEN
SCIENCE AND SPECTACLE, 1890–1940

Daniel J. Sherman

The University of Chicago Press   CHICAGO AND LONDON

The University of Chicago Press, Chicago 60637
The University of Chicago Press, Ltd., London
© 2025 by The University of Chicago
All rights reserved. No part of this book may be used or reproduced in any manner whatsoever without written permission, except in the case of brief quotations in critical articles and reviews. For more information, contact the University of Chicago Press, 1427 E. 60th St., Chicago, IL 60637.
Published 2025
Printed in the United States of America

34 33 32 31 30 29 28 27 26 25   1 2 3 4 5

ISBN-13: 978-0-226-83537-2 (cloth)
ISBN-13: 978-0-226-83538-9 (e-book)
DOI: https://doi.org/10.7208/chicago/9780226835389.001.0001

Library of Congress Cataloging-in-Publication Data

Names: Sherman, Daniel J., author.
Title: Sensations : French archaeology between science and spectacle, 1890–1940 / Daniel J. Sherman.
Description: Chicago : The University of Chicago Press, 2025. | Includes bibliographical references and index.
Identifiers: LCCN 2024038397 | ISBN 9780226835372 (cloth) | ISBN 9780226835389 (ebook)
Subjects: LCSH: Archaeology—France—History—20th century. | Archaeology and history—France—Glozel. | Archaeology and history—Tunisia—Carthage (Extinct city) | Archaeology—Methodology. | Archaeology in mass media. | Communication in archaeology—France.
Classification: LCC CC101.F8 S54 2025 | DDC 930.1—dc23/eng/20241021
LC record available at https://lccn.loc.gov/2024038397

♾ This paper meets the requirements of ANSI/NISO Z39.48-1992 (Permanence of Paper).

*For Alice, Sandrine, and Judy, in friendship
and for Eduardo, always*

Contents

Acknowledgments		ix
INTRODUCTION: From the Archaeological Archive		1
1	"FOR CARTHAGE": Scientific Networks and the Glare of Publicity	15
	Prologue: The Field and the Network	15
	Serious Men (and Their Archives)	21
	"To Save Carthage": Networks, Authority, Publicity	25
	Enter the Americans	33
	Heritage and/as Science	44
2	THE NEWS FROM GLOZEL: Scholars, Media, and the Making of a Scandal	51
	The Glozel Archive: Exit the State	57
	Exchanging News	60
	Blame the Media	69
	Fake News/News of Fakes	75
	Affairs to Remember	82
	Coda	87
3	BODIES AND MINDS: The Work of Archaeology	91
	The Archaeologist's Police File	95
	Bodies at/of Work	101
	Looking Like an Archaeologist	113

4	REALITY EFFECTS: Staging Archaeology	125
	Performing Carthage	129
	Glozel and the Performative	143
5	PICTURING THINGS: Archaeology and the Imagined Past	157
	What Are These Objects?	163
	Who Were These People?	173
	Imagining the Glozelians	179
	EPILOGUE	193
	After Bizerte: 1926–1933–1962	193
	"Glozel For Ever": 1968–1974–2021	196
	Objects, Knowledge, and the Archive: 2022–	201
	Notes	205
	Index	257

Image gallery follows page 138.

Acknowledgments

I am most grateful to the institutions that have supported the research on this book. A six-month fellowship at the Institut d'Études Avancées (IEA), Paris, in 2014 proved decisive in shaping the project. At a later stage, a fellowship at the School of Historical Studies, Institute for Advanced Study, Princeton, in Fall 2016, supported by the Hetty Goldman Membership Fund, and an Edith Bernstein Fellowship at the Institute for Arts and Humanities (IAH) at the University of North Carolina, Chapel Hill, in Spring 2017 helped me transition from research to writing. Most of the writing was accomplished during a fellowship from the National Endowment for the Humanities in 2019–2020, with a final phase supported by a Research and Study Assignment from UNC in Spring 2022, when I was also a Visiting Fellow Commoner at Trinity College, Cambridge University. In addition to the precious time and congenial space all these institutions offered, it is a pleasure to acknowledge those individuals who went out of their way to make fellows' time more pleasant and productive: at the IEA, Gretty Mirdal, Marie-Thérèse Cerf, and Simon Luck; in Princeton, Yve-Alain Bois, Marion Zelazny, and Linda Cooper; at the IAH, Michele Berger and Philip Hollingsworth; at Trinity, Sachiko Kusukawa. I am also grateful for the advice, feedback, and companionship of my fellow fellows, notably Leor Halevi, Andrew Jainchill, Alena Ledeneva, Silvio de Souza Correa, and Nadège Weldwachter in Paris; Roland Betancourt, Nick Cheesman, Thomas Dodman, Emine Fetvaci, Jane Hathaway, and Columba Stewart in Princeton; and Sam Amago, Morgan Pitelka, Lien Truong, and Molly Worthen at the IAH. In late-COVID Cambridge, Andrew Arsan, Susanne Hakenbeck, Jean Khalfa, Peter Mandler, Jonathan Parry, Miri Rubin, Samita Sen, Richard Staley, and Stephen Toope offered hospitality of various kinds, for which many thanks.

I am also pleased to acknowledge the support I have received from my home institution, UNC Chapel Hill, chiefly via the Lineberger

Distinguished Professorship, which I have held since 2014. The professorship has supported research travel, acquisition of books and images, and research assistance from my former students Joshua Smith and Rachel Ozerkevich. In addition, a William C. Friday Arts and Humanities Research Award in Fall 2023 funded invaluable work by my stalwart undergraduate research assistant, Sean Sabye, whose work on images, image descriptions, and permissions I deeply appreciate. An Arts and Humanities Publication Support Grant from the IAH and UNC's Office of Research Development supports the book's illustration program. This is the place to acknowledge with deep gratitude and affection the Department of Art and Art History's accounting technician, Abigail Brooks, for her efficient management of my research fund, and the department manager, Lindsay Fulenwider, for her extraordinary kindness and patience. Many thanks also to former department chair Carol Magee, for her vital support and her friendship.

My research has benefited from the kindness and helpfulness of a number of librarians and archivists, many of whom took time out from other responsibilities to help sort minimally cataloged materials or point me toward treasures I would not otherwise have found. My thanks to Philippe Ferrand at the Bibliothèque Méjanes in Aix-en-Provence; Elisabeth Lacoste at the Auvergne–Rhône–Alpes Direction régionale des affaires culturelles, Service régional de l'archéologie, in Clermont-Ferrand; Soline Morinière and Grégoire Meylan at the Musée d'archéologie nationale in St. Germain-en-Laye; Eleanor Wilkinson and Melanie Norton Hugow at the Cambridge Museum of Archaeology and Anthropology; and Angela Grimshaw at the British Museum. A special shout-out to Sébastien Chauffour, formerly at the Institut national d'histoire de l'art (INHA) in Paris. In January 2014, having just completed the cataloging of the INHA's Fonds Poinssot, and at a time when the library's special collections were accessible only two afternoons per week, he offered me additional research sessions in the none-too-spacious "Mezzanine des conservateurs." Without this access during my IEA fellowship, it is safe to say that this book would have taken a very different form, if it existed at all. In addition, Sébastien invited me to a *journée d'études* about the Poinssot collection and introduced me to scholars with intersecting interests. Many thanks, Sébastien; this was truly the beginning of a beautiful friendship.

On the subject of France and friendship, two concepts that have been happily interconnected in my mind and my life for more than four decades, I am grateful to the French friends and colleagues who have welcomed me to their homes, their seminars, and their conferences over the years. Anne and Alain Cornet-Vernet, Marie-Hélène and Moiffak Hassan, and Marie Petit-Ketoff have known me for most of my life, and keep inviting me back. Martine Joubert was a model landlady and generous host. Laurence Bertrand Dorléac, Michèle Fogel, Rémi Labrusse, Eric Michaud and Maria Stavrinaki, Dominique and Monique Poulot, and Christophe Prochasson have been invaluable interlocutors and good friends. Colette Capdeboscq's extraordinary hospitality gave me the opportunity to visit Glozel on one of its rare open days; many thanks. I'm also very appreciative of the continuing companionship—not only, but frequently, in Paris—of Leora Auslander, Judy Coffin, Nélia Dias, Amy Freund, Philip Nord, M'hamed Oualdi and Augustin Jomier, Sandrine Sanos, Camille Serchuk, Todd Shepard, and Len Smith. It has been a special pleasure to see Paris and France through the eyes of a group of remarkable people I am privileged to have called my students: Davenne Essif, Robin Holmes, Massie Minor, Rachel Ozerkevich, Joshua Smith, and Weixin Zhou. Thanks also to Ridha Moumni for his friendship and advice; his introductions assured me a warm welcome in Tunis from Leila Ben-Gacem, Fatma Kilani, Sihem Lamine, and Zoubeir Mouhli.

As a neophyte, not to say an interloper, in the world of archaeology, I am deeply appreciative of all the advice, assistance, and support I have received from friends and colleagues in this and other fields, including Carla Antonaccio, Zainab Bahrani, William Carruthers, Stephen Driscoll, Susanne Hakenbeck, Arnaud Hurel, Jose Lanzarote Guiral, Jodi Magness, Christina Riggs, Nathan Schlanger, Richard Talbert, and Alexia Yates. Alain Schnapp, the dean of the history of archaeology in France and beyond, has been generous in his support of many kinds, as has his last doctoral student, Rose-Marie Le Rouzic, an invaluable source of information on French archaeology past and present. I am particularly grateful to Charlotte Bigg and Andrée Bergeron for including me in multiple iterations of their international project on the Staging of Science: a conference in 2015, a presentation at their seminar at the Centre Alexandre-Koyré in 2017, and a remarkable workshop at the idyllic Fondation des Treilles in 2018. I am most grateful to those whose invitations gave me the opportunity to try out my ideas in preliminary form: Noit Banai at the University of Vienna (2016); Phil Nord at Princeton University, in 2016 and again at his retirement conference in 2022; Paul Redman at King's College, University of London, in 2017; Will Carruthers, organizer of the conference on Decolonising the Field at the German Historical Institute and University College London in 2018; Herrick Chapman at New York University in 2018; Catherine Clark at MIT in 2019; Cécile Colonna and Néguine Mathieux at the conference on Collecting Antiquities during the Belle Époque at the INHA and the Louvre in 2019; Len Smith at Oberlin

College in 2021; and Matt McCarty and Megan Daniels at the University of British Columbia in 2022.

Thanks go also to my copanelists at various conference sessions, mostly at the Society for French Historical Studies and the Western Society for French History, whose work provided a stimulating context for thinking about my own: Elizabeth Campbell, Will Carruthers, Alice Conklin, Tom Dodman, Bonnie Effros, Sarah Griswold, Suzanne Marchand, Louise McReynolds, John Monroe, and Roxanne Panchasi. The comments of members of the Triangle French History and Culture Seminar and the New York French History Group were particularly helpful. Bonnie Smith saw a book in the various elements of this project before I did; I cannot sufficiently thank her for her decades of support, generosity, and friendship. In addition to the anonymous readers for the University of Chicago Press, I am enormously grateful to the friends and colleagues who spent time wading through drafts of various chapters: Charlotte Bigg, Mirjam Brusius, Will Carruthers, Judy Coffin, Alice Conklin, Alex Csiszar, Tom Dodman, Don Reid, Sandrine Sanos, and Jay Smith. Thanks for all your efforts to lighten and clarify. What remains is of course entirely my responsibility.

This is my third book with the University of Chicago Press but my first acquired by Karen Merikangas Darling, whose enthusiasm for the project and whose calm through the various stages of review and preparation I appreciate in equal measure. Thanks also to my former editor, Susan Bielstein, for putting me in touch with Karen, and to editorial assistant Fabiola Enríquez Flores, for her patience and efficiency in dealing with image and permission questions. I am also grateful to Sophie Berrebi and Jean-François Julliard for assistance with permissions. Portions of the introduction and chapter 4 were previously published by Oxford University Press as "Staging Archaeology: Empire as Reality Effect at the *fêtes de Carthage*, 1906–07," *Classical Receptions Journal* 13, no. 3 (July 2021): 336–67, © 2021 Daniel Sherman. In compliance with French archival regulations, every effort has been made to locate rights-holders for unpublished correspondence. I invite anyone who claims rights to such material to contact me directly.

I honor the memory of cherished people in my life with whom I can share this book only in spirit: my undergraduate mentor Stanley Hoffmann (1929–2015), my doctoral advisor John Merriman (1946–2022), and my beloved parents, Stanley M. Sherman (1922–2019) and Claire Richter Sherman (1930–2023). I cannot sufficiently express the gratitude I feel for the many friends and family members who have sustained me through seasons of loss, among them David Barquist, Martin Berger, Louis Cooper,

Isabel Geffner, Peter Guzzardi, Cary Levine, Louise McReynolds, Don Reid, Anthony Richter, Giles Richter, Holly Russell, Paula Sanders, and Tania String. My husband, Eduardo Douglas, has been the proverbial tower of strength; he makes my life worth living every day, as does our cat Celia (RIP Jacqui, whose colonization of my arm while I was typing somehow seemed to stimulate rather than impede my writing). Judy Coffin, Alice Conklin, and Sandrine Sanos have given me so much—advice, support, knowledge, wisdom, empathy, and not least, laughter—that the word "friendship" does not seem adequate to describe it. Merci mille fois.

Introduction

FROM THE ARCHAEOLOGICAL ARCHIVE

"Carthage Again!" runs the headline in *La Tunisie française*, one of the leading dailies in the French Protectorate of Tunisia, in June 1924. The front-page story, by a journalist called de la Porte, highlights press coverage of excavations in Carthage, a growing suburb of Tunis on the site of the ancient city. De la Porte targets *Lectures pour tous*, a general-interest magazine that misidentified stock photographs of other archaeological sites as views of Carthage. But he also takes to task the illustrated weekly *L'Illustration*, for publishing a polemical article that accused the French archaeological authorities of neglecting Carthage, charges he considers alarmist and unjust.[1] A newspaper article about two other articles, de la Porte's piece makes clear not only the sensitivity of archaeology in a colonial setting but also the inextricable ties between archaeology and its media coverage. Further evidence of this connection comes from an uninspired bit of wordplay in the Paris daily *Journal des débats* a few years later: "Archaeology is a science that screws up as much as it digs up."[2] Thus begins an article on an archaeological controversy written in the form of an imaginary interview with the Egyptian pharaoh Tutankhamun, whose tomb had recently been the object of much publicized excavations.[3] Media coverage of archaeology—here the interview with scholars and excavators—had itself become so familiar that it offered, even to this normally staid daily, a tempting target for mockery, notwithstanding, or perhaps in part because of, the field's status as "science."

This book argues that archaeology as we know it today grew out of a fundamental tension between archaeologists' scientific ambitions and their continuing need for media attention. Although controversies in other scientific fields and other countries also played out in the media, France offers a particularly compelling instance of the ways journalists, editors, and publishers played a constitutive role in legitimating archaeology's claims to scientific status. The book focuses on two long-running controversies that roiled the French archaeological world and its wider public in the first third of the twentieth century. The first involves a dispute about the place of Carthage in state-sponsored excavations in Tunisia from the 1890s through the 1920s: spreading their limited resources among sites throughout the country, the French directors of antiquities faced continual criticism for their "neglect" of Carthage. In the second case, accusations of forgery clouded what seemed to be a stunning Neolithic find at a hamlet called Glozel, in the Auvergne region in central France; the affair, which began in 1924 and reached a climax in late 1927, divided the scholarly community and attracted enormous media attention across Europe and North America.

Though quite different in their particulars, the two cases have much in common, notably an overlapping cast of characters that included some of the most respected archaeologists of the day. Both affairs demonstrated the limits of the state's power to regulate scholarship and debate in archaeology, whether through its own agencies, its colonial relays, or the semi-autonomous but state-funded Académie des inscriptions et belles-lettres.[4] The Academy could not resolve disputes over Carthage, and although the vast majority of its members considered Glozel a hoax, a few ardently endorsed it as authentic. In both cases, international actors furthered the sense that French science was under threat. At Carthage, a well-funded American expedition upset convention, and in so doing called attention to the complex patchwork of public, private, and religiously affiliated digs there. At Glozel, following a practice of transnational cooperation that went back to the beginnings of prehistoric archaeology, an international commission of archaeologists was assembled to adjudicate the authenticity of the site.[5] Considered together, the two controversies provide a new, ground-level view of the formation of archaeology as both scholarly field and media sensation in France.

As the keywords of this study have known many uses, some working definitions are in order. *Archaeology* may be summarily defined as the study of the human past through the systematic examination of its material remains; one group of authors, insisting that the endeavor comprises "an ecology of practices" centered on things, offers a useful reminder that excavation comprises only one of its many and manifold activities.[6] For the archaeologist Jodi Magness, indeed, "the most important part of archaeology is not excavation, which gets all the attention, but rather the process of publication afterwards."[7] The word *science* has a broader meaning in French than in American English, where "science," unless qualified by words like "social," typically refers to the hard or experimental sciences.

In French, by contrast, the adjectival form of the word, *scientifique*, can be translated as "scholarly": the French national research organization, the Centre National de Recherche Scientifique (CNRS), funds research in the humanities as well as the social and natural sciences. The questions that arose in the early twentieth century about archaeology as "science," then, had to do as much with the integrity and reliability of its basic methods and protocols as with its affinities to "harder" sciences. I use *spectacle* roughly in the sense given to it by Guy Debord, as "*capital* accumulated to where it becomes image"; Debord later dated the beginnings of the "society of the spectacle" to 1927, a key moment in the controversies considered in this book.[8] Although the economics of excavations, and indeed of the press, fall largely outside this book's scope, the term serves as a reminder that media attention has more than symbolic value to the field.

The history of archaeology has largely been the province of archaeologists themselves; Nathan Schlanger observes that such histories have traditionally celebrated achievements and legitimated disciplinary claims.[9] Perhaps for this reason, many working archaeologists have doubted the value of studying their own history, deeming it at best an irrelevance and at worst a form of ideological investment in outmoded areas of research.[10] Schlanger has been at the forefront of a movement over the last three or so decades to make the history of archaeology more critical, more reflexive, and more outward-looking, in dialogue with the history and theory of neighboring disciplines. For Marc-Antoine Kaeser, this more critical history stems from "present debates and questionings in the discipline" but finds the rigor it needs in a "historicist methodology" that recognizes the specificity, indeed the difference, of past archaeological practices.[11] Archaeology's implication in the production of colonial and nationalist discourses occupies an important place in this new, more critical history, as do its gender assumptions and practices and its implication in nineteenth-century racial science.[12] As the editors of one anthology put it, "Archaeology is a politicized discipline, for the state needs the remote past to justify its authority and to exercise its rule."[13] More specifically in relation to colonialism, Margarita Diaz-Andreu observes that "archaeology fulfilled a part in the state's strategy of surveillance and observation that gave the imperial powers a perspective on the dominated."[14] These topics have attracted widespread interest both within and beyond academic archaeology as a result of the traffic in and destruction of antiquities in areas of conflict, notably Iraq, Syria, and Afghanistan, and of continuing disputes over ownership, most famously of the Parthenon sculptures in the British Museum.[15] Ongoing debate over these issues reminds us that archaeology always entails acts of dispossession, cultural as well as physical.[16] This dynamic underlies much of the discussion of archaeology in Tunisia throughout the book.

Recent scholarship on archaeology in the media has emphasized books and periodicals aimed at general audiences; in one such work, Elliott Colla explicitly discusses "the tensions between the science and the spectacle" of the Tut excavation and Howard Carter's writings about it.[17] Scholars have considered the role of newspapers and magazines in awakening public interest in archaeology, as well as the representation of archaeology and its finds at worlds' fairs throughout the second half of the nineteenth century. At the 1867 Universal Exposition in Paris, for example, the Egyptian display included a reconstruction of a Pharaonic temple, and prehistoric artifacts were on display for the first time.[18] The illustration of archaeological sites and objects in various media—drawing, engraving, photography, film—has attracted considerable attention, and the visualization and visual rendering of archaeology will be an important element of the analysis that follows.[19] But with few exceptions, previous studies have not treated the relationship between science and media as central to, indeed constitutive of, the archaeological field.[20] That dynamic offers a particularly productive frame for investigating fundamental questions: How have archaeologists developed methods for both excavation and the categorization and analysis of finds? How have scholarly and popular media traced and refracted popular interest in archaeology? How have both formal and informal networks shaped the formation of archaeology as a discipline and profession?

The book brings together a conceptual frame, controversy studies, from the history of science; a historical location in early twentieth-century France and its colonies; and an empirical base in what I am calling the archaeological archive. I will set out each of these in turn. By 1985, Steven Shapin and Simon Schaffer write, studies of controversies had "become set-pieces in sociological studies of the making of modern scientific knowledge." They locate the originality of their contribution to the genre, which they rightly cast as a form of microhistory, in their focus not only on the making of knowledge but on the "'rules of the game'" of how knowledge was made.[21] Along the same lines, in a concise survey of scholarship on scientific controversies, Sheila Jasanoff argues for considering "social controversies as laboratories for studying how science and technology work in society," rather than taking the laboratory "as the site par excellence for studying scientific controversies."[22] This is a particularly valuable insight for archaeology, an area of knowledge based in the field rather than the laboratory. Moreover, the scientific status of archaeology was, in its formative period in the late nineteenth and early twentieth centuries, a matter of contention. Axiomatic in this approach to controversies is Shapin and Schaffer's assertion that "solutions to the problem of knowledge are solutions to the problem of social order."[23] In the field of archaeology, then, evolving positions on controversies about the human past closely track interpreters' assumptions about the present-day social, cultural, and epistemological structures that frame their lives and work.

France in the early twentieth century offers an ideal place and time for a study of this kind. The country had a long tradition of archaeological exploration both at home and abroad, the latter tied up with colonial expansion, not only in North Africa but in the Middle East and later in Southeast Asia.[24] The *Description de l'Egypte* (1809–1829), a massive, multivolume opus presenting the findings of the artists and scientists who accompanied Napoleon on his ill-fated expedition to Egypt in 1798–1799, offered a model for fusing information gathering, the projection of power, and the creation of knowledge (not all of it archaeological) on an imperial stage. French diplomats, clerics, adventurers, and scholars from the early nineteenth century on participated in excavations and the partage of antiquities from the Middle East and North Africa. And not only French discoveries but the hypotheses of French scholars played a pioneering role in the emergence of prehistory as a subfield of archaeology: beginning in 1859, new insights in geology and biology informed the examination of fossil evidence that confirmed the long-standing intuition of the great antiquity of humanity.[25]

For much of the nineteenth century, French archaeology comprised a multitude of sites, practices, and interests. The slow professionalization of the field, with France lagging conspicuously behind its German rival, reflected limited state support for excavations and the lack of postsecondary training.[26] Spectacular finds like those of Paul-Emile Botta at Khorsabad (ancient Dur-Sharrukin) in Mesopotamia in the 1840s were the fruits of private initiative, with little follow-through from the French state.[27] Napoleon III did create several institutions to promote the study of Gallo-Roman antiquities, notably the Commission de topographie des Gaules (1858) and the Musée des antiquités nationales in St. Germain-en-Laye (1867).[28] But only after France's searing defeat in the Franco-Prussian War of 1870 did the government heed the admonitions of scholars like Ernest Renan (1823–1892) and expand higher education to create a university system offering instruction in, among other things, the latest techniques of historical study. Archaeology took its place among the "auxiliary sciences of history," with the creation of chairs in universities and specialized institutions such as the École Pratique des Hautes Études, founded in 1868, and the École du Louvre, which opened in 1882 as a training academy for curators. Most of the scholars named to these early professorships had, like Renan himself, a background in philology, the pattern science for historical study in German higher education at the time. But the classicists sent to organize the French schools in Athens (founded in stages between 1843 and 1850) and Rome (1872) understood them as places where talented students could gain practical training in the evolving methods of archaeology. Both schools operated under the auspices of the Académie des inscriptions et belles-lettres, an association that endowed them with a certain scholarly authority.[29]

In her comprehensive history of French archaeology, Ève Gran-Aymerich has pointed to the importance of the discovery and analysis of preclassical remains in Greece, the Eastern Mediterranean, Egypt, and the Near East in the late nineteenth century. These finds validated not only the idea of prehistory but the methods of archaeology in general: only systematic archaeological study could unlock the mysteries of societies that lacked written records. Yet as Gran-Aymerich acknowledges and Natalie Richard has elaborated, throughout this period and into the first decades of the twentieth century, the prehistory of metropolitan France remained largely in the hands of amateurs, with very little academic attention to the field. The same was true of protohistory as well as Gallo-Roman and early medieval archaeology. Together these specializations constituted "national archaeology," the field concerned with the past civilizations of the metropole. Its practitioners were overwhelmingly amateurs, from doctors and lawyers to priests, landowners, and local officials, their activities centered on provincial learned societies, which devoted the largest share of their meager resources to publications.[30] As Bonnie Effros has observed, a young scholar returning to France after training at one of the classical schools could quickly rise to the top of the world of national archaeology.[31]

Historians have cast the half century covered by this book, from 1890 to 1940, as a period of transition from archaeology's "heroic age" to the era of its professionalization.[32] Although France fits this description, a few particular features of the French archaeological field need to be highlighted. First, the gap between prehistory and the more professionalized fields of classical and Gallo-Roman archaeology persisted in this period, even as it narrowed. Provincial notables and learned societies, jealous guardians of their primacy in national archaeology, successfully lobbied against including archaeology in the 1913 law protecting historic monuments. Tunisia and Algeria instituted laws regulating archaeological excavations and the export of finds in the 1920s, two decades before France, then under the

Vichy regime, finally did the same.[33] This lag offers a reminder of a second feature: archaeology's intimate relationship with colonialism. In archaeology as in so many other domains, the colonial theater served as a kind of laboratory for the metropole. In North Africa, as Gran-Aymerich notes, the French imperium gave its archaeologists a monopoly on archaeology (which post–World War I French mandates in Syria and Lebanon would expand) that it did not enjoy elsewhere in the Mediterranean.[34] Third, the slow process of professionalization made those aspiring to professional status especially insistent on the rigor of their methods. In particular, archaeologists in the first decades of the twentieth century emphasized the need for excavations to attend to distinct layers, whole environments, and fragments that could furnish evidence of the patterns of daily life, not simply spectacular objects. Paradoxically, however, in what Alain Schnapp has identified as a typically French attachment to the written word, many archaeologists chose to write about scientific methods rather than to teach them.[35] Salomon Reinach, who conducted fieldwork only in his youth, in 1911 published an essay called "La méthode en archéologie," but, as Effros observes, field methods seem entirely absent from his lecture plans for the École du Louvre.[36] Overall, the frontier between amateur and professional archaeology remained blurry well into the twentieth century, a large transitional zone rather than a demarcated border.

The period from around 1900 to 1940 also marks both the apex and the beginning of the decline of the French newspaper press. The interwar years saw a stagnation in circulation, a decline of profitability, and only limited innovation: the success in the 1930s of the illustrated daily *Paris-Soir* came at the expense of more established dailies, and the growth of the regional press similarly cut into the circulation of Paris-based national dailies.[37] Moreover, no less an observer than Léon Blum (1872–1950), a journalist before he entered politics, expressed the view that the well-known venality of the financial papers, which enforced a system of regular payoffs from banks and investment entities to prevent the publication of unfavorable stories about them (true or false), had infected the whole of the press.[38] In this context, archaeology offered the press a way of tracing French scholarly achievement on stages both national and international. The press also could shape diverting stories around both the careers of archaeologists as models of Republican meritocracy and, in cases of controversy, the spectacle of learned men (and a few women) behaving badly.

Colonial-era archaeology in Tunisia has received thorough scholarly consideration in recent books by Myriam Bacha and Clémentine Gutron, but in neither work is the dynamic of science and media a central concern.[39] Glozel earns brief mentions in several informative histories of

French archaeology and prehistory, but its sole monographic treatment comes from a historian convinced of the site's authenticity, a position that remains an outlier in the French archaeological profession, which universally treats it as a hoax.[40] Considering the two cases together offers an opportunity to probe the ways colonial and postcolonial situations have shaped archaeology. To the extent that the structures of imperialism conditioned and shaped the methodology and core insights of archaeology, it is important to consider how those shaping influences manifested themselves both in North Africa and in the metropole.

Most core definitions of the *archive* have two components: a place where records are kept, and the records kept at that place. The archaeological archive then refers, obviously, to the sum of records related to and produced by archaeology in all its stages and venues, from planning and excavation to publication, from the dig site to the museum and lecture hall. Beyond that, however, at a conceptual level "the archaeological archive" signifies a profound and active set of connections, theoretical, methodological, and empirical, between archaeological and historical practice. The archival branch of library and information science has for some time mooted an all-encompassing conception of the materials archives collect, one scholar, for example, describing the contents of archives as "objects, artifacts, and documents."[41] Among the materials treated as archival objects in the 2007 collection *Archives, Documentation, and Institutions of Social Memory* are the classic British phone box and presidential limousines.[42] Conversely, historians of an ever-expanding range of periods and places use archaeological evidence as data in their own research, part of a material turn that both complements and competes with sources committed to more customary historical media like paper.[43] Beyond this, the pervasiveness of archaeological metaphors in writing about archives suggests a common imaginary. To take just one of innumerable examples, in their monograph *Processing the Past*, Francis Blouin, an archivist, and William Rosenberg, a historian, describe early archival research in these terms: "the sheer romantic pleasure of opening new historical treasures, engaging the authentic if fragmentary shards of a really lived past."[44]

"Authentic if fragmentary": historians as well as archaeologists engage a poetics of the fragmentary. Florence Bernault probes this connection when she writes of historians, "Our paradox is to use incomplete and fragmentary traces to understand broad patterns and timelines in a past that remains, by definition, unreachable."[45] "Incomplete and fragmentary": scholars of archives make clear that they are always partial records of what happened in the past, assembled for a variety of reasons and organized in ways that reflect the moment of classification and its vision of the future as much as of the past.[46] Historians and archaeologists alike also have to confront the inescapable politics of what we do. The first syllable of both archive and archaeology, *arch*, can be found in such constructions as archduke, archbishop, and archrival: it signifies superiority, lordship, authority, reminding us that the Greek word αρχειον designates the residence of magistrates, a site of power.[47] "Archives are not neutral," the critic Allan Sekula has written: "they embody the power inherent in accumulation, collection and hoarding as well as that power inherent in the command of the lexicon and rules of a language."[48]

The ordering of archives has played a crucial role in the disciplining of history—or, in other words, archives have served as an instrument of the professionalization of history (and vice versa).[49] In parallel, the cataloging, classification, and publication of archaeological materials, all according to emerging scholarly norms, formed an equally indispensable basis for the professionalization of archaeology. If these developments overlapped chronologically, they took place slightly later in archaeology, which has long had an ambivalent relationship with its own archive. Nathan Schlanger, indeed, describes the archaeological archive as "surprisingly undervalued and unproblematized," and notes that it was long considered "trivial, redundant, self-evident"—or, alternatively, too revelatory of archaeologists' human flaws to be shared with outsiders. He is clearly referring to the paper generated *around* archaeological excavations—drawings and photographs, as well as dig journals, correspondence, labor contracts, and other sorts of texts and images.[50] But in a 1989 manifesto, the British archaeologist Christopher Tilley, calling for a new kind of socially conscious excavation, also urged colleagues to spend more time on the vast accumulations of artifacts that remained in storage, largely uncataloged and unpublished.[51] The authors of *Archaeology: The Discipline of Things*, the group who call archaeology an "ecology of practices," similarly argue that archaeologists have a responsibility to care for the objects they have accumulated, including both artifacts and documents.[52]

The understanding of objects as documents has a long tradition in French archaeology: in the preface to his influential *Manuel d'archéologie préhistorique*, originally published in 1908, Joseph Déchelette promised a thorough account of "how persevering research managed to dig up these archives of humanity, and thanks to them, to reconstitute important aspects of our origins."[53] Archaeologists' commitment to acquiring, preserving, and analyzing such documents marks their practice as eminently archival and parallels historians' sense of responsibility to the archival documents, textual and otherwise, from which we construct our narratives and interpretations.[54] The archive, with its multiplicity of materials and ways of storing, classifying, and transmitting them, gives scholars purchase on the construction of our disciplines, a glimpse into how earlier generations understood the media through which they produced and transmitted knowledge.

Notwithstanding these connections, archaeology and history have developed as discrete disciplines, with their own archival protocols and their own claims to scientific status.[55] My own method for investigating controversy owes much to Michel Foucault's genealogy, "gray, meticulous, and patiently documentary."[56] For Foucault the distinct codes, categories, and arrangements of different archives allow us to assert both our connection to and our difference from the past(s) we are studying.[57] Recognizing not only the connections but the differences between archaeology and history enables me, a historian studying archaeology, to strike a position melding empathy with measured distance, a perspective of the sort Schaffer and Shapin recommend for the study of controversies.[58] In my own case, differences in professional culture, affiliation, and various forms of personal identity coexist with the empathy I feel for scholars struggling to find time for research amid their myriad administrative responsibilities: fundraising, supervising sites and personnel, cataloging artifacts and collating data, chasing down images for publication. At times I had to work to find empathy for the characters in the story whose positionality did not so clearly mirror my own.

The archives from which this book emerged bear multiple traces of the processes of discipline formation I have just set out. Even before the establishment of the French Protectorate in Tunisia in 1881, the French government was funding archaeological excavations and other scholarly missions to Algeria's eastern neighbor, and the grantees' reports can be found in the voluminous files of the Ministry of Education's *service des missions* in the French National Archives.[59] By the early 1890s the French had established an Antiquities Service in Tunisia, and traces of it exist in the expected official archives, as well as in the private papers of archaeologists from Ann Arbor to London to Rome.[60] Yet the mother lode of Tunisian archaeological archives exists in none of these places; rather, that distinction belongs to the Poinssot papers, over one hundred linear feet (239 boxes) of archives assembled by three generations of archaeologists who worked in Tunisia between the 1860s and the 1960s.[61] Acquired by the Institut national d'histoire de l'art (INHA) in 2005, the papers opened to researchers in 2014. At the core of this archive sits Louis Poinssot (1879–1967), whose work in Tunisia spanned over forty years beginning in 1899, the last two decades (1920–1942) as director of the Antiquities Service. The Fonds Poinssot also include the papers of Louis Poinssot's father, a lawyer working in Algeria who undertook an archaeological survey of Tunisia in the early 1880s, and those of his youngest son, who excavated in Tunisia in the 1950s.[62] Through various means Poinssot also acquired some of the papers of his two predecessors as director of antiquities, as well as those of the secretary-general of the Protectorate, a noted Arabist. At a workshop marking the opening of the Poinssot archives, the archivists who cataloged them gave a talk entitled "One archive can hide another... and another... and another."[63]

The intricate layering of the Poinssot archive has many of the characteristics of a family archive as described by John Randolph, but it also contains much correspondence Louis Poinssot received in his official capacity as director of antiquities. Its richness and variety confirm Kaeser's argument for the value of private papers as a source for microhistories of scientific controversies.[64] Poinssot's goal in amassing this collection seems to have been to document his and his colleagues' professionalism, seriousness of purpose, and integrity, notably in the face of continuing criticism from well-connected amateur archaeologists who sought to change the orientation of French archaeology in Tunisia. The assemblage of the Poinssot papers thus indexes, among other things, the ongoing conflict between amateur and professional archaeologists in the first third of the twentieth century, even if, as with any archive, its component parts do not all cohere with its founding purpose. This is also a colonial archive par excellence: the correspondence between the director of antiquities and his deputies flows especially during digs and vacation periods, reflecting the imperative that someone from the metropole remain in post at all times.

The accumulation of documents about Glozel also put a premium on private collecting and accumulation, partly for technical reasons. After the international commission issued its verdict in late 1927 that Glozel was not in fact ancient, the state declined to list it as a protected site.[65] At a time when government regulation did not automatically extend to excavations on private land, this decision meant that the state was bowing out of the Glozel affair. Yet archival traces of Glozel abound, with extensive private correspondence of some of the major players housed in repositories from Paris (mostly in the library and archives of the Académie des inscriptions et belles-lettres, a component of the Institut de France) to Aix-en-Provence, and no doubt beyond.[66] Also abundant are collections of newspaper clippings assembled by those with both major and bit parts in the affair. Even in the age of digitization, the convenience of clippings files retains an obvious appeal, but their interest goes beyond their function as a form of information storage.[67] Clipping also seeks to preserve something of the materiality of the past, as both a form of self-archiving and a kind of anticipatory archaeology. If the act of clipping inverts archaeology, taking a whole and turning it into fragments, it does so in service of a fundamentally archaeological epistemology, one that posits a missing whole from the sum of its available parts.[68]

The book comprises five chapters. The first two set out the chronologies of the controversies at issue in the book, Carthage in chapter 1 and Glozel in chapter 2, as well as the main lines of my argument about the mutual constitution of archaeology as science and media spectacle. The remaining chapters bring the two affairs together around common themes: embodiment in chapter 3, performance in chapter 4, imaging and imagination in chapter 5. With background on the beginnings of French archaeology in Tunisia and the creation of the French Antiquities Service in the 1880s, chapter 1 considers the scientific networks that developed around that service, its link to imperial power structures, and the stresses and fractures posed by a press campaign for more excavations at Carthage as well as by well-funded American excavations there in the 1920s. Key figures include the director of antiquities in Tunisia, Louis Poinssot; his main antagonist, a military doctor and accomplished amateur archaeologist, Louis Carton; and a group of American archaeologists with a knack for publicity. Chapter 2 examines the Glozel affair from the perspective both of archaeologists seeking to make sense of it and of the press. It shows how newspapers and magazines served as both relay and loudspeaker in the conflict between those who deemed the Glozel finds Neolithic and those who considered them the work of a talented forger. Drawing on scholarship in media studies and the history of communications, as well as an article by the great French historian Marc Bloch on the spread of "false news" during World War I, the chapter also explores the technology of news distribution, skeptical coverage of archaeology's scientific claims, and the appeal of familiar journalistic tropes, notably those inherited from the Dreyfus affair, in accounts of the Glozel affair.

Informed by literary scholarship on embodiment, the third chapter examines the "work" of archaeology and its manifestation in different kinds of bodies, including those of the laborers doing the physical work of archaeology, as well as those of archaeologists and the various bodies of knowledge they devote themselves to assembling. I argue that in the case of Tunisia, the two friends who successively served as directors of archaeology, Alfred Merlin from 1906 to 1920 and Poinssot from 1920 to 1942, were haunted by the case of their predecessor, Paul Gauckler, who was forced to resign when his same-sex entanglements became too well known. The second part of the chapter considers visual representations of archaeologists in Tunisia in relation to written accounts of their work of all kinds, as well as of the Tunisian workers they supervised. Attitudes toward the physical work of archaeology deploy both racial and classist assumptions, the common currency of colonialism but in evidence wherever digging takes place. The chapter's final section shows how the Glozel affair challenged the image of archaeologists and archaeological work in the wider culture, with particular attention to questions of class and gender. Chapter 4 looks at the performative dimensions of the two controversies, my analysis framed by key concepts in performance theory. The chapter begins with pageants staged in the Carthage ruins in 1906–1907, organized by Carton to alert the public to the importance of the site and the potential contribution of archaeology to imperial rule. I then consider scholarly and popular performances related to Glozel in 1927–1928, archaeology as courtroom drama in the latter phase of the Glozel affair (1928–1932), and finally a 1928 novel about Glozel that treats it as a kind of farcical performance.

Chapter 5 employs insights from the history of photographs and archaeological illustration to explore how archaeology draws on and helps to construct wider cultural imaginaries.[69] This chapter looks closely at the depiction of ancient Mediterranean societies in catalogs of Tunisian archaeological collections and related publications before turning to the visual representation of Glozel. Theories of race and the origins of writing make up the key components of these imaginaries, focused (in both affairs) on the ancient Phoenicians. Finally, via episodes from the 1930s, 1960s, and 1970s, as well as an account of my own encounters with archaeological objects in museums in London and Cambridge, an epilogue offers a sense of the afterlives of the Carthage and Glozel controversies and the changing but persistent dynamics of the constitutive tension in archaeology between science and the media.

"Archaeology does not discover the past as it was," Bjørnar Olsen and his colleagues write: "archaeologists work with what has become of what was; what was, as it is, always becoming."[70] Much the same can be said of the writing of history. Throughout the book, a series of what I call archival moments points to my understanding of archaeology and history as distinct but closely related endeavors rooted in overlapping objectives, common undertakings, and similar epistemological challenges. By "archival moments" I mean instants of discovery in the archives, encounters with documents or objects that change one's sense of a subject and perhaps give new shape to a project. That flash of inspiration also opens new questions, many of them also archival in that they acknowledge, with a mixture of empathy and puzzlement, the complex mental and physical processes, the combination of intent and accident, that put that material in the place it was found. Each archival moment, in other words, constitutes the latest but surely not the last in a chain of such moments, some perhaps routine but all part of the complex process of constructing, conceptualizing, and interpreting the past. Many archives begin in a border zone, as familiar to archaeologists as to historians, somewhere between amateur and professional, evidence and oddity, discipline and passion, science and spectacle. Recapturing the uncertainty, the tension, the challenge, and the excitement of that moment before the fixing of disciplinary norms and boundaries is the goal of this book.

A note on terms: in discussing the non-European population of Tunisia, notably the workers who performed the physical labor of archaeology there, I follow the lead of several thoughtful recent historians of Algeria who eschew the word indigenous (in French *indigène*) because it is indissociable from racist laws, texts, and practices in North Africa under French colonization.[71] Because Europeans and other archaeologists frequently used the word "Arab" in an inexact and pejorative way, I also avoid the ethnic descriptors Arab and Berber unless quoting contemporary usage. Emulating Owen White's eminently sensible choice, I use the term "Tunisians" to refer to people of non-European origin whose progenitors lived in Tunisia prior to the installation of the French Protectorate.[72] Very occasionally the words "native" and "local" are unavoidable; readers should understand them in the most generic senses.

✷ 1 ✷
"For Carthage"

SCIENTIFIC NETWORKS AND
THE GLARE OF PUBLICITY

Prologue: The Field and the Network

Two incidents, two archival moments, encapsulate the dynamics of French archaeology in the era of colonialism. In May 1881, just a week after the signing of the Bardo Treaty establishing a French Protectorate over Tunisia, René Cagnat, a young classicist dispatched by the French government to survey the archaeological terrain in Tunisia, sent a letter to Xavier Charmes, head of the Higher Education Bureau of the Ministry of Education in Paris. Begun in late January, Cagnat's mission had not gone well, hampered by an uncertain military situation that forced an abrupt return to Tunis in April and then a retreat to Algeria. In his letter of 21 May, Cagnat explained why he had not yet set sail for France: an opportunity had come up to visit the Tunisian coastal town of Tabarka, near the Algerian border. He wrote, "Before leaving Algeria, I found a favorable opportunity to retake scientific possession of Tunisia, and I did not want to let it slip."[1] Now on his way back to France, he promised a verbal report on his visit, which he described as "très pénible," a frequently used French word that can describe anything from back pain to an unusually long faculty meeting.

Almost half a century later, in July 1925, an American engineer and dig manager based in Paris, Edward Stoever, wrote a letter to Francis Kelsey, a classicist and founder of the archaeology program at the University of Michigan in Ann Arbor (the university's archaeology museum bears his name to this day). Stoever had just had a visit from the abbé Jean-Baptiste Chabot, a French scholar and member of the Académie des inscriptions et belles-lettres, the governing body and one of the main funders of French archaeology overseas; Chabot had been advising the excavations Stoever and Kelsey were conducting at Carthage, just north of Tunis. "From what he said," Stoever recounted, "our friend in Tunis is so well protected by some one in authority that he is difficult to dislodge, and the abbé seems to think that under the circumstances Poinssot will refuse authorization to send the 500 urns which you have requested."[2] The person referred to, Louis Poinssot, was the director of the Tunisian Antiquities Service, and Chabot's information was accurate: cantankerous and highly sensitive to any breaches of rules or agreements, Poinssot exercised tight control over archaeology in Tunisia. Although unnamed in the letter, his protector was none other than René Cagnat.

These two moments frame the emergence of archaeology in Tunisia within a network of institutions at once scientific and colonial. But the politics of archaeology do not simply replicate on a smaller scale those—complex, shifting, and dynamic as they are—of colonial systems writ large, nor do they pertain only to the space of intersection between colonialism and the knowledges it both generates and relies on. Colonialism takes many forms, and "to think like an empire," in Frederick Cooper's terms, entails different types of mental gymnastics depending on the time and place.[3] Similarly, distinct fields of scientific investigation involve different relationships to colonial authority; if George Trumbull, in his study of ethnography in colonial Algeria, is surely right to identify "methodology and narrative as forms of control," those forms of control vary across disciplines.[4] Of particular importance to archaeology in a colonial setting are the institutions, both in the metrople and the colony, that fund and regulate it.

The emergence of heritage institutions and practices in Tunisia has attracted the interest of a number of scholars in the twenty-first century. Accounts beginning barely a decade before the institution of the Protectorate reveal a growing interest in antiquities among Tunisian elites; in the 1870s, notably, the reformist prime minister Khair-al-Din al-Tunisi began assembling a collection for a planned archaeological museum in Tunis. These plans, however, fell by the wayside after Khair-al-Din's forced resignation in 1877, and the embryonic collection was dispersed. Thus, as Houcine Jaïdi has observed, the networks that emerged around heritage institutions in Tunisia were French: archaeology developed simultaneously as colonial field and as scientific network.[5] Bruno Latour defines a network as a concentration of resources in nodes and knots connected through lines and mesh: "These connections transform the scattered resources into a net that may seem to extend everywhere."[6] Latour's "may seem" offers both a warning and an invitation: scientific networks at once constitute and represent an image of dispersed power, requiring careful study and analysis to understand their workings. In particular, the networks of science and colonial governance are so tightly overlaid that disentangling the two, even when studying only one scientific field, poses considerable analytical, not to say practical, challenges.

In the period of its formation as a scholarly discipline, roughly from the mid-nineteenth century to the eve of World War II, archaeology emerged as two things: first, a discourse and set of practices rooted in, though not limited to, fieldwork; second, a colonial field, in which practices of domination and appropriation coexist, not always comfortably, with a set of higher ideals broadly characterized as "scientific." In Latour's actor-network theory, "centers of calculation" draw together objects, knowledges, and inscriptions—which for Latour betoken not simply the actual inscriptions sought and recorded by epigraphists like Cagnat, but any depiction or description of a phenomenon committed to paper—sent via a network's radiating lines. That transmission or movement in fact creates centers of calculation as the network's nodal points; it forms part of "a cycle of accumulation that allows a point to become a *centre* by acting at a distance on many other points."[7] If the pertinence of these concepts to the study of colonial situations is clear, a number of historians of science have pointed to the ways Latour's understanding of networks reproduces colonial assumptions and omits or glosses over the agency of local actors in their encounters with imperial (European or North American) scientists.[8] Effective use of Latour's schema depends, as Michael Bravo and Simon Schaffer have shown, on applying culturally specific notions of locality and agency to situations of colonial contact.[9] As Schaffer observes, "by attending to science's geographies, the pattern science becomes fieldwork."[10] This patterning offers a model for using Latour's concepts to chart a path through the colonial networks that constituted early French archaeology as a scientific field.

If the chapter focuses on archaeology's operation within a scientific network, examining archaeology in this particular time and place inevitably involves its relationship with the media. Scientific networks, among other things, serve to communicate findings so that, at least in theory, the science in question can progress. Operating chiefly from the centers of calculation they help to constitute as such, the scholarly journals published by academies, museums, and universities represent the principal, approved vectors of such communication. In the field of archaeology, however, these organs never have a monopoly over the diffusion of information, as the period of archaeology's formation as a scientific discipline coincided with that of the emergence of mass media as both a technological and commercial phenomenon. In an important essay on the presentation of archaeology in *National Geographic*, Joan Gero and Dolores Root have argued that the ideological stakes of archaeology as presented in popular media are usually congruent with those of the scholarly discipline.[11] Yet the discipline, as it develops, defines itself in part through its distinction from the forms and procedures of mass media, even as it uses such popular outlets to promote its activities and gain financial support to carry them out. As the case of Cagnat makes clear, this dynamic emerged from the earliest moments of French archaeologists' presence in Tunisia, and intensified in the 1920s.

A word, finally, about why this particular site attracted so much interest in the 1920s. Tunisia offered many sites for excavation, some, like Dougga, extensive and relatively well preserved.[12] But Carthage, known to locals and travelers since its abandonment in the seventh century CE, held a privileged place in the European imagination because of its role as the most formidable rival in Rome's rise to power as a Mediterranean empire. Amateurs acting on behalf of foreign governments and institutions began excavating in the 1830s, but it was the arrival of the White Fathers order of missionaries, founded by Cardinal Charles Lavigerie, the primate of French North Africa, to support French colonization, that led to sustained archaeological activity there.[13] Beginning in 1875, Father Alfred-Louis Delattre, a member of the White Fathers, was the first to carry out sustained excavations on property in Carthage deeded to France in 1840 as a memorial to Louis IX (St. Louis), who died in Carthage in 1270.[14] The White Fathers' extensive landholdings and strategic position on the Byrsa Hill at the heart of the pre-Roman city set the stage for conflict with the Tunisian antiquities service, the Direction des antiquités et des arts (DAA), set up a few years after the establishment of the French Protectorate in 1881. Salomon Reinach, one of a pair of young French archaeologists sent by the government to survey the archaeological terrain in 1884, gave a lecture in Tunis a decade later criticizing the White Fathers' effective monopoly over archaeology at Carthage.[15] Paul Gauckler, who became director of the DAA in 1892, clashed with the church when he attempted to list nineteen Carthage sites as historic monuments, and thus subject to government control, two years later. Although Gauckler respected Delattre's work and deferred to his experience, he regarded the White Fathers, who had a close relationship with the French resident general (the top-ranking French official in the Protectorate) as a rival in the quest for government funding. So the beginning of Gauckler's own Carthage dig in 1899, though ostensibly undertaken in a spirit of cooperation with Delattre, marked a distinct assertion of the authority of the Protectorate government.[16]

In 1904 Gauckler discovered the remains of a Roman theater on a hillside near the Odeon, and by early 1905 he was able to establish the plan of the Roman city as laid out in the second century CE. Like the general public, however, archaeologists were most interested in the pre-Roman city, known variously as Punic, Phoenician, or simply Carthaginian, that the

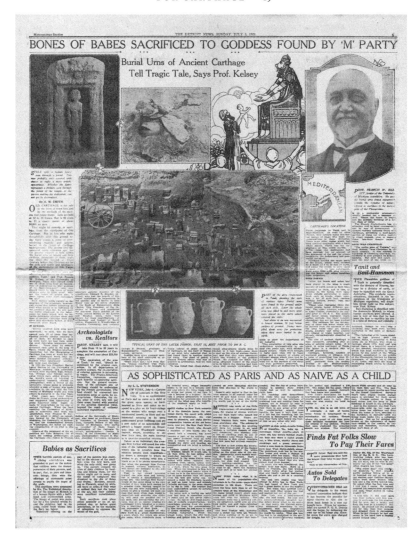

FIG. 1.1. "Bones of Babes Sacrificed to Goddess Found by 'M' Party," *Detroit News*, 5 July 1925. Kelsey Museum of Archaeology Papers, folder 79.5, Clippings and Publicity, Bentley Historical Library, University of Michigan, Ann Arbor. Image courtesy the Bentley Historical Library.

Romans had infamously sacked in 146 BCE. Flaubert's novel *Salammbô* (1862), made into an opera that premiered in 1890, helped to popularize exotic Orientalist tropes about ancient Carthage, notably the practice of child sacrifice.[17] Beginning in the early 1920s, reports of the discovery of urns with what seemed to be human remains began to surface, creating

FIG. 1.2. Father Delattre at the Chapelle St. Louis on the Byrsa Hill, Carthage. Postcard, early twentieth century. Wikimedia Commons.

enormous worldwide attention, as a full-page article from the *Detroit News* (fig. 1.1) testifies.[18] Fruitful excavations at Carthage thus promised to enhance reputations and, perhaps, secure funding for future digs.

A comprehensive account of the excavations and finds at Carthage from the 1890s to the 1920s falls well outside the scope of this book. The site comprises a number of distinct zones, from the hill that dominated the Punic city to the later Roman city and the coast, where archaeologists searched for remains of the famous Punic port. At any one time, distinct excavations could be underway on Punic, Roman, and early Christian levels, to say nothing of unsanctioned recuperation of ancient stones for building materials. Delattre himself alternated among different sites and strata; although best known for his finds in Punic cemeteries, including both monumental funerary architecture and abundant grave goods, he also excavated extensively in early Christian cemeteries and basilicas. Of equal note, Delattre, who received only modest support from the White Fathers for his archaeological work, clearly understood the importance of publicizing it to diverse audiences. Over the half century beginning in 1879 he published nearly eighty reports in the proceedings of the Académie des inscriptions et belles-lettres (AIBL), the venue conferring the greatest scholarly credibility in the field of archaeology.[19] But Delattre also contributed regularly to *Cosmos*, the weekly science supplement of the Catholic newspaper *La Croix*, and many of those well-illustrated articles, which mixed narratives of archaeological seasons with descriptions of the most important find, were reprinted as pamphlets for sale at the White Fathers' museum in Carthage.[20] Beyond his own publications, by the early twentieth century Delattre had become a literary and visual icon of Carthage, appearing as an almost saintly character—equally devoted to the church and to archaeology—in books for which he clearly served as a prime source and even in postcards (fig. 1.2).[21] By this time, however, he also had serious rivals for both scholarly glory and popular attention.

Serious Men (and Their Archives)

Cagnat undertook three additional missions (at the time the French administrative term for funded research) to Tunisia, of varying length, between 1882 and 1888. Just as these expeditions prefigured French archaeological efforts in the Protectorate, the areas of concern Cagnat outlined, in both his reports and his actions, proved remarkably durable. All of these areas, it is worth noting, involved activities temporally distinct from actual excavation: preservation, scholarly publication, and what might be called social relations. Preservation, first: Cagnat's official dossier in the French NationalArchives includes an October 1881 newspaper clipping on the French military's imminent occupation of the holy city of Kairouan, Tunisia. The article calls for a group of scholars to join the troops, along the lines of the scientific delegation that accompanied Napoleon on his expedition to Egypt in 1798–1799. While the author, Marius Vachon, anticipated that French scholars would take possession of artistic treasures in Kairouan, he warned that the French army must at all costs avoid "shameful pillaging" of the sort that had occurred at the summer palace in Beijing.[22] From his first visit to Tunisia, Cagnat was preoccupied with disposing of the objects he found; while inclined to send the most important to France, he did so only in consultation with the Protectorate administration, and assiduously followed instructions to deposit materials with local authorities in anticipation of the opening of an archaeological museum in the Bardo Palace.[23]

For Matthew McCarty, French archaeology was seeking above all to demonstrate something material evidence could not prove, the biological descent of French colonizers from their Roman predecessors; thus archaeological sites offered only a stage set or backdrop for performances of this imagined history.[24] Unearthed Christian structures, which became the priority for Delattre's excavations early in the twentieth century, offered a different kind of direct connection, since the faithful could in theory observe the same rites as their ancestors. In contrast, McCarty argues, prior to the 1921 discovery of evidence of Carthaginian infant sacrifice, Punic archaeology held little appeal, since, in the words of Jacques Alexandropoulos, it could not "directly serve to valorize the colonial enterprise."[25] Yet as early as 1882, Cagnat followed the prescriptions of his mentors and correspondents in the metropole, who wanted him to go beyond already excavated sites to new terrain, particularly that of known Punic (pre-Roman) cities; this emphasis differed from the preoccupation with Rome that dominated French archaeology in Algeria in the first decades after the conquest.[26] Cagnat readily accepted such instructions because they promised to help secure his own scholarly reputation. For this, timing was crucial. Since the early nineteenth century, as Alex Csiszar has shown, scientists closely linked discovery with publication, so publication offered archaeologists the opportunity to associate themsleves with recognized scientific practice.[27] Writing to request offprints of a forthcoming report in the scholarly missions service's own journal, Cagnat observed, "As the documents I'm publishing there will be available to all, I have considerable interest in making them known as soon as possible so as to stake my own prior claim."[28]

Finally, in an uncertain political and military situation, discretion and the observance of proprieties became the precondition for success in the scholarly realm. To this end Cagnat sought the advice of Delattre; requested authorization from religious authorities before entering mosques; and praised the army officers who accompanied him in 1883 as "serious men seeking something other than publicity (*la réclame*)."[29] In contrast to his own eminently respectable plans for publication, the use of the word *réclame* suggests a degree of puffery and of commercial calculation. The *Petit Robert* gives this example for a now-dated sense of the word: "this writer's celebrity is based on *réclame*."[30] Already in 1883, then, disinterested science constitutes itself in opposition to commercial publicity. It is important to recognize, however, that this binary was at once aspirational and artificial; it should not be taken at face value.

Cagnat arguably owed his brilliant career not only to his actual excavations and scholarship—in 1892 he published what Bonnie Effros calls the first "holistic study of Lambaesis," a Roman military site in Algeria—but to his reputation as a similarly "serious man" with a sense of mission beyond mere publicity.[31] By the first decade of the twentieth century he controlled all the levers of power in the world of archaeology: holder of a chair in epigraphy at the very prestigious Collège de France, member and from 1916 until his death in 1937 perpetual secretary of the Académie des inscriptions et belles-lettres, and secretary of the North Africa Committee of the Comité des travaux scientifiques et historiques (Committee for Historical and Scholarly Works), which controlled research funding within the Ministry of Education.[32] In the latter capacity Cagnat served as de facto editor of an impressive series of scholarly catalogs of North African archaeological collections that began to appear in 1890. But he also authored texts that clearly aimed at the general public, including *Carthage, Timgad, Tébessa*, a volume in the generously illustrated *Villes d'art célebres* series that went through three editions beginning in 1909.[33]

More generally, Cagnat's missions show that even at the very beginning of organized, state-supported archaeology there, fieldwork in Tunisia was always connected to the constitution of archaeology as a field or discipline in the broader sense elaborated by Pierre Bourdieu. Bourdieu understands a discipline as an area of practice constituted by a set of structurally determined relationships among agents with varying amounts of symbolic capital (and actual resources).[34] And the archaeological field always existed in tandem with a larger colonial apparatus. The Tunisian Antiquities Service, or DAA, was set up in 1885; in 1896 it became an autonomous directorate under the authority of a single administrator within the Protectorate government.[35] Gauckler, technically the second director but really the institution-builder, and his two successors, Alfred Merlin and Louis Poinssot, shared certain key credentials, including a degree in epigraphy or a related field from a prestigious Paris institution and a stint at the École française de Rome, which along with its counterpart in Athens served as de facto training grounds for budding French archaeologists. Merlin had the most illustrious career of the three: after leaving Tunisia at the end of 1919, he spent a year teaching in Lille before taking up a curatorial position at the Louvre; he was elected to the Académie des inscriptions in 1928, and eventually became its perpetual secretary—in this as in other respects following in the footsteps of Cagnat, who was his father-in-law.[36] Poinssot, just three years younger than Merlin, began excavating at Dougga in 1899 and served as Merlin's deputy for over a decade before succeeding him in 1920. Merlin and Poinssot had known each other since their days at the Lycée Louis le Grand in Paris, and the hundreds of letters from Merlin to Poinssot, available since the 2014 opening of the Poinssot family papers to researchers at the Institut national d'histoire de l'art (INHA) in Paris, are written in the intimate second-person singular, with frequent inquiries and observations about personal and family health and other such matters.[37]

As noted in the introduction, the Poinssot papers, which encompass other partial archives, including correspondence of and related to Gauckler, provide an extraordinary complement to existing official archives. Besides offering backstage glimpses of incidents only partially, if at all, reported in the press, these archives illuminate the workings of colonial science in at least two distinct ways. First, the correspondence between Poinssot and Gauckler, between Poinssot and Merlin in the years they worked together, and then between Poinssot and Raymond Lantier, his deputy from 1920 to 1925, flows in tides regulated by the structures and assumptions of colonial authority, which required someone from the metropole to supervise Tunisian workers at all times. During the field season, letters move between Tunis and the excavation that either the director or his deputy, who had the title of inspector, was supervising; they follow the rhythms of a dig, its discoveries and challenges, from bad weather to labor disputes. Merlin and Poinssot took turns supervising the work at Mahdia from 1907 to 1913; one of the first major underwater excavations, it brought to light significant works of art and architectural elements from a Greek vessel that foundered off the coast of Tunisia in the first century BCE. Their letters from the coast combine, in almost equal measure, excitement at the treasures being discovered and concern over practical constraints involving unfamiliar equipment and trained Greek divers who were paid by the minute.[38] It is hard to imagine correspondence more fully embodying, and illuminating, James Secord's idea of science as "a form of communicative action."[39]

Letters also moved between the director and his deputy during vacation periods, which cumulatively extended from late June or early July into late October but were staggered so that one or the other was always on post in Tunis. Here a second, slightly different relationship to colonialism and scientific networks becomes apparent. Holiday letters from France pulsate with the bustle of networking, as one archaeologist made the rounds of the Academy, museums, libraries, and bookstores before heading to his country retreat, while the colleague still in Tunisia complained of the heat and hoped to have nothing dramatic to report. The letter writers, while clearly aligned with the metropolitan center, are thus not completely of one or the other world, and do not fully adopt center/periphery models in navigating between them. Like the Tunisians they employ and work with, they operate more as cultural translators or intermediaries, the importance of whom Kapil Raj has demonstrated.[40] But administrators never held a monopoly over cultural translation, and sometimes faced direct competition in that role.

"To Save Carthage": Networks, Authority, Publicity

The gap between seriousness and the publicity, or *réclame*, to which Cagnat referred clearly privileged the scholarly elite, whose organs could ensure that claims to "priority" like Cagnat's would not be confused with shameless self-promotion. The question of publicity, and of priority, runs throughout the career of another figure granted official archaeological missions in the early years of the Protectorate, a military doctor and archaeologist by avocation named Louis Carton. For the administrators of archaeology in Tunisia, Carton became the archetype of a practitioner subservient to *réclame*. Just a decade into the new colonial regime, digging at a defined locale—Dougga, the Roman Thugga, with strata extending from the pre-Punic to the Roman period—rather than surveying potential sites, Carton's concerns differed markedly from those of Cagnat. In his initial request for a formal grant from the French government, Carton argued that establishing Dougga as an official archaeological site would promote tourism, an important part of his vision for the colony. Such use would also "protect" the site from the imminent threat of depredation by its contemporary inhabitants and colonial development. The receptivity to his proposal in Paris stemmed from more pragmatic considerations: the ministry anticipated supporting Carton's petition for leave from his military duties while providing him with a subvention to support his work. At the same time it would delegate to the Tunisian government the purchase, from its own revenue stream, of the land required for the dig. In this way Carton's excavation would provide a testing ground for the mechanisms of the Protectorate administration, although officials stressed that local authorities would have to be consulted on protecting the site.[41] Apart from, indeed long before, interpretive questions about past achievements and conquest, archaeology's sensitivity within the colonial theater comes from the inescapable fact that it entails physical, and not just scientific, possession of land, and often the dispossession of its prior inhabitants. Subsequent correspondence between a frustrated Carton, the Ministry of Education, the Foreign Ministry, and the French Resident's office in Tunis conveys something of the complexity, and the slowness, such actions entailed.[42]

For Carton, archaeology not only promised to provide evidence of the success of earlier European colonization of Tunisia, but itself offered a kind of model for an imperial project of reclamation and development.[43] His lack of scholarly credentials and dependence on both the government and his military superiors for access to archaeological sites and resources placed him at a structural disadvantage in the emerging field, but Carton still aimed for prestige and recognition, publishing his finds when and as he could. In a letter to Gauckler in 1896, he thanked him for constructive criticism of a published article, complaining that too often his work was simply dismissed.[44] Wherever he was posted in Tunisia, from Sousse to Tunis, Carton founded archaeological societies, often with their own publications; for a decade beginning in 1903, he published an annual "Chronique d'archéologie nord-africaine" in the organ of the Institut de Carthage, a combination learned society and booster club for the Carthage archaeological site. In this self-appointed role, Carton praised his fellow amateurs as "modest savants" whose "disinterest, devotion, and love of knowledge is neither a profession nor a leg up [*marchepied*]."[45] Carton had no choice but to celebrate amateurs: his official archaeological missions had ended in 1899, when the ministry turned down his application for renewal on the grounds that all its resources would henceforth be directed to digs sponsored by the DAA.[46] His attempts to promote his own vision inevitably led him into conflict, both personal and ideological, with the people actually entrusted with authority over French colonial archaeology.

Gauckler and his two successors had difficult relations with Carton. Despite the doctor's assiduous flattery, Gauckler found his self-importance unbearable, and preferred to entrust excavation work to credentialed young men from the metropole, notably Poinssot.[47] Merlin, the most emollient of the three, got along best with Carton, but his correspondence with Poinssot shows that he did not trust the doctor. By 1911, Merlin could dismiss as typical of Carton an article in the *Revue de Paris* calling for more intensive efforts to excavate at Carthage and preserve the site from development.[48] The mutual antipathy reached new heights when Poinssot became director in 1920. Around this time, Carton, by now retired from the military and a full-time resident of a Tunis suburb, relaunched his public campaign in North Africa and the metropole to remedy what he considered the DAA's scandalous neglect of Carthage. In Gauckler's case, Myriam Bacha sees the controversy as simply a quarrel of egotistical men seeking to affirm their authority over a prestigious if highly contentious field, and she is not entirely wrong.[49] But something more than personal interest was at stake: nothing less than the definition of the archaeological field and its role in colonial governance.

In June 1920 Carton's view of state-supported archaeology in Tunisia—essentially, that it was in crisis—found an echo on the front page of an establishment daily, the *Journal des débats*, which published an article by the influential writer and colonial propagandist Louis Bertrand, a longtime friend and confidant of Carton's.[50] Taking the form of an open letter to the minister of education and entitled "Pour les ruines antiques de l'Afrique du Nord," the piece concisely made the case for the importance of ancient ruins in North Africa. "Besides the aesthetic and scientific interest attached to them," Bertrand wrote, "these vestiges of a civilization of which we are the heirs represent, in reconquered Africa, our titles of nobility and our rights as occupiers."[51] Bertrand then launched into a detailed critique of the present archaeological administration. After noting that in contrast to Algeria, which lacked an archaeological service, Tunisia at least had one, he criticized the DAA for its too limited conception of its responsibilities. "This service," Bertrand opined, "seems to limit its efforts to excavating, describing the ruins, making an inventory of the objects discovered . . . and classifying them in museum displays."[52] This might seem a sufficient, even ambitious, range of activities for an agency that was, as Bertrand acknowledged, chronically short-staffed and underfunded. But for Bertrand the agency needed to engage not only in study and collecting but in the *mise en valeur*—that familiar colonial term—of the ruins, safeguarding them from depredation and making them into tourist attractions with modern signage and interpretive tours.[53] This view neatly incorporates Carton's advocacy of transforming archaeological sites into tourist attractions to boost the Protectorate's economy and attractiveness to settlers.[54]

Just as significant as this broad conception of what an imperial heritage policy should look like is Bertrand's jaundiced view of professional science. He took pains to praise Merlin personally, as "not only an eminent archaeologist but—and this doesn't hurt anything—a man of letters and a *galant homme*." But the sentence following the description of his agency's "limited efforts" runs as follows: "Its goal is the search for the 'scientific' document. It is concerned only with the progress of science, science as a kind of mystique, science with a capital S."[55] Bertrand expressed his support for science, calling for excavations to be expanded, before setting out his larger program. Yet if archaeology in itself was admirable, Bertrand criticized archaeologists for losing interest in their finds once they had published them or reported on them to the Academy: "These excavations in the end have no other purpose than to expedite a civil servant's promotion." And he even charged that some archaeologists secretly hoped the ruins would disappear, leaving their scholarship as the only remaining trace of them. In contrast, Bertrand praised in lyrical terms those who followed "the religion of the past and of our great memories, all those who exult at the mere sight of a column or of a temple rising from the splendid barrenness of the African desert."[56] It would be difficult to find a more comprehensive expression of Carton's case for the amateur gentleman scholar. Endorsed by the *Journal des débats*—a header expressed its confidence that "our readers will want to join in [*s'associer*]" Bertrand's position—excerpts from the article were republished soon after in a Tunis daily reputed to be close to the French Residence.[57]

Bertrand's article did not appear in isolation, but formed part of a multipronged media campaign, orchestrated by Carton, that began before the war and reached a crescendo in the 1920s. If Carton's original lament about the neglect of Carthage focused on what he called "modern vandals," as the years went by his criticism increasingly targeted French officials, which after 1920 meant Poinssot. This was *réclame* at its most advanced: numerous articles in the press, from local newspapers to the national weekly *L'Illustration*; the founding of several associations dedicated to saving Carthage, including one, the Comité des dames amies de Carthage (Carthage Ladies' Friends Committee, or CDAC), presided over by Carton's wife; and collaboration with the Touring Club de France and several publishers to produce information panels and guidebooks.[58] The latter invariably included polemical prefaces (at least) commenting on the deplorable lack of care and decline of the site.[59] In 1924 the CDAC was transformed into an association with a broader compass, the Société des amis de Carthage, with Bertrand as its president. Antonella Mezzolani Andreose has argued that the new association marginalized the women who had run the CDAC and its chapters all over Europe, and as a result the Société soon foundered.[60]

In December 1922 Carton delivered a lecture in Paris with the title "Pour sauver Carthage," with, according to a ticket and notice in the Tunisian Protectorate archives in Nantes, the late addition of Bertrand to the program.[61] One imagines that the content of this lecture was quite similar to that of a pamphlet promoting the CDAC that Carton had published earlier that year. The pamphlet offered a justification for devoting resources to excavating and preserving the site at a time when "there are so many other, more costly ruins to raise": Carton noted that the cost would be minor in comparison, but he also compared the site to pleasures, like sports, the cinema, or dessert, of which his listeners would not deprive themselves even while supporting worthy charities.[62] Even more revealing is the way Carton casts himself as a sort of seer. On many occasions, he writes, even skeptical visitors who believe nothing to be left at Carthage, "after three hours' walk with me, looking at the ruins *with other eyes*, they have always changed their minds and proclaimed their enthusiasm." This was by way of justifying his claim that "these ruins should be judged not as they appear, but *as they would be if we had the will*."[63]

With few exceptions, Carton's backers comprised mostly politicians, society ladies, and literary celebrities such as Bertrand. Although he had supporters among the academicians, and he was elected a correspondent of the AIBL in 1911, Carton always felt like an outsider among them. In a 1910 letter to Salomon Reinach, by this time a well-established curator and scholar with whom he had a long-standing epistolary relationship, Carton conveyed his chagrin at having unwittingly offended one of Reinach's fellow academicians, in fact the perpetual secretary, the elderly Hellenist Georges Perrot:

> Please don't forget that I live far from the metropole, and that having never studied archaeology at a university or special school, and having no connection to a member of the Academy, things that for others are routine are unknown to me, as a consequence of which, when I present myself, either in person or in writing, in this world where I'm at best a modest auxiliary, but to which I don't belong, I must be considered a bit like the peasant in the living room, who can't be blamed for his poor manners.[64]

One suspects that Carton here is playing up his outsider status to gain his interlocutor's sympathy, and that he had some success at this tactic: two years later Merlin wrote Poinssot that Perrot, apparently "moved by the tears [Carton] shed in his office in October," had written him a "*very pressing letter*" urging Merlin to grant Carton an excavation permit, and the director thought it best to comply.[65] Nonetheless, Carton's was not primarily a scientific network, which makes archaeologists' defensive response to the doctor all the more telling.

Merlin's presence in Paris after leaving Tunisia in 1919 allowed him to keep Poinssot informed of what they called Carton's "intrigue" (*agissements*) or "*agitation*." In late 1921, Carton sent a letter to the Academy and the Education Ministry's North Africa Committee asserting that lax oversight at Carthage had resulted in damage to monuments, looting, and new construction that impinged on parts of the site. Carton excoriated the DAA for its poor planning, misuse of funds, scorn for scholars, and refusal to work with volunteer groups such as his wife's. Attributing such failures to "vanity, stubbornness, or personal animosity," Carton called for intervention from Paris, not only more funds but an independent administration of excavations at Carthage.[66] If at the Academy Cagnat could simply deflect Carton's report, Merlin sensed danger in a meeting of the committee in which several members reacted favorably to it. He thus intervened to state forcefully that the question was one of money, but that with the necessary resources the DAA was best placed to "do what is necessary."[67] At the committee's next meeting, Merlin and Cagnat were able to formulate a resolution (*voeu*), to be sent to the minister of education and then to the French resident general, calling for more funds to be allocated to excavations at Carthage but specifying that the DAA be in charge of spending them. "I saw nothing but benefits," Merlin blandly added, "to this indication."[68]

Poinssot and his supporters could not simply ignore Carton's public campaign, but they had their own media relays. In July 1923, a jurist, historian of Roman law, and civil servant in Tunis, Charles Saumagne, published an article in a Tunis daily entitled "Le problème de Carthage."[69] Drawing on Saumagne's own experience as a respected amateur archaeologist, the article deplored the recent "agitation" about supposed official neglect of the ruins. The idea that Carthage concealed untold treasures Saumagne cast as a "recurring vertigo, a collective hallucination" going back to the Roman era.[70] Over a half century of excavations had thus far produced little—Saumagne detailed the finds—making it difficult to allocate to it the disproportionate resources that would be needed to preserve Carthage from further development. Certainly parts of the site were worth preserving as an archaeological park, individual excavations needed supervision, and architectural ruins should be protected, but all of these policies, the article concluded, were well within the competence of the existing Antiquities Service. Over the next year, several Tunis dailies picked up some of the themes of the article, notably concerning the considerable cost of excavations in a place with high real estate values, and often explicitly defended the DAA.[71]

Saumagne's article provoked a furious reply from Carton in the form of a lengthy handwritten letter to the Ministry of Education in Paris; although the article did not mention Carton by name, Carton wrote that the author's tone and scholarly assertions did not permit him to respond publicly. Countering his colleague's assessment of the Carthage excavations as having produced only modest, aesthetically uninteresting results, Carton took the opportunity to name-check other archaeologists whose finds Saumagne had minimized or neglected to mention. Yet the wounded, defensive tone undercut this clever rhetorical strategy; Carton took particular umbrage at the phrase "tourist commercialism" (*mercantilisme touristique*) that Saumagne used to describe the spirit of the open-air performances he, Carton, had organized at the Roman theater.[72] Carton noted that Saumagne himself had helped design the set and costumes for a performance in 1907, and cited both Cagnat (a passage from a picture book he had published on the Roman ruins of North Africa) and Bertrand in his own defense.[73] The detailed, ad hominem mode of argument was typical of Carton. Indiscriminately citing other archaeologists, men of letters like Bertrand, politicians, and organizations like the tourist board and automobile association, Carton presents greater investment in Carthage excavations as a self-justifying goal. For Carton the imbrication of archaeology and empire operated primarily in the realm of tourism and settlement rather than in that of science. Although, he does, as Clémentine Gutron argues, anticipate some of the strategies outlined in the 1972 UNESCO plan for safeguarding Tunis and Carthage, so do many ideas about historic preservation and the promotion of tourism already circulating widely before World War I and that grew more insistent afterward.[74] Moreover, Carton's vision of "Tunisia in the Year 2000" notably fails to anticipate one signal development: independence. This was part of a larger gap in his vision, which included the Tunisian population only as local color animating the Tunis Medina. Such blindness to the colonial situation, and more particularly to the budgetary and political constraints that impinged on colonial science, inevitably brought Carton into conflict with officials like Poinssot, who could not afford to ignore them.

In mid-September 1923, the head of the Higher Education Bureau in the Ministry of Education forwarded copies of Saumagne's article and Carton's letter to Poinssot and asked the director to return them with his thoughts.[75] Observing that he had already had to "refute" some of Carton's views several times over the past few months, Poinssot chose to contest only the charge that Saumagne's article had "indubitably been inspired by the antiquities service or written in complete agreement with them," an assertion Poinssot called "absolutely false (as Mr. Carton knows perfectly well)."[76] In a few succinct paragraphs, Poinssot observed that he had no need or desire to ask Saumagne to defend the work of previous directors, "the merits of which I am better placed than anyone to appreciate," and intimated that he too found Saumagne's assessment "a little severe," while making clear his respect for Saumagne's own archaeological work. Poinssot's tone, though firm, remains dignified and somewhat distanced until the penultimate sentence, where he writes, "Like my predecessors, I attach no great importance to Mr. Carton's words, since, like all those who know him well, I consider him at once a dishonest man and a crazy one."[77]

This exchange illustrates Poinssot's mastery of the communication channels afforded him as a man of science in the French colonial hierarchy. In gently distancing himself from Saumagne, he nevertheless praises the latter for the "patient research" along the coast of Carthage that had produced "very interesting results." Poinssot thus makes clear that in his worldview credibility comes from scholarly research, that is, from science, not from whatever publicity might be attached to it. (It is these "results" that Poinssot points to as the source of Carton's hostility to Saumagne.) The twin references to his predecessors, first—because obviously most important—to their own scientific work, and then in passing to their assessments of Carton, place him in a lineage that undergirds his authority: "I am better placed than anyone." At the same time, this lineage seamlessly conjoins the authority of the directors' position with that of their research. Up until the last paragraph, the report contrives to cast its author as an independent agent within a scientific network, taking for granted the communicative act—the request from Paris for "your thoughts"—that prompted it. And even the closing, which immediately follows the dismissal of "a dishonest man and a crazy one," comprises a knowing rhetorical gesture: "I only regret that because of him [Carton] I lose precious time I would have taken much pleasure in devoting to defending and exploring the ruins of Tunisia."[78]

The summer of 1923 actually saw Poinssot fighting several public relations fires at once. In June the conservative Paris daily *Le Matin* published a front-page article with the ominous title "Sacreligious Devastation: The Ruins of Carthage are Being Pillaged," ostensibly inspired by an article in the weekly *L'Illustration* the previous year by a young associate of Carton called J. Jaubert de Benac. A clipping of the article made its way to the French Residence in Tunis, and the Resident's staff considered it significant enough to draft a letter to two senators known to be supportive of Carton, assuring them that the article was largely hype motivated by local intrigue, and that "I [the Resident] am doing for Carthage everything that is possible with limited means." The letter concludes, "Peaceful minds are, in this as in everything else, an essential condition for useful and productive work in the science of archaeology."[79] But in August *L'Illustration* published a second, equally incendiary article by Jaubert de Benac about the DAA's neglect of Carthage, in which a series of before and after photos purported to show serious neglect of the site.[80]

In his detailed official (but private) response to the article for the Ministry of Education, produced at more or less the same time he was responding to Carton's report, Poinssot pointed out a variety of photographic tricks. In one pair of photos of columns in an underground basilica, for example, the photographer's change of position, effectively a form of cropping, suggested decay where none had occurred. In another pair, the DAA itself had removed the votive stelae and altars visible in one picture and absent from the other. It took this step as a preservation measure after the amateur archaeologists who found them ignored Poinssot's plea to protect them and the site; no looting had occurred. A photograph of men in local dress carting off architectural fragments actually showed people working for Carton, whose illegal and deceptive collecting practices Poinssot noted he had already denounced. "This explains," Poinssot wrote, "their rather conventional poses"—they felt they had nothing to fear.[81] Equally important as this thorough debunking, on receipt of these two reports—on the magazine article and on Carton's response to Saumagne—the director of higher education, in preparing to present them to the North Africa Committee, sent them for comment to someone he considered especially qualified to assess the situation: Merlin. Merlin predictably responded as he had in the earlier committee meeting, emphasizing that the question was, above all, financial and affirming the DAA's competence and efficacy.[82] Thus the center of calculation continued to work within a framework that recognized the authority of the outlying points in its network. At the same time, however, Poinssot and the authorities in Paris had to confront a new type of archaeological practice, closely attuned to and calculated to feed media interest, that they would have more difficulty mastering. Perhaps unsurprisingly, it was American in origin.

Enter the Americans

A signed studio photograph in the Poinssot papers (fig. 1.3) testifies to the first encounter between Louis Poinssot and a young American calling himself Byron Khun (sometimes Kuhn) de Prorok. Both the photograph and its presence in the archive have their puzzling aspects. Although the accoutrements and the backdrop suggest a garden setting, the angle of the shot makes the illusion visible. The informal pose and casual attire of the sitter—tieless, legs crossed, socks in need of pulling up—coupled with his intent but somewhat uncertain gaze have the effect of emphasizing his youth. The affect is that of an intelligent but unworldly student, not someone one would necessarily entrust with a sensitive archaeological excavation. And what to make of the signature? The idea of associating a calling card with a studio photograph by this time had a venerable tradition behind it, but inscribing a full-size photograph with one's name smacks more of the celebrity culture then beginning to emerge in Hollywood.[83] If the gesture seems naïve and at least slightly maladroit, Prorok's talents as a

FIG. 1.3. Signed photograph of Byron Khun de Prorok, May 1922. Photographer unknown. Bibliothèque de l'Institut national d'histoire de l'art, Collections Jacques Doucet, archives 106, box 33. Published with permission.

networker belied his innocent appearance: he came to Tunis with a letter of introduction from none other than Merlin, whom he had visited in Paris a few months before. Merlin's letter to Poinssot about that meeting makes it sound more like a one-way conversation than a dialogue, with the experienced archaeologist setting out the lay of the land and making clear that Poinssot's approval would be a precondition for any archaeological work at Carthage; he also took the opportunity to warn Prorok off any dealings with Carton.[84] Merlin made no judgment of Prorok personally—"We've already seen many of his ilk," he wrote—but clearly felt that his promise of American funds could not simply be dismissed.[85] Although Prorok's visit to Tunisia in 1922 left few other traces in the archives, by the fall he was back in the United States lecturing on his adventures and, so he assured Poinssot, raising funds for a future excavation season.[86]

Poinssot was sufficiently impressed with Prorok's energy to express satisfaction at the prospect of his leading further excavations in 1923. This statement—drafted in the name of the French resident general in Tunis, Lucien Saint—came as part of a flurry of correspondence prompted by a report from the French ambassador in Washington, Jules Jusserand, that Robert Woods Bliss, a wealthy American diplomat and philanthropist, was willing to support such a dig.[87] The initial identification of Bliss only as a "Harvard alumnus" created confusion, leading the minister and others to suppose that the actual dig would be carried out under Harvard's auspices. The arrival of the American team in Tunis thus caused some consternation, as Saint wrote to Paris that they had no mandate from Harvard and, moreover, that Prorok seemed inexperienced and unaware of the financial and logistical difficulties of the project he was ostensibly leading.[88]

Thus emerged a deeper problem with the signed photo Prorok had given Poinssot the previous year, what photography scholars might call its problematic indexicality. To put it more simply, who was Byron Khun de Prorok? Was he who he said he was? Was this unlikely name really his? In response to queries, Jusserand dismissed Prorok's claims that he was the adoptive son of a Polish count (Polish authorities could find no trace of any such count, and further stated that Prorok was not a Polish name) and deemed the distant relationship he had intimated with the poet Lord Byron "at the very least improbable." Whereas to the French authorities Prorok vaunted his fictive Harvard connections, during his lecture tour in the winter of 1923 he presented himself to his American audiences—Jusserand enclosed a note from the director of Boston's Museum of Fine Arts inquiring about this—as "commissioner of the French Government" to carry out work in Carthage.[89] Clearly vexed by this array of false claims, Jusserand followed up by writing the president of Yale to inquire about

Prorok's supposed teaching position there, which he was told the young man had never held.[90] As an educated man, moreover, Jusserand could tell when Prorok was blowing smoke about his archaeological work, for example his suggestion that he would find the treasure of the Jews that Alaric, the king of the Visigoths who sacked Rome in 410 CE, had supposedly brought to Carthage. "As Alaric was never in either Jerusalem or Carthage," the ambassador drily observed, "that would clearly be the rarest of finds."[91]

Certainly these accounts led to doubt and hesitation on the part of some French authorities: informed of Jusserand's conclusions about Prorok, the minister of education called for greater prudence in relations with American scholars and universities.[92] The obvious question then arises: how could a young man of uncertain nationality, at best uncredentialed and at worst a con artist, succeed in carrying out excavations at Carthage not only for the 1923 season, but for the next two years as well? That he obtained the essential support of credible American academics slightly modifies the question but does not fundamentally alter it. The answers say much about the emergence of archaeology as a science within a complex concatenation of money, publicity, and imperial authority.

Formal American involvement in excavating Carthage came about through a consortium of the Archaeological Association of Washington, the University of Michigan, and Prorok acting as an independent agent. The association, a chartered corporation in the capital, enjoyed the patronage of both a former secretary of state, Robert Lansing, and the president of one of the Carnegie foundations.[93] The University of Michigan offered some funding but above all the scholarly prestige of Francis Kelsey, a distinguished classical archaeologist. As for Prorok, a proposal for a 1924 campaign assumes he would bring in half of the anticipated budget of $20,000 (approximately $290,000 today).[94] Prorok's letters to Poinssot and others imply that he raised money through his extensive lecture tours, which certainly received widespread publicity.[95] But records in the Kelsey papers in Ann Arbor associate Prorok with a New York businessman named William F. Kenny, who was Prorok's father-in-law. In 1925 Kenny gave $18,000 for the purchase of land at Carthage, a sum close to that allocated (separately) for the whole excavation campaign that year.[96] Surely this alone was a sufficient reason for including his son-in-law, whatever his actual name and credentials, in the dig. Since the documents specify that Prorok would receive no salary for his work at Carthage,[97] it is likely Kenny was also covering his travel and expenses.

The proposed plan of operations for the Americans' 1925 dig included the following provision: "No publicity is to be permitted without the unanimous vote of the Staff, and, in important cases, only after communication by cable or letter with the Chairman of the Committee in Washington." Taking responsibility for the clause, Kelsey wrote Stoever,

> I have added the note in regard to publicity, which I fully believe in; had Mr. Carter kept out of the papers instead of exploiting by publicity the tomb of Tutankahmen [sic] there would have been no trouble in Egypt of a sort either to upset his own arrangements for excavation or to disturb the arrangements of other excavators in that country.[98]

Restricted publicity thus operates both openly and behind the scenes as a pledge of scientific responsibility, attesting that those bound by it know the difference between scholarly reputations, enhanced through publication in appropriate venues, and mere *réclame*, the word Jusserand used to describe Prorok's self-advertising.[99] Yet the Kelsey papers are full of clippings, from local and national newspapers like the *Detroit Free Press* and the *New York Times*, as well as nonscholarly art magazines, describing his finds and their significance. Perhaps he followed his own rules and sought the authorization of his colleagues to give interviews and speak to the press.[100] Kelsey might also have made a distinction between publicizing the results of a campaign and talking to the press during the dig season, or even before it began. But the *réclame* at which Prorok excelled also served a purpose for Poinssot and his supporters.

Perhaps because his penchant for autofiction demanded it, Prorok proved something of a quick study. In May 1923 he gave a lecture at the Sorbonne accompanied by film of the Carthage excavations. Merlin, who was in attendance, noted Prorok's care to credit French scholars for their work, naming himself and Delattre but not Carton.[101] Later that summer Prorok was writing Poinssot of the "disgust" that Jaubert de Benac's article in *L'Illustration* had produced in his entourage, but added reassuringly, "patience, dear sir, you will see that time will prove things very different from the words of Doctor Carton!"[102] Coming in the same letter that announces his forthcoming engagements in England, this seemingly casual reference cast him as a defender of the DAA. In the fall he wrote that his lectures made clear how much work the service had accomplished, as much as Carton's own digs, and framed his requests for photographs as serving the cause: "I can use everything you can send me, discreetly, to counter *his* press campaign."[103] Even if Prorok was at least in part flattering Poinssot's susceptibilities to secure materials he needed for his own *réclame*, the director could hardly have been indifferent to a propagandist as talented and prolific as Carton producing for public consumption words and images defending the service. Of particular value was the Americans' use of the latest technology, notably film, which suggested the extent of their resources and brought a breath of the modern to the Carthage excavations. When Prorok presented his 1923 film and lecture in Tunis, with Poinssot and many luminaries of the French establishment in attendance, he did so in a cinema.[104]

Yet in narrating his North African excavations, Prorok is at pains to demonstrate his knowledge of the operating assumptions and techniques of scientific archaeology, for example, contrasting the old practice of "collection of specimens" with the contemporary goal of "complete reconstruction of the civilisation or conditions of life." To underline that even the smallest fragments can serve this goal, at several points he describes the exhaustive sieving of earth before its final dumping.[105] Prorok also points to the expertise brought to bear on dating and structural analysis, and the importance of careful measurement and recording of sites and objects. A later description of the team's housing, "an old Arab palace at Sidi Bou Saïd," a picturesque village not far from Carthage, emphasizes early hours and posted charts of daily responsibilities.[106] In contrast, Prorok's brief account of lectures in Paris and the US verges on the picaresque, recounting mishaps like a stolen suitcase, technical difficulties, and a locked lecture hall; throughout, he proclaims his preference for scholarly meetings over popular events.[107] It as though he is trying to slough off his reputation as a charming publicity hound and earn that of a serious man.

Ultimately, however, whatever his reputation, Prorok's public relations campaign helped Poinssot only to the extent that it could portray continuing activity, thus countering charges that the DAA was neglecting Carthage. Here the problem remained what it always had been: land. Although Cardinal Lavigerie's plans to construct a new town on church-acquired land in Carthage were never realized, the church's sale of land for development, and its construction of a large basilica at the top of the Byrsa hill, both of which made excavation impossible, exacerbated tensions with the DAA. As a seaside suburb, moreover, Carthage was the object of intense real estate speculation beginning in the 1890s, and the arrival of a streetcar line from Tunis in 1907 marked its effective incorporation into a larger conurbation.[108] As development continued in the 1920s, the high cost of real estate made the acquisition of land for excavations prohibitive.

In their contacts with the Americans, Poinssot and his supporters had always stressed the importance of land acquisition. Recounting his first conversation with Prorok, Merlin told Poinssot that he had said, "The essential thing, if you want to save Carthage, is not to excavate but to buy land, and that's the first purpose to which you should put the funds you will gather; the Antiquities Service will be able to tell you how best to use these funds, and later, if you have enough money, you can excavate."[109] A memorandum drafted by the Washington Archaeological Society's attorney in February 1924 raised the question of whether it could legally acquire land in Tunisia; the question remained unresolved that year, but the plan of operations for the 1925 season included a list of properties that would need to be acquired. Somewhat ominously, given the size of William Kenny's gift of $18,000 for land acquisition, the total funds required for this acquisition came to over $52,000, with no indication of how the additional sums would be raised.[110]

Prorok makes only a few passing mentions of the issue of land in his memoir, preoccupied as he was with actual digging and with the twin acts of recording and publicizing it.[111] Photography and film occupied a major place in the activities of the "American mission." In the spring of 1923 an article in the *Dépêche tunisienne* entitled "Film américain à Carthage" informed readers that they could (already) find photographs of the dig in the windows of its Tunis office, though the excavation was still in its preparatory stages.[112] This press coverage was misleading in at least two ways: by suggesting that the Americans were undertaking "major projects" (*grands travaux*) when the extent of the dig was still highly uncertain, and by crediting the photography program with much greater comprehensiveness than it actually had. A confidential memo for the Washington Archaeological Society listed "lack of proper photographic records, not compensated by moving pictures" as one of the main defects of the 1924 campaign.[113] The fall of 1924 found Prorok, his father-in-law, and Stoever in Tunis trying to negotiate land deals without calling too much attention to themselves, as they feared their known presence would drive up prices.[114] Passing through Paris on his way back to America, Prorok told Merlin that a group of his friends (including Kenny) were contributing a million francs (the approximate equivalent of $52,000) to land purchases, which were being carried out through the intermediary of a prominent French artist living outside Tunis, Rodolphe d'Erlanger. "The project seems to be holding together a little better," Merlin wrote cautiously.[115] In fact it was teetering on the brink.

1924 had been a complicated year for Poinssot. In January, Carton finally achieved one of this long-standing goals: the Foreign Ministry, to which the resident general reported, commissioned an independent inquiry into archaeology at Carthage.[116] Poinssot took this inquiry, which threatened his jurisdiction over the site, as both a professional and a personal affront, an attempt by his adversaries to undermine his position. He had reason to be suspicious: for several years Merlin had been writing Poinssot from Paris about Carton's maneuvers to have himself named director of excavations at Carthage, with his own budget and full autonomy from the DAA.[117] At the same time, one of Carton's outlets, the Carthage Ladies' Friends Committee, or CDAC, tried to claim credit for the American excavations in Carthage and to bring Prorok into their orbit.[118] But the inquiry did not take the exact form that Carton's supporters had proposed, that is a commission of three members of the Académie des inscriptions. Cagnat having discreetly used his influence to quash this idea—as perpetual secretary, he pleaded the press of business—the inquiry was in the end entrusted to one well-known academician, the classicist Stéphane Gsell.[119]

As a former director of the Musée des antiquités algériennes (Algerian Antiquities Museum), Gsell had a general familiarity with archaeology in North Africa, but he had never lived or excavated in Tunisia. On the ministerial North Africa Committee, he had shown himself open to Carton's proposals—he was a friend of both Carton and Bertrand—which had forced Merlin to refute them with more firmness than came naturally to him.[120] Poinssot thus greeted Gsell with something less than collegial warmth: responding to a letter informing him of Gsell's impending arrival, the director wrote, "far from being able to do anything useful here, your coming will make an already deplorable situation even more difficult."[121] Gsell's report, unsurprisingly, put Carton and Poinssot on the same level, calling them "two archaeologists who are furiously over-excited and, to put it plainly, unbearable [*insupportables*]." While recommending that Carton be granted authorization to excavate at Carthage, he dismissed the criticisms leveled at Poinssot and did not endorse the idea of an autonomous excavation director.[122] The director of higher education concurred, writing the minister's chief of staff that, if Poinssot had a difficult character, he had a strong scholarly reputation and the esteem of the AIBL, whereas Carton is "a man of disordered and chaotic mind, rather vindictive, and a mediocre scholar."[123]

Poinssot thus emerged from Gsell's inquiry with his authority more or less intact, and as Gsell's report garnered little to no attention in the press, it cannot have done Poinssot much harm. But given the director's irascibility and sensitivity to criticism, the tenor of the report could hardly have pleased him. True, Gsell acknowledged, as a component of his critique of the DAA's lack of initiative, that Poinssot had inherited an untenable situation from his predecessors, who had failed to acquire land at the site in the 1880s, when it had still been affordable. But even this concession would have been of little consolation to Poinssot, as it would surely have reminded him of the sensitivity of issues of land acquisition and transfer in the Protectorate. Since shortly after their arrival in Tunisia in the early

1880s, French authorities had been working to simplify and Westernize the complex system of landownership in Tunisia, which drew on multiple forms of Muslim law and practice. The French had notably introduced a system of *immatriculation*, or registration, that supposedly shifted disputes over registered land to French courts. But the registration system failed to resolve matters, which were further complicated by French commitments to respect the rights of non-French Europeans in Tunisia to have land cases judged under their own legal systems.[124] In 1911, French attempts to register a historic cemetery, the Djellaz on the outskirts of Tunis, had led to riots, and just over a decade later a newly created, already formidable nationalist party, the Destour, prompted a crisis in relations between the French and the nominal Tunisian ruler, the Bey.[125]

Poinssot's actions in the spring of 1925 make clear his increased sensitivity to questions of property. The 1925 American season—cloaked in the diplomatically correct appellation "Franco-American Mission"— promised to be the most impressive yet; not coincidentally, the addition of Kelsey as general director and Stoever as dig director relegated Prorok to a secondary position. The Americans were also bringing a French academician, the abbé Chabot, a specialist in Punic inscriptions, to assist them.[126] Far from being reassured, however, Poinssot found it necessary, in sending Prorok the official excavation permits, to remind him of the terms of the agreement: "that this is much more about preserving than discovering Carthage," that excavations were to take place only on purchased land, systematically, with restoration and consolidation to occur progressively as part of the work, and that all the earth removed was to be disposed of "*outside* the territory of the ancient city."[127] The director apologized for recalling all this to Prorok, whose good will he did not doubt, but observed that his own previous support for the Americans had "earned me only problems of every sort . . . and the near unanimous criticism of my compatriots." He implored Prorok and other members of the American team to respect local sensitivities and "not to appear to want to monopolize all of Tunisia or even the whole territory of Carthage."[128]

Within a few weeks of the drafting of this letter, the American excavations in Carthage ended in a morass of mutual recrimination. Stoever had clearly riled Poinssot with what he regarded as discourteous behavior, and Kelsey infuriated him by altering a letter the director had written (among other things to excise a derogatory mention of Carton) before sending it on to the Archaeological Society in Washington.[129] But ultimately the dispute came down to land: when Poinssot learned that Stoever had been negotiating to acquire land through promissory lease arrangements (the promise of purchase being conditional on preliminary finds), he accused the Americans of reneging on their agreement to work only on land they had purchased outright and canceled their permits. Kelsey observed that Poinssot's conditions exceeded the requirements of the Tunisian antiquities decree of 1920 and declared that "we came here with no other thought than to support French science in a disinterested and cordially collaborative manner," with no claim to objects for American museums: "we seek no benefit but that to science." Moreover, he warned that their departure might poison Franco-American relations and deter American tourists from coming to North Africa.[130] Rejecting Kelsey's assertion that the different views of land sales resulted from a misunderstanding, Poinssot wrote simply, "Commitments were made that have not been kept." And, he added coldly, "it is impossible for me to accept the 'veiled threats' in your last letter."[131]

The failure of the American mission to Carthage does not lack for signs of cross-cultural misunderstanding. An article in the *New York Times Magazine* the month after the break bears the title "Saving Carthage from Tunis Babbitts." This strenuous exercise in both cultural translation and contextual updating—Sinclair Lewis's novel, the title character of which was a real estate developer, was published in 1922—suggests that a hoard of cynical real estate speculators had despoiled a great archaeological site, a proposition of which neither term holds up.[132] (Prorok later used the trope in his memoir, noting that Carthage was at risk of becoming "Babbitville" and vaunting the Americans' efforts at preservation.[133]) A subsequent confidential report from Kelsey attributes Poinssot's "hostility" to jealousy at the efficiency of American operations.[134] A more fundamental dynamic emerges from the invocation of science in Kelsey's letter, which occurs in two successive sentences. Perhaps meant to recall an earlier cordiality, the repetition suggests the extent of the American's discomfiture. A line from a letter Poinssot wrote before receiving Kelsey's, warning him of the unacceptable behavior ("l'attitude singulière") of his compatriots, makes clear that from the Frenchman's point of view true scientific dialogue was no longer possible: "Whatever disdain they may have for our 'old civilization,' they should have respected certain proprieties and not believed themselves to be in 'conquered territory.'"[135] The blame for injecting national differences into the internationalism of scientific exchange in this view lies firmly with the intruders. Yet this turn of phrase not only seamlessly elides the "old civilization" of the French with that of the native inhabitants of Tunisia, and perhaps of their Semitic predecessors, it also, through an act of rhetorical erasure, obscures the *French* conquest of this very *pays* less than half a century before.

Kelsey would soon regain his equilibrium, writing in his confidential report that "in the relations with Mr. Poinssot . . . he has found a source of amusement rather than of chagrin."[136] In the fall, however, when Prorok, who had expressed his dismay about the conflict to both sides, raised the possibility of approaching Poinssot again, Kelsey responded very firmly that the director "is not a man whose word can be trusted," and no Americans could safely work in Tunisia while he was in charge.[137] But in putting the blame on Poinssot's character, Kelsey glosses over one final element that had widened the gap between the two sides and that Poinssot and his supporters saw as quintessentially American: publicity. In mid-June Chabot, who had maintained polite relations with Poinssot but counseled Kelsey throughout the spring, took the extraordinary step of sending several Tunis dailies a letter critical of the director. The letter, which Chabot had sent privately to Poinssot a few weeks before, broke with the conventions of scientific discourse in several ways, notably by accusing Poinssot of having lied about the Americans' commitments and having projected onto the Americans his own "incredible insolence and discourtesy." Chabot, himself known for a rather brusque manner, claimed to have heard the director say that he wished to have "no Americans at Carthage, and no tourists in Tunisia."[138] A week later, Poinssot availed himself of his legal right of response. Though he admitted to having used the word *muflerie* (boorishness) in a letter to Chabot, he said the abbé was taking it out of context; he otherwise refuted Chabot's letter point by point and accused him, in turn, of lying.[139] Although he usually preferred to rely on surrogates to convey the DAA's position, Poinssot here demonstrates his ability to engage directly with the media when the situation demanded it.

Back in Paris, a flabbergasted Merlin, with whom Poinssot had shared the original letter, could not get over Chabot's last act. "The letter alone was incredible; what can one say about its publication? I am unnerved by such events, which I could never have imagined." He described Prorok, who had come to see him, as shaken and deeply regretful.[140] A week later Merlin wrote in a more reassuring vein: the latest meeting of the North Africa Committee had gone off without incident, a sign that their network was still working. Poinssot had sent him two copies of his published reply to Chabot's letter, and Merlin had passed one on to Cagnat, who was keeping it in case of further ministerial inquiries. But, Merlin wrote, while "a few people in our milieu have some idea" of the problems between Chabot and Poinssot, none had seen or was even aware of the published letters, and it was better to keep it that way. Poinssot would be better off countering whatever impact the incident had had on the resident general, perhaps by having some leftist politicians put in a word.[141] A paragraph followed with advice on the kinds of behind-the-scenes maneuvering through which archaeologists had customarily negotiated with their political masters.

Yet both Poinssot and Merlin must have realized that Chabot's recourse to the press marked a decisive shift in the operation of their networks, one that blurred the line between true scientists and outsiders. In 1923 Poinssot's deputy Raymond Lantier found himself the subject of scurrilous attacks spread by Jaubert de Benac. Among other things the reports charged that he was permitting looting at Carthage, claiming credit for the work of others, and allowing illicit construction on the site through the agency of a corrupt guard. Significantly, such rumors had been circulating in the Tunisian press for a year before they reached the well-connected Salomon Reinach in Paris, who forwarded them to Lantier.[142] Lantier understood this attack as an attempt to provoke a split between himself and Poinssot, but declared that their adversaries would not succeed. All these responses reveal at once the power and the limits of scientific networks in an age of public controversy. "I thought that in our profession," a furious Lantier wrote, "there could be no quarrels other than on scientific grounds. I see that I was badly mistaken."[143] If scientific authority retained considerable respect among the politicians and administrators to whom the archaeologists reported, neither the scientists nor the officials could henceforth ignore public pressure relayed through the press. Poinssot's own experience, moreover, whether behind the scenes or himself writing to the press, makes clear his awareness that publicity constituted an integral part of his archaeological work.

Heritage and/as Science

For Louis Poinssot the events of 1925 meant, practically and one suspects also psychologically, that the Carton nightmare was not over, even though the doctor himself had died unexpectedly in late December 1924. Chabot's letter included the charge that Poinssot had defamed both Carton and his widow, and Poinssot's response, which began by questioning whether Chabot had actually written the letter published over his name, implicitly casting the abbé as a tool of his old adversary, though without naming the latter. (Lantier, who was working at Dougga at the time, wrote him knowingly that Poinssot's measured yet decisive tone would surely not please the Carton faction.[144]) The conflict would take a bizarre turn in March 1926 when the widow, Marie Carton, filed suit against Poinssot on charges that he violently grabbed her wrist on the sidewalk after a heated public meeting about plans for performances in Carthage's Roman theater.[145] The charge led to Poinssot's suspension as director for three months and an administrative inquiry carried out by his fellow directors (essentially the French cabinet of the Protectorate). The administrators having deferred to the court, when Marie Carton's charges were dismissed, Poinssot was reinstated. The press carried few reports of this episode, which lasted less than two months and will be discussed further in chapter 4, but it made clear the depths of the animosity to which Poinssot was subject, and the risks he incurred when venturing into a public arena.

But if Carthage unleashed the most heated and public phase of the quarrel between Carton and Poinssot, their antagonism had other causes and deeper stakes. Carton accused Poinssot, for example, of having covered over or removed labels in the Bardo Museum identifying the doctor as the donor of certain antiquities.[146] The complaint might seem as trivial as the ostensible action; why, apart from personal pettiness, would Poinssot have done such a thing? Here we may look to the signal 1920 decree protecting archaeological heritage in Tunisia, of which Merlin identified Poinssot, then his deputy, as the primary author.[147] Two decades before the equivalent regulations in the metropole, the decree, in nine chapters and seventy-seven articles, declared all artifacts dating from before the Arab conquests, found or yet to be discovered, the property of the state. It also banned any trade in antiquities not declared to the government, and made the director the absolute arbiter of excavation permits (article 33).[148]

A document of sweeping colonial power, framed by the royal we of the Bey at the beginning and the signature of the French resident general at the end, the 1920 heritage code gave the director of antiquities enormous authority over excavations, ruins, and the preservation of finds on private property. Notably, the DAA had the right to intervene on any ancient site at any moment (article 6) and to forbid any construction or renovation work on or in proximity to ancient remains (article 9). An entire chapter (title 4) is devoted to chance discoveries, which were to be declared to the Antiquities Service within two weeks, and for which the finder had the right to a reward amounting to a quarter of the objects' value (article 30; in case of dispute, a curator from a Tunisian museum would arbitrate). Throughout, the decree specified monetary fines in set amounts for infractions of particular provisions.

Significantly, however, the code did not exempt the DAA from the normal rules of expropriation (article 4). The decree was thus calculated to affront colonists more than Tunisian landholders, who were already at a disadvantage in relation to the colonial authorities, and it clearly did so. In May 1923 a French landowner and amateur archaeologist called Paul

Gielly published an article in the *Dépêche tunisienne* with the suggestive title "Carthage and Science Upside Down."[149] Ostensibly inspired by rumored plans to create an oceanographic institute in Carthage, the article charged that "in Tunisia no one has any interest in science, and indeed private initiatives are discouraged in every possible way." Gielly, an amateur archaeologist, owner of the land on which the first funerary urns with human remains were discovered (the so-called sanctuary of Tanit), and an acolyte of Carton, specifically referenced the antiquities decree, which he charged was counterproductive and discouraged reporting.[150]

But the main object of contention between Poinssot and Carton (besides the question of excavation permits) involved the expansive title 6 (articles 39–62), which banned the illicit commerce in antiquities. The doctor was known to have a large collection of antiquities and was even rumored to have obtained some objects he did not himself excavate by passing himself off as an official of the Protectorate government.[151] At the very least, Carton seems to have assumed that most Tunisian workers would not question the authority of a Frenchman in uniform. Even if most of his acquisitions predated the 1920 decree, which was not retroactive, they violated regulations dating back to 1886, as well as norms of archaeological practice; even Prorok in his memoirs decried looting and the informal trade in archaeological "souvenirs."[152] For Poinssot, then, Carton was no more the "donor" of the objects in the Bardo than he was president of the French Republic. And the doctor's collecting practices as much as his ideology account for Poinssot's unusually pronounced personal distaste for him, conveyed in adjectives like *répugnante* and *immonde* (which literally means, among other things, tainted or impure).[153] That Carton was even willing to have himself depicted surrounded by his collection (fig. 1.4), an image imbued with the visual tropes of colonial possession, and one his widow described as very true to life, must have shocked Poinssot deeply.[154]

In his 1924 report on Carthage, Gsell touched on Carton's collection, a matter Poinssot had obviously brought up, and acknowledged that "Dr. Carton could not be the legal owner of the objects he found, even before 1920," so that, "from a strictly legal point of view, M. Carton is in the wrong." But Gsell minimized the value of the objects in Carton's collection, and noted that "these misguided ways, however regrettable, have for quite some time been current practice in Africa." Indeed, Carton's illicit collecting, though well known, had not kept the doctor from receiving official subventions for his excavations.[155] Poinssot, however, conscious of the duty of a department head to set a good example, was not inclined to make exceptions. As one of his defenders wrote of Poinssot at the time

FIG. 1.4. Dr. Louis Carton at his home outside Tunis. Postcard, ca. 1920. Centre des Archives diplomatiques de Nantes, 1 TU1 2079A. Published with permission.

of the Chabot controversy, "In truth this affable man, who is moreover a highly intelligent man, is a lot more than a 'grumpy bureaucrat.' His position as director of antiquities sometimes requires him to put a stop to the machinations of a bunch of profiteers for whom archaeology is only a pretext, and he is obliged to defend himself against their attacks."[156]

What, though, accounts for the differences between Gsell and Poinssot, particularly on the question of strictly enforcing the heritage code? The gap could certainly be attributed to the different locations of the two men within colonial scientific networks. Gsell was speaking from the lofty vantage point of the imperial center of calculation (although he had spent many years in Algeria), whereas Poinssot was preoccupied with conditions in the field. A draft letter from Poinssot to Cagnat thanking him for news of his election as an associate member of the Academy of Inscriptions in 1934 is framed in uncharacteristically humble terms: clearly he recognized that his own reputation was bound up with his performance as a colonial functionary.[157] Poinssot's general attitude may also evince some of the distrust, even contempt, toward those whose work takes place only on paper, which Latour describes as characteristic of scientists in the field.[158] Although Poinssot never begrudged Cagnat or Merlin the prestigious positions in the metropole their earlier fieldwork had earned them, his own position was firmly in the field.

Yet one could also argue that Poinssot had the more expansive sense of archaeology as colonial science, in which the comprehensiveness of the data predefines the terrain of calculation. As science, first: Latour stresses the vital role of scientists within networks who "make traces of all sorts circulate better by increasing their mobility, their speed, their reliability, their ability to combine with one another."[159] In this regard the creation of massive inventories and catalogs, such as the multivolume series of catalogs of North African archaeological collections shepherded by Cagnat, played a role as important as the collections themselves. Indeed, in Latour's terms the publications' importance would arguably be even greater than that of the collections, since a center's position of dominance depends on its capacity to absorb information in a form abstracted from the physical objects or phenomena it describes.[160] Even so, in archaeology, contrary to so-called universal sciences based on replicable experiments, the pertinence of the objects does not disappear once they have been tabulated and recorded. Hence the importance of defined channels of acquisition and storage, under the control of a supposedly dispassionate authority acting in the public interest. Exercising authority in the realm of archaeology involves three steps: defining, according to criteria often extrinsic to the society in which it takes place, a domain worthy of protection; establishing a legal and administrative regime that exerts control over objects within this domain; and surmounting the opposition that such regimes invariably provoke. The value to imperial rule of enforced regulations in archaeology, a domain involving disruption of the colonized terrain and of its native inhabitants, lies precisely in the appearance of impartiality and legality that authorities seek to project.

This appearance served to legitimate professional authority, but it came at a cost, which many scholars will find familiar. Poinssot's correspondents, notably Merlin and Lantier, contrasted the real *work* of archaeology with the minutiae of administrative routine, often characterized as *besogne* or drudgery. Whatever bodily discomforts, labor disputes, and other problems fieldwork may entail (to be discussed in chapter 3), archaeologists never sound as happy as when they are in or adjacent to the field. They acceded with fairly good grace to demands by scientific networks that they regard as necessary and that offer real rewards, such as preparing oral reports to the Academy or to committees with grant-making authority. But directors (Lantier, on his return from Tunis, became a curator and eventually director of the National Antiquities Museum in St. Germain-en-Laye) also had other time-consuming duties, as much at the margins as on the front lines: negotiating budgets, staffing basic operations and seasonal digs, escorting visiting dignitaries to excavations, obtaining medals for support staff, and attending endless meetings. In other words, the emergent "ecology of practices" of archaeology, to use a term with some theoretical currency, includes not only field and museum practices, such as site photography, stratigraphic analysis, and cataloging, but bureaucratic procedures that make the rest possible and occupy a prominent place in the archaeological archive.[161] The distinction between the two, and the strong preference for fieldwork over the rest, itself becomes a characteristic of the archaeological habitus, Bourdieu's influential term for an informal system of perception inculcated alongside formal training. Poinssot's frustration with *besogne* spilled over into a long-running dispute with his immediate superior, the Protectorate's director of public instruction, Henri Doliveux, whom he regarded as a brainless mediocrity—and who was responsible for Poinssot's suspension and administrative review in 1926.

The distinction archaeologists made between work and drudgery should not be taken at face value any more than that between science and publicity. Recalling the seasonal patterns of correspondence discussed earlier, Latour observes that what we call information is a compromise between presence and absence.[162] The tradeoffs scholars make in publishing their work, sometimes sacrificing length and the quality of illustrations for speed, or vice versa, form part of the habitus as well. Most striking, however, and most decisive in the entanglement to which I referred at the beginning, is the melding, over and above such distinctions, of scientific and colonial authority. Part of Cagnat's and, after he returned to the metropole, Merlin's credibility as defenders of Poinssot came from their familiarity with the exercise of imperial power, circumscribed as it inevitably was. Their *inclination* to defend him, though, came from the habitus of archaeology they shared, shot through with the conviction that their true authority came from their scientific vocation and the ability to balance it with the demands of a government position. For them, this complex and delicate concatenation of authority distinguished them at once from Carton, who in the 1920s, after his retirement, had no public responsibilities; from the Americans, whose ample funding appeared to the French largely free of constraints; and from their administrative superiors, who were not real men of science.

Latour notes that the extension of science to new geographic locations is in some ways deceptive: one of the purposes of scientific networks is to convey a false sense of omnipresence, for "scientists travel within narrow and fragile networks, resembling the galleries termites build to link their nests to their feeding sites."[163] The directors of antiquities in Protectorate Tunisia understood the fragility of their authority, taking it as simply a means to a scientific end, that of understanding the past through its systematically discovered material remains, and thus one that could find proper employ only in the hands of people like themselves. A conscientious regard for what was framed as the public interest counted, on this view, for more than sometimes messy practice; the distastefulness of this precarious equilibrium perhaps explains the indulgence of Poinssot's supporters for his most intemperate displays. Carton and his defenders they saw as inverting the relationship, making science an instrument of colonialism. The complex legacy of the 1920s Carthage controversy for archaeology and for heritage fields in general lay not so much in the exercise of colonial authority but in the scientific principles that made such authority appear more reasonable, embodied and modeled in a certain professional discretion. As the next chapter will show, the omnipresence of the press and other forms of publicity both put a premium on that discretion and made it increasingly difficult to come by.

✶ 2 ✶

The News from Glozel

SCHOLARS, MEDIA, AND THE
MAKING OF A SCANDAL

Two men stand in a pastoral landscape, a rolling hillside visible in the background (fig. 2.1). Their baggy jackets and trousers, their wooden shoes—reminiscent, along with their downward gaze, of Jean-Francois Millet's celebrated *Angelus* (1855–1857)—clearly identify them as farmers. To dispel any doubt, *L'Humanité*, the Communist party daily that published this press agency photograph in early January 1928, refers to them as "peasants" (*paysans*). Readers of *Paris Times*, a short-lived newspaper catering to the city's Anglophone population, which published the same photo, would learn that the two figures, Emile Fradin and his grandfather Claude, were examining an object found on their property in Glozel, a remote hamlet in the Auvergne. The object came from an archaeological excavation that, according to *Paris Times*, had prompted "the controversy which seems about to seek legal solution."[1] *L'Humanité* left no doubt as to its position on the controversy, joining those who believed that the many objects unearthed there, far from being Neolithic, as some scholars maintained, were forgeries. Both the headline and the caption point out that the Fradin family had opened a museum to house the finds, the latter mocking the object they contemplate by comparing it to a potato or a "giant truffle" and noting the admission fee of four francs.[2]

The reference in *Paris Times* to "the controversy," rather than simply *a* controversy, suggest that its readers would already be familiar with Glozel. In the spring of 1924, the two men shown in the photograph were plowing their land when they accidentally uncovered what appeared to be a burial site of great antiquity. After initial consultations with local teachers, the Fradins rented the site to Antonin Morlet, a doctor in the nearby spa town of Vichy and amateur archaeologist, who in late 1925 began publishing the finds. As news of the discoveries spread, Glozel became controversial, with a number of distinguished scholars casting doubt on the site's authenticity; the controversy prompted extensive press coverage, not just

FIG. 2.1. Claude and Emile Fradin examining an object at Glozel, late 1927 or early 1928. Agence Meurisse photo. Source: gallica.bnf.fr/BnF.

in France but internationally. Most of the objects unearthed at Glozel seemed to belong to the late Neolithic period, perhaps twelve thousand years BP (before the present). Yet Morlet's finds also included two types of objects that archaeologists had until then found only in much later sites, typically from the Bronze Age (ca. 6000 BP) or later: ceramics and, most stunningly, small stones with what looked like a form of alphabetic writing.[3] Although the object the Fradins are holding in the press photograph is hard to discern, it is likely one of those engraved stones. If authentic, the Glozel finds would revolutionize archaeologists' understanding of the beginnings of the alphabet and would make a plausible case that Western civilization emerged not where scholars had long placed it, in the Fertile Crescent, but millennia earlier in central France.

But the stakes of the controversy extended well beyond the realm of prehistory. Even more than Carthage, Glozel tested the ability of scholars and institutions to manage disagreements in a civil and credible way, and of various media, from scientific journals to the general-interest press, to relay those disagreements accurately and without sensationalism. These were not new issues for archaeologists or for the scholarly community more generally; Alex Csiszar has shown how conflicts over the manners and venues of publishing divided the Paris scientific community in the 1820s.[4] Yet issues of civility and media coverage loomed large in the years following the Great War, as France faced a difficult reconstruction and doubts about its ability to meet the challenges of the postwar world. As the controversy intensified, scholars and journalists alike worried that it was discrediting French science in the eyes of the world. Archaeologists, their attitudes toward the press a mixture of grudging respect, resentment, and opportunism, feared becoming a laughingstock. When in the fall of 1927 Morlet and the Fradin family agreed to a verification dig to be carried out by an international group of archaeologists, Morlet wrote Auguste Audollent, dean of the Faculty of Letters at the University of Clermont-Ferrand, saying,"I have insisted that the press be present, to make sure that everything takes place in broad daylight [*au grand jour*]." Morlet had received warnings from his supporter Salomon Reinach that some commission members were convinced skeptics of Glozel, but he expressed confidence that, in open proceedings, "it is impossible for truth not to triumph over the cabal."[5]

A few glosses make clear both the similarities and some notable differences between Glozel and Carthage. Morlet, like Carton, was both a medical doctor and an enterprising amateur archaeologist. Reinach, who appeared in passing in the previous chapter as a young archaeologist in Tunisia, by this time directed the National Antiquities Museum in St. Germain-en-Laye, just outside of Paris, and as a member of the Académie des inscriptions et belles-lettres (AIBL) had emerged as Glozel's most prominent defender.[6] Even the idea of a cabal could be seen as an echo of Francis Kelsey's belief that Poinssot owed his job to powerful protectors, though as Glozel was a private excavation without formal institutional relays, the power dynamics of the two cases diverged substantially. The most striking difference, however, lies in the idea of the press, at least rhetorically the bane of serious archaeological scholarship in Carthage, as the guarantor of transparency at Glozel. From the first, Morlet had been publishing his finds not in scholarly journals but in a high-end literary

FIG. 2.2. Salomon Reinach at Glozel. Postcard, April 1928. Public domain.

and political magazine, the *Mercure de France*, and his correspondence manifests a constant concern with press coverage of the excavations. As the controversy developed, the press would become its central arena.

The invention of the telegraph in the mid-nineteenth century and subsequent founding of international wire services radically transformed the dissemination of news, above all by separating news transmission from physical transport.[7] Other technological advances, such as the introduction of halftone and rotogravure processes to facilitate and improve the quality of visual images, also led to significant innovations in the media.[8] In a recent study of news in colonial Algeria, Arthur Asseraf observes that new media do not immediately displace the old: rather, when newspapers arrive, they add to existing forms of circulation, many of them oral, such as gossip, rumor, manuscript, and song.[9] News of the Glozel discoveries spread first by oral report to schoolteachers and a small-town mayor, then by letter to members of the local learned society, from its bulletin on to wider regional networks, and then, once Morlet had published his first fascicule of results, to regional, then national and international newspapers. Originally published in the Paris daily *L'Intransigeant*, the image of Salomon Reinach on a cart pulled by cows (fig. 2.2) indexes the continued importance of a slower means of communication, notably what we now know as snail mail, alongside more rapid transmission of the news.[10]

Jason Hill and Vanessa Schwartz observe that, "whatever its subject, once generally known, news is thought to make an impact on those who learn about it—the news is a business, but at its best it is also a catalyst generating discourse, dialog, and debate about what matters in the present."[11] The news about Glozel certainly generated considerable debate; one of the questions this chapter considers is not only why but how, beyond the scholarly community most concerned with it, it came to matter in 1920s France. Pointing to the business of the news, in a period scholars have identified as one of relative stagnation for the French press,[12] only redirects the question: what was it about Glozel that made editors and publishers see it as a salable story over several years, a long time in their business?

Part of the answer has to do with the relatively new field of prehistory, which had attracted much attention in France in the relatively brief period since its initial elaboration in the mid-nineteenth century.[13] Scholars and antiquarians in a number of fields had long had the intuition that the earth was much older, and human habitation much more ancient, than biblical exegetes had calculated. But prehistory as an archaeological subfield can date its origins to 1859. In that year, which also saw the publication of Darwin's *Origin of Species*, British geologists validated the claims of a French customs inspector, Jacques Boucher de Perthes, to have found animal remains, pottery, and stone tools from what is now called the Paleolithic period seven meters below the town walls of Abbeville, in northern France.[14] Important discoveries would follow apace throughout Europe, with the Dordogne department in south central France one of the focal points (the Cro-Magnon skull, unearthed by Louis Lartet in 1868, takes its name from a cave in the Dordogne village of Les Eyzies). As in 1859, and as in other fields of archaeology, prehistory would become the object of international competition as well as cooperation, although differences of interpretation and chronology did not always follow national lines.[15] New and unprecedented discoveries, such as that of the paintings in the caves of Altamira, Spain, in the late 1860s, often prompted considerable skepticism and even charges of forgery, as they would continue to do through and beyond the discovery of the Lascaux cave paintings in 1940.[16] Many of these developments, ranging from scholarly disagreements to public controversies, garnered considerable media attention. Prehistory, indeed, seems to have been born and developed on the highly fungible border between science and spectacle.

In terms of the transmission of scientific knowledge about prehistory, the creation of specialized journals, which could subsequently be summarized or repackaged in more general-interest periodicals, played a key role.[17] But prehistory also regularly attracted attention in a range of popular media, including novels, paintings, and the graphic arts, as well as the newspaper press. Prehistorians, moreover, did not always make a clear distinction between scholarly and popular outlets.[18] Archaeology was not the only field where this was the case: the historian of science Joshua Nall has persuasively argued that debates over the techniques and findings of Mars observation in the late nineteenth century were intimately tied up with debates over the appropriate forms and venues of scientific communication. Nall has coined the term "event astronomy" to describe how Mars's predictable but discontinuous periods of proximity to earth not only facilitated observation but lent themselves to newspaper coverage and encouraged close, mutually beneficial ties between astronomers and print outlets.[19] Archaeological discoveries did not follow the same predictable rhythms as those in astronomy, but a case like Glozel posed similar dilemmas to those faced by astronomers a generation earlier. For archaeologists and astronomers, the questions included how to move from local discoveries to larger insights and how to present these observations in the public sphere, while journalists debated how to report controversy in a manner at once accurate and interesting to the public. In archaeology as in astronomy, the interweaving of technology and discourse, and the limits they placed on each other, played a crucial role not only in transmitting but in constructing knowledge.[20]

In the Glozel controversy, the processes of transmission played out in an unusually open and self-conscious fashion, as scholars, journalists, and interested members of the public hotly debated not only the different versions of the past that divided them, but the channels and language they chose for their dissemination. Within a year or two of the Carthage controversy, Glozel presents another case of scientific networks trying to cope with the glare of publicity, but with several key differences. First, the strangeness of the Glozel finds led to the involvement of scholars in a number of fields, hence "networks" in the plural: prehistory, in France largely associated with anthropology, but also classical archaeology, epigraphy, and even geology and forensic chemistry. Although intersecting at points and via individuals, each of these fields occupied its own habitus, differing in methods, epistemology, and traditions. Second, the disruptive potential of Glozel, which threatened to overturn the chronology of human cultural development as then understood, invited discussion of even larger issues than those raised in Tunisia. Certainly Glozel offered archaeologists an opportunity to make a case for the scientificity of their work, and to debate their methods on a very public stage. But discussions of Glozel also entertained, implicitly but also often explicitly, questions about the nature of scientific learning, the course of civilization—whether it moved from east to west or vice versa—and popular understandings of both. The scope of these questions offers an indication of the enormous stakes of the controversy, which—a final difference from Carthage—ultimately overwhelmed the ability of scientific networks to contain it.

The Glozel Archive: Exit the State

Two documents frame the extremely limited state involvement in the Glozel controversy, limits that in turn had important consequences for the shape of its archive. The first, an official circular from the minister of education and fine arts, Léon Bérard, asked that public schoolteachers "collaborate in the search for and study of prehistoric monuments." Dated 11 January 1924, the circular contained a questionnaire that all primary schoolteachers were asked to fill out and return to their regional inspector, beginning with the most general questions ("Does the commune have any *grottoes* or *caverns*?") and moving on to the more specific ("Are there *prehistoric sites* [*stations*] offering stone tools? What are the physical forms of these tools?"). Just over two months after its publication, the primary schoolteacher in the Auvergne village of Ferrières-sur-Sichon, Adrienne Picandet, referenced the circular in a letter to the regional inspector announcing the first Glozel finds.[21] The initial excavations were then carried out by the young Emile Fradin, the member of the family responsible for first finding the site, in concert with a teacher from the local middle school, Monsieur Clément.

The second document, a handwritten note on the external folder or *bordereau* containing the file on the Glozel affair in the Archives Nationales de France reads as follows: "M. de Bar/Glozel/dossier to be filed [*à classer*]/keep only the clippings."[22] De Bar was the head of the Higher Education Bureau at the Ministry of Education, in charge of the government's research programs; the note instructed him, or perhaps relayed his instruction, to file away the report, indicating that the affair no longer required the attention of the state. This is an archival moment of dismissal, an almost semaphoric message to the future to look elsewhere. Although it is undated, this notation probably was made after the ministry's Commission on Prehistoric Monuments, source of the 1924 circular, decided not to list—another sense of *classer*—Glozel in January 1928.[23] At this point the state officially had no further interest in Glozel, a private excavation taking place on private land; as one official had noted earlier, "is it really up to the government to seek and proclaim an official truth in a scholarly discussion? I do not think so."[24] Newspaper reports presented the government decision as a blow to Glozel; *L'Oeuvre* titled its story "Glozel Does Not Merit Listing." But a few months earlier Morlet had publicly protested the prospect of government listing; though he claimed he would welcome the listing itself, he feared being removed from the "scientific headship" of the excavations. When the ministry's decision was announced, he privately exulted, saying in a letter to Reinach that "happily for us, hatred blinded the members of the Prehistoric Monuments Commission!"[25] As a consequence of this decision, both the national archives and the departmental archives of the Allier in Moulins house only slim official dossiers on Glozel. The archives of the Historic Monuments Service of what is now the Ministry of Culture have a somewhat more comprehensive file on Glozel, but it too terminates with the decision against listing, for which this agency was responsible.[26] But what about that notation "keep only the clippings"? Why clippings? Why *only* clippings?

In another unexpected archival moment, I originally found Morlet's public statement not in a folder of clippings but cited in a letter between two academicians, both skeptics of claims that Glozel was Neolithic. Writing the classicist Camille Jullian at his country retreat near Bordeaux, his younger colleague René Dussaud, at the time just finishing a short book charging that Glozel was a hoax, copied out a paragraph from Morlet's letter to the *Petit niçois*. For whatever reason, Dussaud preferred to keep the article for himself rather than enclosing a clipping of it, a widespread practice at the time. The sending of clippings clearly belongs among the "processes of storage and transmission" that have prompted the so-called material turn in media theory and science studies.[27] These are practices of archival accumulation in which archaeologists, among others, create layers of knowledge that, whether intentionally or not, reproduce the sedimentation of their objects of study.

In October 1927, Dussaud took out a subscription to the Argus de la Presse, a commercial clipping service.[28] Dussaud, then in his late fifties, was a busy man: curator of ancient Near Eastern art at the Louvre and a professor at the École du Louvre, like Reinach and Jullian a member of the AIBL, and editor of the leading journal in his field, *Syria*. Like many clients of the Argus he was using it to keep track of his own achievements, timing his subscription to coincide with the publication of his long essay in pamphlet form on the Glozel controversy. But the scope of the selection went well beyond his own publication to include any article with the keyword "Glozel." Within three weeks, Dussaud wrote Jullian, he had received 728 clippings.[29] By the time he let his subscription lapse, at the end of February 1928, the pile of clippings numbered nearly fifteen hundred individual items, from newspapers all over France, Europe, and the French colonies, in nine languages besides French. In addition, Audollent's papers in Clermont-Ferrand, the gift of his family to the Regional Archaeology

Service, and the deposited archive of the Société d'Emulation du Bourbonnais in the departmental archives of the Allier in Moulins, contain considerable quantities of privately amassed clippings.[30]

The sheer bulk of the assemblages, as well as their retention for decades in cupboards and attics, records the importance, the weight of the affair as it unspooled, justifying or excusing the clipper's compulsion as a small part of a national obsession. Whatever the present whereabouts of clippings and other printed matter, they clearly constitute one version of the documentary genre as Lisa Gitelman has described it: "ongoing and changeable practices of expression and reception" recognizable "at once and also across time."[31] Here, the masses of paper seem to say, is what it was like to live through Glozel, to try to keep up with it, to master its surprising twists and turns while also following the general news. More simply, here—somewhere, buried in these piles—is the truth, as well as, simply as a logical proposition, much fiction.

Among the clippers, only Dussaud's motives can be identified with any clarity: his letter to Jullian suggests that he was tracking the public response to Glozel as an ongoing battle between two sides, and he expressed satisfaction when the tide seemed to be turning in the anti-Glozelians' favor. Others shared clippings as a form of information exchange. Reinach's correspondents in the provinces, for example, made sure he saw what they considered significant articles or letters in the local press, while he occasionally shared with them clippings from low-circulation establishment newspapers, notably the *Journal des débats*, that might be difficult for them to find. Like attachments or hyperlinks, the clipping could serve several purposes beyond the merely informative. Emile Espérandieu, by late 1927 one of only two other members of the AIBL to share Reinach's belief in the authenticity of Glozel, wrote him from Nîmes with a clipping from a Toulouse paper that he thought would boost Reinach's morale. "What a fine apostle!" he wrote of its author: the clipping as exclamation point.[32] In contrast, Denis Peyrony, director of what is now the national museum of prehistory in Les Eyzies (Dordogne) and a member of the international commission, wrote Reinach to complain about defamatory comments made by Morlet. His brief letter begins by asking sarcastically, "Do you not read the *Dépêche de Vichy* or *La rumeur*? One must have very little to support one's argument to get to that point."[33] If Peyrony here seems to be disparaging the provincial press and by extension the clipper's obsession with it, he also enclosed a clipping, from the *Courrier du Centre*, designed to demonstrate Morlet's lack of credibility. This is the clipping as exhortation, paralleling Peyrony's pleas to Reinach to listen to other voices.

The circulation of clippings thus illustrates several dimensions of the ways scholarly networks functioned in and around the Glozel affair. Ultimately, as the next sections will show, archaeologists and other scholars blamed the media they so assiduously consumed for the breakdown of those networks. But an examination of their correspondence through a diachronic lens reveals a more complicated picture. At the beginning, and for some scholars throughout the affair, scholarly networks, including both personal letters and limited-circulation publications, offered what they always had: a means of seeking and transmitting information at a time when it was still very unequally distributed. As the controversy grew, however, not only the range of publications but the functioning of those networks changed substantially.

Exchanging News

"I was one of the first to go to the site, I've followed the matter as closely as possible, and I am very perplexed."[34] So wrote a lawyer and officer of the Société d'émulation du Bourbonnais, Joseph Viple, in May 1926 in response to a request for information from Camille Jullian. Viple enclosed a copy of an article he had published in the society's journal. Many people responded with perplexity to the first fascicule published by Morlet in the fall of 1925, which contained both the kinds of stone tools archaeologists would expect to find in a Neolithic site and many strange objects, including what he described as a "neolithic idol representing a rudimentary human figure atop the phallic organs." But the "bricks with alphabetic signs," of which the publication illustrated four, plus several fragments, generated the greatest confusion. In his conclusion, Morlet argued that the finds could only date from the early Neolithic, "most likely between the 10th and 8th millennium [BCE]."[35] This was millennia before the time, around 1300 BCE, that most scholars believed had seen the emergence of alphabetic writing in the eastern Mediterranean.[36]

Jullian's central position in the scholarly networks that would become enmeshed in the Glozel controversy stemmed from several factors, beginning with his scholarly distinction and celebrity as a public intellectual. From a modest background in the south, he exemplified the meritocracy of the Third Republic, moving from public schools in Marseille to the École normale supérieure, then coming in first in the national *agrégation* competition in history and geography in 1880, at the age of twenty-one. Having studied in Berlin with Theodor Mommsen, Jullian was committed to the rigorous practice of history as a science. Although for most of his career not an active archaeologist, he was at pains to include the latest archaeological finds, including those from prehistoric and protohistoric sites, in his magisterial eight-volume *History of Gaul*, of which the final volume appeared in 1926. In 1905 Jullian took up a chair in "national history and antiquities" at the Collège de France, traditionally regarded as the apex of the French university system. In addition to election to the AIBL in 1908, routine for someone of his reputation and career stage, in 1924 he was also elected to the Académie française, a more unusual honor for an academic.[37]

It was therefore entirely natural that Morlet should inform Jullian of his finds and seek his assessment quite early in the excavations. Though Audollent, another distinguished epigrapher, was much closer to hand in Clermont-Ferrand, around sixty kilometers from Vichy, Morlet dispatched his initial letter to him, with a copy of the first fascicule, only on Jullian's instigation.[38] Yet Jullian owed his centrality not only to his eminence but to his determination to approach Glozel as a serious scholarly problem, one demanding wide reading and consultation. His abundant correspondence makes clear that Glozel quickly became something of an obsession for Jullian. In a letter to the president of the Société d'émulation thanking him for the contacts he was providing, Jullian wrote, "I'm going to do everything possible and impossible to shed light on this affair of the Glozel bricks, so mysterious—and so important, if one could believe in it without reservation."[39] Jullian shared readings of the "mysterious" inscriptions published in Morlet's fascicules and in the *Mercure de France* with his academic colleagues, trying out ideas and soliciting their suggestions. Letters sent to him often contained sketches, facsimiles of the Glozel signs, and references to scholarly publications; in some instances, Jullian seems to have asked for specific citations of works unavailable in Bordeaux, the city nearest his rural retreat. Albert Grenier, a younger scholar who considered himself Jullian's disciple, offered research assistance from his post in Strasbourg, where he was able to consult books and articles of German philologists in that city's fine university library. Jullian sent photographs of the Glozel tablets, his own readings of the alphabet, and letters from other scholars; Grenier sent these materials back with his own readings, generally supportive of the master's.[40]

Alone among the protagonists in the controversy, Jullian corresponded extensively with major figures on both sides: Morlet, Reinach, and Espérandieu among the Glozelians, Dussaud among the critics. Writing from Nîmes, where he was curator of the archaeological museum, in September 1926, Espérandieu enclosed a sketch of the most recently discovered tablets and speculated that unpierced beads found at the site might be "analogous to the cowry shells of the backward peoples of our time."[41] In the fall of 1925 Morlet had been sending Jullian photographs and sketches, and even a fragment of a floor tile (*dalle*) from the site, though he planned to recuperate the latter; he mentioned that he had also shared a fragment of one of the tablets with the noted prehistorian Henri (abbé) Breuil.[42] A year later, however, Espérandieu could not satisfy Jullian's desire for more samples: he had received only a few objects on his own visit to Glozel, and had given several, including a "phallic idol broken in several pieces," to the Nîmes museum of natural history. A tablet Espérandieu had himself unearthed at Glozel had crumbled to dust in the course of a bicycle ride to Vichy. But Morlet had sent Espérandieu another one, which he had in front of him while writing, and "there is no doubt whatever of its authenticity."[43] A few weeks later Audollent wrote Jullian in similar terms, and with more detailed explanations: "This excavation is perfectly unadulterated; neither deceit nor tampering."[44]

These comments indicate another reason for the breadth of Jullian's correspondence: his colleagues were lobbying him. Soon after learning of the discoveries, and certainly by the summer of 1926, according to a letter he sent to Morlet, Jullian had concluded that some of the tablets were likely fakes, and that those that were authentic dated to the Gallo-Roman period, much later than the Neolithic.[45] In November 1926, Jullian revealed his position in various forums: a letter to the establishment daily *Journal des débats*, public meetings of the AIBL, and a cover article in a new general-interest weekly, *Les Nouvelles littéraires*.[46] Glozel, he believed, had been the den of a Gallo-Roman sorcerer, and the inscribed tablets, which he insisted were in Latin cursive, were spells. A postscript in the *Nouvelles littéraires* dispatched contrary arguments in a few sentences.[47] A few weeks later, just after New Year's, a member of the Société d'émulation, Count de Lacarelle, reported to his cousin, the society's president de Brinon, about a long conversation he had had with Jullian the previous evening in Paris. Jullian greeted his interlocutor warmly, but the latter wrote, "on the question of whether one can have a conversation with him, I would say it's very difficult, because he has much to say on the subject and is always *totally positive*." Over the course of their conversation, however, Lacarelle found Jullian's explanations less than convincing, occasionally almost laughable.[48]

From the beginning the Glozel finds had aroused skepticism. In his May 1926 letter to Jullian, the local scholar Viple wrote, "The discovery of the bricks would need to be clarified. First one brick with inscriptions is found, then a second, then a third, and now they're saying thirty! Exactly how many have been found and how have they been found?" And he went on to point the finger: "It appears that it's always the young Fradin, by himself, who finds them."[49] Morlet would later argue that Viple's doubts sprang from the Société d'émulation's bitterness toward the Fradins, who refused to accept association members' claims to have discovered the site. This dispute, which constituted the first phase of the Glozel controversy but was not publicized until later, culminated in Morlet's resignation from the society in March 1926.[50] In fact, in the first year or so of excavations, local antiquarians tended to defend the authenticity of the finds against the skepticism of outsiders. Responding in August 1925 to a letter in which Espérandieu had expressed doubts about the inscriptions, de Brinon wrote that "deception appears to me completely impossible."[51]

Even Salomon Reinach, who would become Glozel's most prominent defender, at first expressed skepticism.[52] Although in the summer of 1926 he found in Jullian's preliminary readings evidence that the tablets were authentic, he wrote Jullian a few weeks later, "Cutting one's losses is impossible: *you have to accept it whole or leave it.*" While others he named had made a decision, "*I oscillate*, which is unfortunate."[53] Clearly ready to be persuaded, in late August Reinach visited the excavation site himself and was, he wrote Jullian again, unalterably convinced of its authenticity.[54] Yet at the same time, a friend who accompanied him, the historian and Egyptologist Seymour de Ricci, came to the opposite conclusion, declaring Glozel a fraud.[55] Little more than a week later, before Reinach could report to the AIBL, the Paris daily *Le Journal* published an extended article by a reporter on the scene, Lucien Chassaigne, who told the story from the beginning and framed it as "a major scientific controversy."[56] Then, on 9 September 1926, a prehistorian from Toulouse, Henri Bégouën, published an open letter to Reinach in the *Journal des débats* calling for an international inquiry into the authenticity of the Glozel site and finds. The Glozel affair was well and truly launched.[57]

On the surface archaeology's scientific networks were operating within prescribed norms in the fall of 1926, with the AIBL at the center of what was increasingly becoming an international discussion. The local teacher Clément, who aided Emile Fradin with the excavations from the time of the first discovery in 1924 until Morlet's arrival a year later, had wanted nothing more than to report on the finds to the Academy. First he imagined himself giving the report, then entrusting the task to the prehistorian Louis Capitan, a professor at the Collège de France, before having to give up the idea entirely in the wake of the dispute between the Fradins and the Société d'émulation.[58] Members of the Academy, their instinctive reactions not yet transformed into positions, wrote each other politely, bids for support for the most part just below the surface. Edmond Pottier, an academician and well-named curator of ancient ceramics at the Louvre, wrote Reinach in mid-September 1926 that while he was inclined to accept the attestations of those who had seen the finds (which he, Pottier, had not), when it came to dating them, it would be impossible to find an answer. For this he used an old French idiom for staying mum: "we give our tongues to the cat."[59] Although a longtime friend of Reinach and a ceramologist of towering reputation, Pottier would remain one of the rare agnostics for the duration of the controversy. His silence reflected a commitment to the civility long considered essential to the functioning of scholarly networks.[60] In the fall of 1926 that civility still held, but it and the networks it undergirded were beginning to fray.

Even those with firmer convictions strove to maintain this civil tone. Returning from a trip to the Iberian Peninsula in May 1927, Bégouën noticed that in his absence Audollent had published a response to an article on Glozel that Bégouën had published in La vie catholique. Although the two were not acquainted, Bégouën wrote Audollent a letter, recounting his conversations about Glozel while abroad, and expressing mild surprise that Audollent had not accepted Jullian's views. "You will excuse me," Bégouën wrote, "for having contacted you directly, but it seems to me that it is the fairest [le plus loyal] and most courteous means of discussion."[61] In his reply Audollent said that he had, in responding to Bégouën's article, been mainly concerned to refute the charge that Morlet was allowing access to the dig only to his friends. Bégouën, in response, wrote that "In any discussion it is important to me to remain polite, above all when dealing with learned men like you, whose knowledge and character I hold in regard." After detailing his concerns about access to Glozel, he concluded, "Let us continue to discuss the *recognized* facts in all fairness and without passion."[62] That Bégouën found it necessary to articulate this point, generally an unspoken subtext of scientific exchange, hints at the upheaval Glozel was already causing.

Several of the substantive aspects of this exchange stand out, beginning with Bégouën's lecture tour of Spain and Portugal, where he had expressed his doubts about Glozel's authenticity. French archaeologists on both sides of the Glozel controversy vied for the support of their Iberian counterparts, whose work excavating and interpreting objects and sites of recognized antiquity in the peninsula had earned them a strong reputation in prehistory. In his initial letter to Audollent, Bégouën noted that he had been the first French scholar to examine inscribed prehistoric tablets from the Alvão caverns in northeastern Portugal: "They are really curious and odd, but they have no relationship to those from Glozel." Bégouën claimed that five leading scholars based in Spain and Portugal shared his views; only José Leite de Vasconcelos, the dean of Portuguese anthropology, refused to express an opinion, saying that more methodical excavations would be necessary.[63] The Portuguese anthropologist António Mendes Corrêa wrote Jullian around the same time that the Alvão inscriptions were probably later than Neolithic and could in any case not be used to date those from Glozel without archaeological context.[64] At the time of Bégouën's visit, Mendes Corrêa was persuaded by his French colleague's lectures, but after examining the finds in situ a few months later he changed his mind and accepted Glozel's authenticity.[65] Supporters of Glozel used chemical tests carried out by Mendes Corrêa, a physician with knowledge of biological anthropology, as well as by colleagues in Norway and Sweden, to argue for the antiquity of the finds, but opponents dismissed them as inconclusive.[66]

Writing Jullian—in fluent French—to inform him of his rejection of the "Gallo-Roman sorcery thesis," Mendes Corrêa characterized his opinion as motivated only by "devotion to scientific truth."[67] Yet the Portuguese scholar was participating fully not only in privately networked discussions of the controversy, with clippings moving back and forth, but in its media coverage, publishing articles in a Portuguese newspaper about the similarities and differences between the Alvão and Glozel inscriptions; he also sent Jullian an article he hoped his "very eminent colleague" would submit to a French publication, perhaps the *Nouvelles littéraires*, if he found it worthy.[68] Mendes Corrêa clearly took an interest in the media coverage beyond the copies of articles his French colleagues might send him. Although this interest did not prevent Mendes Corrêa from continuing to participate in scholarly exchange, he did so discreetly, telling Audollent "*very confidentially*" that he had just sent Reinach a photograph of an unpublished Alvão inscription that demonstrated the existence of some of the Glozel signs elsewhere. By this point (January 1928), Mendes Corrêa wrote regretfully, "Glozel is enveloped in a thick fog in which dogmatism and insults are replacing arguments."[69] Few would have disagreed with him, but some might have pointed to his own actions—substituting private communication for the normal processes of scholarly publication—as itself both symptom and cause of the problem.

Accusations of breaching norms of openness, transparency, and access flew in both directions. Bégouën responded to Audollent's pointed observation that most of the critics of Glozel had not visited the site by saying that Morlet was not permitting access to those who had publicly expressed any doubts. Abbé Breuil, who had visited once, in fall 1926, later recounted that "the toxic atmosphere, the absence of method and of true supervision . . . were enough to keep me away from this setting, to which I prefer the serenity of my caves."[70] Pottier gently rebuffed Reinach's invitation to visit Glozel largely on the grounds of expense, but he also observed that "the conditions of examination, for those who do not have a firm opinion and who moreover wish to retain complete independence, are rather awkward: one necessarily becomes Dr. Morlet's guest and under his obligation."[71] The pro-Glozelian side riposted by accusing Dussaud in September 1927 of a serious breach of scholarly propriety: he chose first to present his charges of forgery in a closed meeting of the AIBL, the so-called *comité secret* normally reserved for sensitive personnel matters, and then (the Glozelians charged) leaked his statement to the press. Certainly the account in *Le Journal* two days after the ostensibly secret meeting was extremely detailed and included lengthy quotes.[72]

Less than a week after this meeting, Reinach penned a four-page letter to Dussaud on his personal stationery. It began with the customary salutation for a member of the same body, "Mon cher Confrère" (a plummier version of "colleague"); relayed the protests of Joseph Loth and Espérandieu, neither of whom had been present; and called Dussaud's decision to speak in a closed session "very regrettable." The bulk of the letter consists of a ten-point refutation of Dussaud's report. Reinach began with the size of the Glozel trove, over a thousand objects found in a relatively small space, which he deemed perfectly natural; moved on to specific points relating to the alphabet; and defended the views of scholars Dussaud had dismissed.[73] A few weeks later, Reinach sent Dussaud a short note thanking him for the copy of his essay on Glozel. The collegial proprieties were still functioning, but barely. Reinach used the occasion to relay a joke contained in a note from Loth, and it had a clear undertone of menace: "Soon Dussaud will have only one way of extricating himself: to bring the Phoenicians to Glozel and install them for a long stay." In his own voice Reinach added drily, "I doubt that would be a solution."[74] Even before the publication of his pamphlet, Dussaud, outraged by the pro-Glozel faction's attacks on him in the media, told Jullian he was breaking off relations with Reinach, which dated back to 1895. Revealingly, in reference to Pottier's attempt to make peace, he wrote, "He does not realize what a hellish line of attack the Glozelians are leading. One has to receive all the clippings, from newspapers I'd never heard of before now, to realize the full measure of these attacks."[75] Noting Reinach's influence with the prominent daily *Le Figaro*, Dussaud charged that he had "led the whole campaign at one end of the Paris-Vichy telephone line."[76]

Whatever the validity of Dussaud's suspicions, the dispute clearly took on another dimension when it moved to the hushed halls of the Academy. A clearly alarmed René Cagnat, the ultimate institutionalist both by instinct and as the AIBL's perpetual secretary, quickly took steps to limit the damage. In early September 1927, before the closed meeting, Cagnat issued a statement making clear that, notwithstanding Morlet's claim to the contrary, the AIBL had never sent one of its members as an official delegate to Glozel and had no intention of doing so. After the meeting, he wrote Reinach urging patience and moderation. The Academy, Cagnat insisted, "has no need to consider journalists' talk; if we did we would spend all of our meetings denying what the press writes about us." And, he went on, "I remain absolutely opposed to any official involvement of the Academy in this quarrel." Cagnat noted that if Morlet had wanted the AIBL to arbitrate the dispute, he could easily have asked; instead, he relied on personal supporters, "up to the moment where he falsely [*mensongèrement*] inserted into an article that which I protested against two weeks ago." "Mensongèrement": this was strong language for a man of his discretion. Cagnat allowed that, after the end of the summer holidays, the AIBL as a whole could vote to become involved if it wished.[77] But Reinach must have realized that, given Cagnat's attitude, this was highly unlikely.

Avoiding the subject of Glozel in the Academy deprived both sides of a platform, and in a letter to Jullian in late October Dussaud grumbled that Cagnat regarded Glozel as a nuisance "because it wrests him from his sweet tranquility" and threatened to upend the next academician elections. But, as he and Cagnat well knew, with the press eagerly printing whatever scholars had to say about Glozel, neither they nor any other party to the quarrel needed the AIBL to get their views across. Dussaud could take satisfaction in the fact that "in our milieus the negative view is rapidly gaining." Methodically he ran through the subgroups of academicians: "the Chartists [graduates of the École des Chartes, archivists and specialists in medieval paleography] are clearly hostile to Glozel, the epigraphists as well, even the Athenians."[78] The plural "milieus" is significant, since Dussaud was including his colleagues in the national museums (where Reinach also held a post) and in the government education and arts bureaucracies, though he knew the latter had to remain discreet. By new year's 1928 he could report that of the fifty-nine members of the Academy, only three—Reinach, Espérandieu, and Loth—viewed Glozel as authentic, leaving fifty-five in Dussaud's camp and Jullian, to Dussaud's great annoyance, pursuing his own hypothesis of a mix of genuine Gallo-Roman artifacts and fakes.[79]

Rather than feeling isolated, however, Jullian expressed "absolute confidence in the future" as he published his views on Glozel serially in the scholarly journal he coedited, the *Revue des études anciennes*, taking nearly two years to do so.[80] Although he allowed himself the occasional bit of snark—"the Ferrières excavation," he wrote in the journal in 1926, "has been much talked about, much too much"—and could rage in letters to Dussaud about the brazenness of Morlet and his supporters, the classicist took the high road in his public pronouncements.[81] "As far as the personal goes," he wrote Dussaud, "I do not want to concern myself with it; too much space has been devoted to matters of individuals."[82] This tone earned the approbation of Jullian's colleagues: Pottier, for example, pointedly told Reinach, "I must give credit to Jullian: he, alone, has made it a principle to never name individuals and to contest only theories or interpretations of facts."[83] In contrast, in his letter to Reinach after Dussaud's report to the AIBL in September, Cagnat warned him against entering into "questions of persons."[84]

This was a warning Reinach did not heed (nor, for that matter, did Dussaud). At an earlier stage in the controversy, when he thought Dussaud still persuadable, Reinach sent his colleague, whom he addressed as "My dear friend," a list of inscriptions ostensibly similar to those at Glozel that had appeared in various works by prehistorians. On its face, this was a classic example of scholarly networking. To the left of his signature, however, Reinach included a postscript about some declared skeptics: "*Between us*, Vayson is a drunk, Breuil is jealous, Bégouen is dumb, Jullian is too learned."[85] Little could better exemplify the breakdown of scholarly proprieties, and of the networks they safeguarded, than this succinct catalog of insults. Networks, of course, consist of people, and one could say that Reinach here, or Dussaud when he dismissed Reinach as Morlet's "lieutenant," or Loth in referring to an opponent as "dumb as a carp," was simply acting as humans do when attacked.[86] It was easy to blame Morlet for making insult the rhetorical currency of his camp, even if Dussaud had earlier cast Reinach as the chief of the Glozelian campaign. At one point Morlet wrote Audollent conspiratorially that he had solid information that Dussaud was a Freemason, information he hoped to use to discredit him.[87] Peyrony, in his letters to Reinach, implored him to restrain Morlet, and around the same time a supporter of the Glozelian camp suggested withholding information from the doctor, fearing he would overreact.[88]

But the rhetoric of blame only confirmed Cagnat's and Jullian's worst fears. If people comprised networks, the ties that bound scholars into networks came from a joint commitment to an endeavor understood as collective, with its own interest and reputation greater than that of individuals. Such networks have their own protocols: as Dussaud, who could be discerning even in his most polemical moments, wrote Jullian at one point, the scandal of Glozel lay not in making mistakes—"we all make mistakes; that is of no importance"—but in the personalization of scientific conversation.[89] By the end of 1927, the bitterness of the Glozel controversy had exposed the tenuousness of the understanding of science as a civilly collective project. As Peyrony wrote Reinach:

> It's not a little surprising that you, a learned man, owner of a highly respected journal, have not used that organ to discuss the truly scientific questions around Glozel, instead of writing things in the daily press [*la presse politique*] that amuse only the public.
>
> Men of learning, what kind of spectacle are you presenting the whole world? Do you think you are raising the reputation of the Institut? If you knew, dear professor, how badly people are speaking of you, you and your colleagues would cease your polemics and would be content to carry on courteously in your journals and your lecture halls.[90]

Apparently unmoved, Reinach used some blank space on the last page of the letter to scrawl a note disparaging Peyrony's motives and calling him a tool of the "Begouen clique." In French a *clique*, according to the *Petit Robert*, is a "coterie, a group of people of small repute."[91] It bears roughly the same relationship to a scholarly network as a faction to a republic. Scholars could blame the press for displacing lecture halls and journals as the main channel for circulating scholarly disagreements, but they could equally have blamed themselves.

Blame the Media

Journalists and archaeologists could agree, at least rhetorically, on one thing in 1927: scientific controversy sold newspapers, but media coverage did science—specifically the scientific claims of prehistory—no favors. In a long overview of the controversy, a reporter for *L'Avenir* ran through the views of various scholars, including Jullian. Then, under the rubric "Polémiques," the reporter clearly assigned responsibility: "The press got involved, publishing the sometimes acerbic statements of one and all, and began taking sides. In short, the case was turning into a scandal."[92] The abbé Breuil grew so disgusted with the media coverage that he refused to comment on the release of the commission's report in December 1927. But this refusal, as multiple newspapers recorded, did not quite amount to silence: a story in *Le Quotidien* cited Breuil's statement to a different daily that criticized journalists' role in turning the affair into a "crude polemic."[93] Some accounts conveyed a certain cynicism about the coverage, one story in *Paris-Soir* commenting that Glozel was fast becoming a *fait divers*, a dependable type of story, the rough equivalent of what in English we sometimes call "news of the weird," calculated to sell newspapers.[94]

Certainly archaeologists at the time had a tendency to blame the media for enflaming the controversy. Wrote one of Dussaud's correspondents, "I think that these hoaxes are a law of nature and that it's also one of the effects of this overabundance of illustrated newspapers and cheap brochures to grow prehistorians and archaeologists out of nowhere. If we excite rural types about flints, fibulas and [illegible], we give rise to imitators. What could be more fun than fooling the bourgeois?"[95] More directly, André Vayson de Pradenne charged in the journal of the society of French prehistorians that the affair was entirely the product of the *Mercure de France* and its regular chronicler of prehistory, Arnold van Gennep, who gave Morlet as much space as he wanted to write up his finds.[96] On another note, in a letter to *Le Temps*, Bégouën complained that the "swarm of reporters" at Glozel had prevented the commission from working in the "atmosphere of calm contemplation" appropriate to its task. Reinach replied that, on the contrary, the presence of journalists had been indispensable: "A surplus of eyes, when it comes to observing facts, is never pointless."[97]

Bégouën did not in fact disagree about the value of press coverage of archaeological finds. Although by the 1920s he devoted himself, as both a museum director and university lecturer, largely to prehistory, as a younger man he had been a journalist, serving for a few years as editor of a Toulouse daily, and he never gave up those ties.[98] In a lengthy letter to the editor of the *Mercure de France* in July 1927, responding to accusations against him leveled by Morlet, Bégouën stressed the importance of openness in, and wide publicity for, archaeological excavations, advocating for a public conference at Glozel where archaeologists could gather to examine the site and their finds. Expanding on a point he had made in the *Journal des débats* the previous fall, Bégouen cited numerous examples of archaeologists encouraging open debate: he recalled that fifteen years before, Peyrony had halted a potentially important dig until he could assemble ten of his colleagues to excavate with him and share their views. "Peyrony's scientific authority was not diminished by this verification, quite the contrary."[99]

Archaeologists were quick to denounce their colleagues' reliance on, and pandering to, the press, but both sides plotted media strategy and used whatever contacts they had to try to influence coverage. If the wealthy and prominent Reinach family, had, as Dussaud observed, influence over *Le Figaro*, a letter from *Le Journal*'s Paul Bringuier—sent by Paris's speedy *pneumatique* system to announce a visit in a matter of hours—suggests that Dussaud served as a confidential source for that reporter.[100] Knowing of *Le Journal*'s skepticism of Glozel, Morlet decided against sending a statement by Reinach to the Havas agency, at the time under the same ownership as *Le Journal*.[101] Since many newspapers took sides in the controversy, both sides prized space in those that did not, notably *Le Temps*, along with *Le Figaro* and the *Journal des débats* one of the establishment newspapers of record, their influence far exceeding their circulation. When, in mid-November 1927, the paper solicited a short opinion piece from Audollent, the latter had already received a tip-off from Reinach that *Le Temps* "has decided to play ball."[102] Just six weeks later, however, Dussaud told Jullian that he was planning to launch his latest anti-Glozel campaign in *Le Temps*.[103] *Le Temps*, indeed, received so many communications from figures on both sides of the controversy that, while publishing them serially under the heading "Autour d'un controverse scientifique" (On a scholarly controversy), it felt the need to recall certain limits. Writing that it would publish only "objective opinions" that avoided personal attacks, the paper added, "Letter-writers must also keep in mind that no newspaper can give unlimited space to any one topic, and this may keep us from publishing some overly prolix statements."[104]

Auguste Audollent's correspondence offers several revealing glimpses of the dilemmas newspapers and magazines faced. Asking Audollent for 100–150 lines about his excavations at Glozel, the editor at *Le Temps* observed, "We would be happy to publish your statement to which, we are quite sure, you will not give a polemical character."[105] The editor of *Le Correspondant*, however, the general-interest weekly in which a long essay by Audollent appeared around the same time, was not so sanguine. In several letters the editor pleaded with Audollent to take into account some of the objections that had surfaced about Glozel. After several publication delays Audollent, refusing to make changes, had to agree to an extraordinary editor's note saying, among other things, that the periodical "would have wished that the very distinguished Dean took account [of the controversy]" and that it was printing this version on his insistence and sole responsibility.[106]

Among the archaeologists, Morlet in particular acquired a reputation for litigiousness and inveterate self-promotion. Prefacing a letter that appeared in early February 1928, the *Journal des débats* wrote somewhat wryly, "Doctor Morlet is very fond of invoking his right of reply. We are glad to grant it to him, and his letter below will confirm for readers the eternal validity of the parable of 'the mote and the beam'" (from the Sermon on the Mount).[107] If from early on scholars like Peyrony and the paleontologist Marcelin Boule fretted that all the publicity was discrediting French science, some journalists and commentators blamed the scientists themselves for their intemperance.[108] "Science!" wrote an exasperated Victor Méric, in terms recalling Bertrand's rant about archaeology in

North Africa earlier in the decade. "What a horrible joke! We've got here a dozen augurs who can't come to any kind of agreement. And yet the tests are fairly simple."[109] Calling attention to the uncertainties and gaps in archaeological knowledge as a whole, the *Guardian* noted that prehistory was still in its early stages, inviting prudence.[110]

It would, though, be unfair to characterize the media's role in the controversy as simply a megaphone, or to view it as purely venal. Newspapers that published communications from scholars felt it their duty to do so, even if they suspected that articles about Glozel would not hurt sales. Others took it upon themselves to act as interpreters: an article in the Belgian newspaper *La Meuse* explained key terms in archaeology such as *mobilier*, referring to objects or artifacts; detailed the types of objects found at Glozel; listed the main proponents of both sides of the controversy; and briefly set out the accomplishments of the members of the international commission.[111] Newspapers also frequently published explanations—some summary, some much longer—of the larger significance of the affair. Over a photograph of the commission members at work, a headline in the staunchly Glozelian *Le Matin* announced, "The Glozel excavations could turn the chronology of prehistory completely upside down," promising a new cultural epoch between those currently known.[112] Even more, as the journalist Marcel Sauvage put it in a thorough account of the controversy, "But here the real problem that Glozel raises gradually reveals itself in full: in what direction did civilization progress, from west to east in the view of Salomon Reinach, or from east to west as Mr. Dussaud maintains."[113] Of course some accounts personalized the controversy, but this was another way of conveying the stakes: in an interview with the *Petit méridional*, Espérandieu intimated that Dussaud, whom he did not name, needed Glozel to be fake because its authenticity would invalidate central aspects of his scholarship. On the other side, Eugène Marsan wrote in *Le Figaro* that Reinach "knows how high the stakes are at Glozel: nothing less than the history of civilization. If Glozel is a fraud, nothing works: civilization passed from East to West: everything came from the Garden, between the Tigris and the Euphrates. Which Salomon Reinach will not admit at any price."[114] The journalist here presents a bullet-point version, accurate as far as it went, of the most celebrated of Reinach's hundreds of articles, "The Oriental Mirage," first published in 1895.[115]

In his own controversial work *Black Athena*, Martin Bernal situates Reinach's long essay among a plethora of fin-de-siècle works challenging, and ultimately dismissing, the influence of the Phoenicians, and thus of an "Asiatic" east, on the early development of European culture. Bernal speculates that Reinach's motive for this work, which rested largely on philological rather than archaeological evidence, involved a mix of antiracism, since one of his targets was the idea of an Aryan race, and his desire as an assimilated, nonobservant Jew to rid himself of "Semitic cultural baggage." The article does acknowledge a cultural influence of the Phoenicians dating to the beginnings of Phoenician trade across the Mediterranean around 1300 BCE, a fact confirmed by a wide range of ancient texts.[116] Dussaud made his name as a Semitic epigraphist and specialist in Phoenician, recognized as the oldest surviving alphabetic writing; a few years before the Glozel affair he had deciphered the earliest known Phoenician inscription, found at Byblos in Lebanon in 1923. He used his knowledge to denounce the Glozel tablets as forgeries based on publications showing *later* Phoenician alphabets, as the forger would not have had access to the 1923 discoveries.[117] "Would you ever have believed," Dussaud wrote Jullian wonderingly, "that the study of Phoenician could be of interest in such a brawl?"[118]

The use of the word "brawl" (*bagarre*) conveys the way the dispute over Glozel, magnified by the press, had exceeded the boundaries of scholarly debate, leaving in tatters scholarly networks to which Dussaud and Reinach had long belonged. To the charge that his theories blinded him to all the factors suggesting fraud at Glozel, Reinach gave, in his letter to Dussaud in September 1927, what we might these days call a nondenial denial: "Yes, you are fighting for the theories you learned from your professors, and there is nothing unfortunate about that. I am not fighting for my old theory of the Oriental mirage, but I am not upset that it is supported by these huge discoveries."[119] But a year earlier, just after his first visit to Glozel, Reinach had written Jullian in less moderate terms: "The Phoenicians are done for! [Victor] Bérard is sunk!" referring to a close colleague of Dussaud's who argued for Phoenician influence on Greek mythology.[120] Morlet, who had cited "The Oriental Mirage" several times in the first Glozel fascicule, was clearly aiming for the big fish, and he hooked him.

Marsan's observation about the "Oriental mirage" in *Le Figaro* actually came in an article on a much earlier controversy involving Reinach, when as a young curator he had recommended the Louvre's purchase of the so-called tiara of Saitaphernes. The museum duly bought the crown, supposedly that of a Scythian king of the third century BCE, in 1896, only to face decisive evidence a few years later that it was a modern forgery.[121] So many accounts of the Saitaphernes affair, including the caricature reproduced here (fig. 2.3), appeared in the press that when a group of anti-Glozelians tried to shout down a lecture by Loth, the pro-Glozel academician and professor at the Collège de France, they repeatedly chanted the name "Saïta-pharnès," making it into a kind of rhythmic refrain.[122]

FIG. 2.3. *The Judgment of Hamlet Salomon Reinach on the Glozelian Field*, published in the weekly *Aux Ecoutes*, 11 February 1928. Artist unknown. Public domain. Source: Direction des affaires culturelles, Région Auvergne–Rhône Alpes, Service régional d'archéologie, Site de Clermont–Ferrand.

Reinach's colleagues were well aware of this episode and happy to use it to their advantage: "The fortunate tiara has been of great service to our side," Dussaud told Jullian as the controversy grew more heated, adding, "the Louvre really didn't pay too much for it."[123] For Vayson de Pradenne, who devoted a lengthy chapter to the tiara in a 1932 book on archaeological frauds, its true significance lay in the fact that "it was thanks to the press and public opinion that the truth could come to light."[124]

Pottier had attended the curatorial committee meeting on the purchase and, three decades later, recalled Reinach's hesitations and warnings to senior curators about the object. He wrote Reinach several times of his willingness to defend him, and did so in the press as late as October 1928.[125] But, as the cartoon shows, the incident remained engraved in the public imagination as channeled by the media. The veiled anti-Semitism, with the mixing up of stories of the Judgment of Solomon and Belshazar's feast, was typical of the media treatment of Reinach, although it was explicit only in far-right journals like *Action française*, where Glozel fueled tangled, obsessive antipositivist tirades by Léon Daudet.[126] Loth, for his part, cast his controversial lecture series as a way of avoiding overreliance on the media: "I insist on saying what I know in public and do not in any way want to start a new press campaign."[127] But Loth issued this statement, of course, in the media, specifically a newspaper interview clearly intended as a teaser for the lectures; courses at the Collège de France had a long history of prompting media attention and public clamor. Nothing could have made clearer the symbiotic relationship between archaeology, its traditional academic relays, and the very media that archaeologists at once engaged with and blamed for distorting and tarnishing scholarly debates.

Fake News/News of Fakes

Recalling the dictum—which itself goes back to Aristotle—that there is nothing new under the sun, in 1921 the historian Marc Bloch, fresh from his experience as a sergeant during World War I, published an article in the *Revue de synthèse historique* entitled "Reflections of a historian on the false news of the war." Part review essay, part research agenda, part memoir, the article probes the psychological dimensions of the waves of disinformation that had spread from the war zone (though they usually originated along the support lines to the rear) to the home front throughout the war.[128] In the *Revue de synthèse* essay, the only scholarly article on the Great War Bloch published during his lifetime, he treats false news as a phenomenon of collective psychology, something that propagates only if it finds "a favorable breeding ground." For Bloch, "false news is the mirror wherein 'the collective consciousness' contemplates its own features."[129] As Leonard Smith has observed, the perspicacity about experience and narration that Bloch demonstrates not only in this article but in his war diaries and correspondence anticipates questions historians would not begin to grapple with for decades.[130] But the fact that, in 1921, he already had several books on false news he could reflect on makes clear that the phenomenon itself had some purchase on "la conscience collective" in 1920s France. Indeed, after the publication of the commission report in late December 1927 several newspaper articles suggested that Glozel would be a fit subject for "studies of abnormal psychology" or that Freud might have illuminating things to say about this syndrome, which the author deemed a "brain irritation from which no one is immune and which we could call *glozelitis*."[131]

The study of false news begins, as Bloch observes, with the historian treating error not as something to be eliminated but as "an object of study." In such a study, Bloch observes, even minor typographical errors may have some significance; he offers as a telling example a false report that involved the homophonic relationship between Braisne, the town where he was stationed in the Chemin-des-Dames sector in 1917, and Brême, the French spelling of the German city of Bremen.[132] Accounts of Glozel were riddled with errors, notably but not exclusively in the spelling of commission members' names: many reports referred to Dorothy Garrod, the youngest and only British member of the commission, as Miss Garrot, and one also rendered the eminent Spanish prehistorian Pere Bosch-Gimpera as "Blosh Guidini."[133] At the very least such errors prompt skepticism about other "facts" proffered by that publication.

I have so far translated the "fausses nouvelles" of Bloch's title as "false news." In a timely recent collection, a number of French scholars have drawn a careful distinction between the French term *fausses nouvelles* as used by Bloch and the more recent but widespread American coinage "fake news." *Fausses nouvelles* forms a triad with the older terms *bruits* (noises) and *rumeurs* (rumors), which generally refer to "the circulation in the public sphere of information, of which the contours are difficult to discern, of which the goal is—directly or indirectly, voluntarily or not, with criminal or honorable intention—that of disinformation."[134] By contrast, "fake news," as Vincent Michelot observes, designates a situation of bitter political division in which one powerful political faction dismisses as "fake" any information that does not correspond to its highly skewed vision of current events, and indeed of the world.[135] The term rejects and delegitimates every rendering of reality coming from an antagonist. Arguably the polemic over the authenticity of Glozel produced "fake news" in precisely our contemporary sense, nearly a century *avant la lettre*.

In some instances, indeed, the intensity of media coverage provides a glimpse into the production and distribution of fake news almost in real time. One such instance, the international commission's verification dig at Glozel in November 1927, is of particular interest because it involves reports on the archaeological process. Conscious of the media frenzy surrounding their work—Garrod later recalled the close proximity of journalists as "rather trying"—commission members tried to avoid communicating while digging, then issued a brief statement at the end of the dig saying they would not comment on press reports.[136] Even before this statement, newspapers reported on the commission's refusal to issue daily bulletins, the *Loire républicain* noting with a mix of understanding and disappointment that the group had decided to say nothing until it completed its report.[137] But this reticence did not prevent journalists from speculating and, indeed, from telling readers that the commission, based on finds made during three days of digging at the site, had determined that Glozel was authentic. In the Dussaud clipping files, the *Loire républicain* story is pasted onto a sheet between two others, one with a headline declaring that the Glozel discoveries "might be [*seraient*] authentiques," another saying that the site "is indeed [*est bien*] authentic," leaving no doubt. The headline in *Le Matin* after the second day of the dig went even further: "The authenticity of the Glozel deposit is unanimously recognized by the international commission." French publications were not the only ones fooled; a headline in the *Daily Mail* described "ancient history rewritten."[138] How could they have been so wrong?

For one thing, newspapers devoted considerable space to describing the objects unearthed by the commission; for *Le Quotidien*, again in a headline, "A reindeer head engraved on a pebble attests to the authenticity of the Glozel deposit."[139] In this epistemology, seeing was believing, in the sense described by Reinach in a letter to Jullian the previous August: "Under my eyes, so close I could touch them, using hand tools in the untouched soil, an inscribed tablet and a clay idol of the lingam–yoni type came out of the ground. It's surprising, it's disconcerting, but it's undeniable."[140] A day later, Reinach added, "Despite the objections, I cannot give up on making public the evidence of my eyes."[141] But the eyes cannot testify without the use of some other senses and faculties. The *Daily Mail* report contains this revealing sentence, which follows a description of some of the discoveries: "The scientists vouch that the finds were made in soil which had not previously been disturbed and could not have been placed there by a modern hand." Yet "the scientists"—the members of the commission, that is—were making no declarations, so this assertion reflects some level of inference. The day after it asserted that Glozel's authenticity "is recognized," *Le Matin* provided a detailed account of commission members' expressions and gestures on the last day of the verification dig; it even offered (in paraphrase) the comments of another archaeologist observing the dig, who interpreted those gestures as confirming the radical importance of the finds.[142] Rumor and the overheard conversational fragment could also find their way into such assurances, even though more prudent journalists pointed out the risks of attaching too much importance to them.[143] Other admonitions against the speculation, such as Bégouën's in a Toulouse newspaper,[144] could be dismissed as partisan, this kind of delegitimation being a central aspect of the fake news syndrome.

In the event, the commission's report, published as a supplement to the *Revue anthropologique* just before Christmas, validated Bégouen's warning: it concluded, on the basis chiefly of careful study of the soil surrounding the finds, that the objects found at Glozel were not ancient.[145] With the Glozelians vehemently rejecting this assessment and attempting to discredit it, the press could and did keep busy with articles on the many responses to the reports, including, early in January, Loth's lecture series at the Collège de France. For *Le Matin*, however, the mass circulation daily that had most heavily staked its reputation on Glozel's authenticity, the prospect of the Glozel controversy receding from the headlines presented too much of a threat to its often shaky bottom line. Taking the story literally into its own hands, the newspaper decided to send one of its editors and another journalist to carry out and report on their *own* excavations in the vicinity of Glozel, where their close relationship with the Fradin family presumably facilitated access. "Les fouilles du *Matin* à Glozel," ran the front-page headline on 6 January 1928, less than two weeks after the publication of the commission report, over photographs of the editor, Pierre Guitet-Vauquelin, and his colleague holding up ther finds in what looks like a snowy field (fig. 2.4).[146] The headline continues with details on the location of the dig, and finally this revealing phrase, "The Fradins don't want to be treated as forgers." Similar discoveries nearby, however, did nothing to exculpate them.

The Dussaud clippings include a number of reactions from *Le Matin*'s competitors. Writing in *Paris-Soir*, Maurice Verne praised the newspaper's initiative, but said its efforts, in bracketing the controversy over authenticity, were insufficient.[147] The Communist daily *L'Humanité* took a far harsher view. In an article entitled "Toujours Glozel" that mentioned

FIG. 2.4. "Les fouilles du 'Matin' à Glozel" (*Le Matin*'s dig at Glozel), *Le Matin*, 7 January 1928, with a picture of the journalist leading the dig at lower left. A subheadline translates, "the Fradins don't want to be treated as forgers." Source: gallica.bnf.fr/BnF.

neither *Le Matin* nor its correspondent by name, *L'Huma* dismissed the finds as well as Guitet-Vauquelin's "bright idea" of examining the clay in a nearby grotto, with its profoundly different geology. Even the most fervent Glozelians, the article goes on, must realize that more spectacular finds would not help their cause, and would probably discredit it in the only circle that mattered for the question of authenticity, international science. Attempting to reassert the boundary between science and the news, *L'Humanité* declared that Parisian journalists could not invalidate the unanimous report of reputable scientists: "We insist that the problem of Glozel is purely scientific. Laymen have no role there."[148]

But probably the most damning response to *Le Matin*'s gambit came from the satirical weekly *Le Canard enchaîné*, then just over a decade old. Founded in 1915, the *Canard*, one of the few French newspaper titles to persist across the divide of World War II, remains today a unique hybrid, a satirical weekly with a strongly subversive political voice.[149] The *Canard* took every advantage of the opportunities Glozel offered to mock the many types of people caught up in the affair: learned men (*savants*), local amateurs, peasants—and their fellow journalists. It published, for example, a detailed if imaginary account of the arrival of the international commission in Vichy and the many festivities put on to welcome them, including a "neolithic ball": the account has the fictional Dutch delegate Van Putzeboom mutter suggestively to his fictionalized English counterpart, Miss Plunkett, who replies with the inspired double entendre, "Go dig yourself" (*Tu peux te fouiller*; *fouiller* can mean either to excavate or to pat down, as in a security check).[150] The *Canard* had a field day with objects published by Morlet in the *Mercure* known as the "hermaphrodite idols" (*idoles bisexuées*). Dussaud had noted drily that one seemed to have its breasts crossed over each other; the *Canard*, in an article entitled "Of a Neolithic Fashion," observed that such breasts must have been remarkably supple, postulating that it was perhaps this very suppleness that Neolithic men valued above all else in their wives' bosoms.[151]

But the *Canard* reacted with particular verve when *Le Matin* began its own excavations. A few days later, the *Canard* published an article claiming that it had itself carried out excavations at what it called "the Glozel subdivision [*lotissement*]." The article consists of a supposed dig journal in which the journalist spends all his time observing the *other* journalists who are trying to excavate at Glozel, sometimes coming to blows, as well as scouting out decent places to eat. Readers are treated to the details of his wardrobe, modeled on those of the *Matin* editor as recorded in that front-page photograph, but after three days at Glozel the reporter has yet to do any digging.[152] For the rare reader who might be wondering if this piece of fake news actually represented real investigative journalism, something for which the *Canard* is now famous but which it did not undertake until much later in its history, the newspaper published several cartoons. Figure 2.5 offers a schematic map of the new "subdivision," with streets named after the main Glozelians as well as a "Prehistory Circle," conveying something of the repetitiveness of the story. Another cartoon presents a rendering of Boulevard Morlet, suggesting that such venerable institutions as the Institut de France and major newspapers had been reduced to carnival barkers.

FIG. 2.5. Henri Guilac, *Development of Glozel* and *Glozel Subdivision*. Cartoons from *Le Canard enchaîné*, 11 January 1928. Public domain.

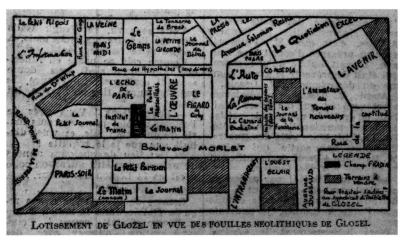

Le Canard enchaîné's satire of its counterparts in the press could be regarded as a kind of safety valve, a way for the press to inoculate itself against the widespread criticism of its tactics and profit-minded opportunism. This criticism started very early in the affair, with Viple writing to Jullian in May 1926 that Morlet, though acting in good faith, had gone too far in "monetizing" the discoveries; several months later, he offered the same characterization of Emile Fradin.[153] As visits to the site became popular, many reports cited the hefty four-franc fee the Fradins charged for admission to the small museum they had opened on their property.[154] Rumors flew that companies controlling the Vichy spas, or even the municipal tourist office, were seeking a financial interest in Glozel, perhaps by buying some of the discoveries.[155] Although Espérandieu told Reinach privately that he had warned the Fradins against charging admission, Morlet, typically, chose to go on the offensive.[156] Through various intermediaries he revived charges that Peyrony, the French member of the international commission and a noted prehistorian, had decades before sold antiquities to a notoriously shady Swiss archaeologist and dealer called Otto Hauser. This commercial venture, resulting in the export of many prehistoric objects to Germany, had caused a well-publicized commotion between 1907 and 1910.[157] And yet, for all that, the press remained the main object of archaeologists' suspicion. The Belgian archaeologist St. Just Péquart wrote Dussaud in November 1927, "Obviously the primary cause of the scandal is the press, which knows how much the public goes crazy over sheep with five paws, ducks with multiple beaks, and other teratological animals."[158] Apart from the obvious slippage from the consumer to the supplier of scandalous news, Péquart's remark also leaves out the many ways archaeologists took advantage of, and benefited from, the media they loved to blame.

Affairs to Remember

By focusing on the commercial aspect of Glozel, *Le Canard enchaîné* cast competition as part of a larger problem with the journalism of the day. But beyond this point of consensus, Glozel hardly provided a reliable metric for diagnosing problems with the press. As should be evident by now, newspapers' attitudes toward Glozel did not break down along conventional political lines: if the royalist *L'Action française* was anti-Glozelian, so was *L'Humanité* and an independent leftist daily, *L'Oeuvre*. The far-right *Le Matin* was Glozelian, but so was the Socialist party daily *Le populaire* and the generally left-leaning *Le Quotidien*.[159] In this disconcerting media landscape, newspapers sought out familiar reference points with which to orient their readers. Doing so extended their role as interpreters and translators, offering an authoritative version of Glozel that would make sense of the unfamiliar and frame archaeology and science in general in terms their readers could understand.

Immediately upon publication of the commission report, Reinach, Loth, and Espérandieu, the three members of the AIBL who accepted the authenticity of Glozel, issued a solemn statement. Referring to the group as "la commission Bégouën," a shorthand for the charge that the anti-Glozelians had unduly influenced the commission's composition, the three wrote:

> [Until now] the admirable Glozel discovery was lacking the highest consecration, the one with which the Roman Inquisition honored Galileo's genius. In this respect, the Bégouen commission has done well by science, and soldiers for the right owe it their thanks.
>
> As for the commission itself and its Toulousain inspirer, they will share with the judges of 1633 the only immortality within their reach, that of ridicule.[160]

Widely reprinted, the three academicians' statement provoked a vociferous, often mocking response. A few newspapers of a Catholic bent picked up on the religious overtones of the statement, taking offense at what they saw as disparagement of the church; at the same time, some leftist papers charged that "a clerical campaign against prehistory itself has been grafted onto the Glozel affair."[161] The right-wing *La Liberté* mildly called the comparison of Morlet to Galileo "perhaps excessive," but a columnist for *Le Temps* scoffed at the reference, saying that no sensible spectator would compare the commission members to Torquemada, and that no one wanted to burn Morlet—surely a lesser genius than "the illustrious Tuscan"—alive.[162] *La Nation* chimed in on the theme of civility, calling the statement "discourtesy toward eminent foreign scholars," and evidence of a lack of "that serenity which is always as necessary to scientific method as it is in the conduct of public affairs."[163] Here the commentary at least returns to the original frame of reference, science. Attempts to find Catholic or anti-Catholic motivation for any position on Glozel, such as *La Rumeur*'s suggestion in February that Bégouën's career had been shaped, and stymied, by his reactionary Catholic upbringing, founder on the multiplicity of available references and connotations. The leftist *Le Quotidien*, in a group interview with the Fradins conducted on New Year's Day 1928, likened Morlet to Joan of Arc and Jesus in his humility, but the devotion here is much more to Glozel than to the church.[164] Morlet's letters to Reinach contain occasional anti-Catholic asides, but whether these reflect Morlet's genuine views or simply an effort to cater to Reinach's would be difficult to determine.[165]

If the Galileo reference in the end fizzled in its own hyperbole, another, more spontaneous comparison suffused discussions of Glozel both in scholarly exchanges and in the media. For many, the divide over Glozel recalled the Dreyfus affair of a quarter century earlier, another moment when the authenticity of evidence was vigorously and divisively debated in public. Several parallels and cross-references exist. Salomon Reinach was the younger brother of Joseph Reinach, a politician and publisher of *L'Aurore*, who was one of the first members of parliament to protest Alfred Dreyfus's conviction and sentence to deportation. Like his brother, Salomon was an early and prominent Dreyfusard.[166] Journalism played an important role in the affair, notably in the form of Emile Zola's celebrated *J'accuse*, published in *L'Aurore* in 1898, which detailed the ways the army and the government had colluded in the suppression of exculpatory evidence. This publishing coup eventually led to a new trial for Dreyfus at Rennes. Finally, the Dreyfus affair notoriously provoked bitter splits among (former) friends and family members who took different positions on Dreyfus's culpability. Just as the word "affair" became emblematic of the Dreyfus affair, a two-frame cartoon by Caran d'Arche, first published in *Le Figaro* a month or so after *J'accuse*, became metonymic of the social divisions it caused. Thirty years later, a reference to one of Caran d'Arche's captions, "Ils en ont parlé" (They spoke of it) in a description of a bitter argument between ladies at a tea party suffices to convey the parallel between the Dreyfus affair and Glozel; a cartoon makes a similar point (fig. 2.6).[167]

Most references to the Dreyfus affair appeared under a similar sign of levity.[168] At least two newspapers published an item about an orchestra in a town not far from Glozel (the town's name was not given, suggesting that the story could be apocryphal) whose musicians were so divided on the matter that they could not play together.[169] Occasionally, however, the evocation of Dreyfus turned into an invocation, as a protagonist and his press relay went off the rails. "J'Accuse Emile Fradin, déclare M. Peyrony," proclaimed a headline in *Le Journal* in early January 1928.[170] The *Journal* article proved less dramatic than its title: Peyrony was accusing Emile Fradin, the youngest member of the family and the one most active in the excavations, of having lied to him on several occasions about his familiarity with certain texts about early alphabets. The books were significant because they could have furnished Fradin with signs and images to use in fabricating inscriptions, and the young man had by this time emerged as the likeliest perpetrator of a forgery.[171] Although Peyrony was not "accusing" Fradin directly, the article, which was reprinted immediately in the *Journal des débats*, accurately conveyed the weight of the Dreyfus affair as precedent in the curator's mind. Peyrony began his futile attempt to reconcile with Reinach with a long disquisition on the impact the Dreyfus affair had made on him as a young man. A little later, one of his letters included a long allegory of truth, of the sort much published and illustrated dur-

FIG. 2.6. "No, really, come dine with us . . . We won't speak of Glozel!" Cartoon from *L'Oeuvre*, 5 January 1928. Artist unknown. Source: gallica.bnf.fr/BnF.

ing the earlier affair.[172] At a time many newspapers were facing financial difficulties, the media understood the commercial and recursive power of references to the Dreyfus affair.

Unsurprisingly, some participants reacted indignantly to the Dreyfus references. "Someone has said it's a Dreyfus affair," Dussaud wrote Jullian at one point. "That is giving it too much credit: it's just an Esterhazy affair."[173] The leftist newspaper *L'Oeuvre*, which in another kind of journalistic gamesmanship claimed to be the only Paris paper to have been anti-Glozelian from the beginning, also took a dim view of such allusions.[174] In an article that appeared a few days after Peyrony's "J'accuse" (which it did not mention specifically), *L'Oeuvre* accused the press of having "unleashed the Glozel affair—and Salomon Reinach." Scientists detested each other, the journalist "D" observed, yet normally got along in private. But, the article went on, this was not a Dreyfus affair. "[Then] we were fighting for a man and for an idea; we will not fight for some shards. Truth must represent something higher and more noble, something clearer than a scholar's opinion!"[175] The media's responsibility—the blame it deserved—came from giving undue publicity to normal scholarly disagreement, and thus not ony enflaming the dispute but harming the reputation of "a science that respectful citizens could have imagined more exact." For *L'Oeuvre*, though, the problem lay less with the science than with science's need, indeed craving, for publicity.

The commission report, for the most part dry and understated, views truth as something patiently and dispassionately arrived at; it expresses regret that the debate had "deviated completely from the terrain on which it should have stayed: that of a scientific discussion carried out in scientific calm, by men for whom scientific discipline is a daily moral obligation."[176] In an article published forty years later, Dorothy Garrod recalls that while the commission unanimously viewed the site as a hoax, it was the younger members, herself included, who insisted on the uncompromising verdict that Glozel was "not ancient."[177] Without the authority to order chemical testing of the objects, the visitors focused more on what they could say about the site itself. The report begins with a rehearsal of the members' credentials as field archaeologists, something Garrod later made clear Reinach was not.[178] It carefully narrates archaeology as *process*, emphasizing the commissioners' meticulous examination not just of the objects but of their placement in the soil, and the nature of the soil itself. Particularly damning evidence came from what appeared to be soil packets surrounding the objects unearthed, distinct for an inch or two from the composition of the earth the commission observed in this sector of the site.[179]

The lesson many archaeologists took away from Glozel thus involved the precarious balance between science and public attention. Commission members insisted that when they arrived at Glozel, they put aside all the impressions that articles on either side could have left, "determined to be guided only by the observations [*constatations*] they would make."[180] But the word *constatations* implies something that goes beyond seeing to involve a certain level of understanding. Such understanding itself derived not only from habits of mind but from learning, *la science* in the broadest sense. The report's concluding section begins on a philosophical note, which suggests a professional culture or habitus that archaeologists shared:

The Commission is obliged to recall that the history of archaeology, like that of other sciences, at every historical moment, has recorded many misadventures (some paleolithic carvings, the age of ivory in Switzerland, vases and statuettes from Spiennes, Moabite vases, etc.). That is why it had the duty to take every possible precaution.[181]

To avoid "misadventures," one must have in place all possible "precautions," what might now be called disciplinary protocols. Though the commission members did not offer any specifics, looming on the horizon of this contrast are the emerging norms and constraints of professional standards, and beyond them of scientific discipline, of the sort that few amateur archaeologists would be able to master. Yet whatever archaeologists' suspicion of the press, the emergent profession could not undo the imbrication of science and media that had made Glozel a spectacle.

Coda

On both sides of the Glozel controversy, the protagonists had the habit of predicting that the next intervention—a report to the AIBL, an article in a scholarly journal, an interview in an influential newspaper—would mark the end of the affair. Grenier, Jullian's protégé in Strasbourg, confidently made that prediction in October 1926, Dussaud a year later.[182] Hopes ran high at the end of 1927 that the commission report would, finally, resolve the matter, though many on the anti-Glozel side correctly anticipated that the Glozelians would never accept a conclusion of forgery.[183] The question then becomes, not when the controversy ended—some people still believe in the authenticity of Glozel—but when it ceased to be a media spectacle.

If for many observers the commission report did sound the death knell of Glozel, the winter of 1928 marked something of a cultural highwater mark, with a carnival float and ball on Glozelian themes, a parody in a student ritual in Paris, a Glozel sketch in a variety show, and a three-dimensional display at the annual cartoonists' salon.[184] In late February, acting on a formal complaint for fraud from the Société préhistorique de France, police conducted a search of the Fradins' property and carted off not only objects from the museum, but pebbles and carving tools stored in a barn that they took as evidence of forgery. Led by Morlet, the Glozelians naturally cried foul, noting that anyone visiting the museum could have accessed the barn, which housed a toilet for visitors, and argued that the tablets, obviously forged, had been planted.[185] In April, Morlet organized a second verification dig, with a group of archaeologists he chose for their open-mindedness—in other words, their acceptance of Glozel's authenticity. This was the occasion of Reinach's visit in the cattle-drawn cart and, though it attracted media attention, it did not succeed in reopening the question of Glozel's authenticity either among scholars or in the wider circle of public opinion.

A number of legal cases—the fraud case brought by the Société préhistorique, the countersuit brought by the Fradin family against Dussaud, and some minor suits brought by the litigious Morlet—provided periodic reminders to the public that Glozel still preoccupied some scholars and (increasingly) lawyers. But the real impact of what might be called Glozel's judicial phase, which began with the filing of suits in early 1928 and extended through the conclusion of the fraud trial in 1932, came from the scientific testing that accompanied it. Of these the most notable was the test the Moulins prosecutor's office entrusted to Edmond Bayle, the respected forensic scientist for the Paris police (he was a protégé of the founder of the forensics lab, the celebrated Alphonse Bertillon). In October 1928, *Le Journal*'s Paul Bringuier quoted Bayle as saying that he had concluded that the samples he tested were not ancient.[186] The actual report, issued in May 1929 and based on fairly rudimentary procedures of physical chemistry, confirmed this, again producing furious rebuttals from the Glozelian side.[187] In response, the Société préhistorique, through its attorney Maurice Garçon, sought an indictment of Emile Fradin.[188]

Already in the spring of 1928, Morlet wrote Reinach that he was having trouble getting his Glozel articles published. Loth, fresh from his well-publicized lectures at the College de France, had similar difficulties. By the fall, Morlet attributed to corruption the Paris dailies' refusal to publish his open letter to Bayle; a few months later, in January 1929, he mused about sending articles to the Vichy papers as a last resort, perhaps anticipating that clipping services would find them wherever they appeared.[189] Not long after, Morlet transmitted a report from his publisher that sales of Glozel literature had more or less ceased after the leak of Bayle's report. By the spring of 1929, the inveterate propagandist was seriously contemplating self-publishing.[190] The opening of a criminal complaint against Emile Fradin for fraud (*escroquerie*) in early June 1929 seemed not to have much of an effect on the dwindling public interest in the affair, even though Fradin was eventually acquitted.

In January 1929, a rising star in French classical studies exchanged a few letters with Salomon Reinach. At forty-seven, with several stints as interim director of the École française de Rome under his belt, Jérôme Carcopino, professor of Roman history at the Sorbonne, was preparing his candidacy for membership in the AIBL, a process that by tradition entailed personal meetings with all the voting members.¹⁹¹ Carcopino thanked his elder for sending a copy of a pro-Glozel report, then remarked that, while a student at the École normale supérieure in 1899, he had had the opportunity to attend a session of the Rennes retrial of Alfred Dreyfus at which several experts, including the celebrated Bertillon (testifying about handwriting), appeared. "I had the impression that all this expertise would lead to nothing," Carcopino wrote. "My conviction came from a simple, commonsense idea: if there had been material proof of the accused's guilt, it would not have been necessary to fabricate some." These remarks clearly did not please Reinach, and a few days later Carcopino wrote him again, apologizing for his apparent "stubbornness" and acknowledging Reinach's point that "chemistry is a science and graphology is not." Should the geological report Reinach had sent him not meet with any contradiction, Carcopino said he would yield to it without reservation, but should the scientific dispute continue, as he clearly expected, he would have trouble accepting claims that would overturn everything he had learned about the beginnings of human civilization.[192]

Only the passage of time would make clear the remarkable aspects of this exchange. A scholar known for his style as well as his historical prowess, Carcopino, like Jullian before him, would eventually be elected to the Académie française as well as the AIBL. But although a skilled and energetic administrator, Carcopino was never known for his moral courage: in the year he served as minister of education under the Vichy regime, he faithfully executed anti-Semitic purges of French professors. Though he was arrested after the Liberation, charges against him were dropped, and he was fully rehabilitated in 1951. In his lengthy memoir of the war years he avoids polemic but never apologizes for his actions.[193] This later history makes the boldness of his letters to Reinach, at a moment when Reinach had some say over his career trajectory, noteworthy, even if one concession Carcopino made in the second letter expresses a sentiment at the core of modern archaeology: "The best of reasonings does not hold up against a duly established fact."[194] And what establishes a fact? "Material observations," *constatations*, that keyword used by the Glozel commission.

Fueled as much by belief in the importance of public debate in science as by a clash of egos, Glozel pitted against each other two different visions of archaeology, one a structured discipline with recognizable subfields and areas of expertise, the other a more informal enterprise in which amateurs could still make important discoveries. After Glozel, only a stricter set of rules delimiting professional practice could secure archaeology's scientific status; under this new regime, as Laurent Olivier has put it, "It is the dig and no longer collecting that produces scientific data."[195] Such rules, as Dussaud argued in a lecture in Moulins shortly after the report's publication, would have to be enforced by state regulation of excavations, the norm, he observed, in most countries besides France.[196] In both visions, however, publicity would play an important role. The legislation that brought France into conformity with international norms came into effect in 1942, just two years after the accidental discovery of the Lascaux cave paintings. The new laws imposed state regulation on all archaeological excavations, even those on private property, and restricted the sale of unearthed artifacts.[197] They sit, still, at the heart of the relationship between archaeology, the state, and the public sphere in France. To this day, even with subsequent modifications, and notwithstanding conflicting claims to authorship, they bear the name of the minister responsible for them, Jérôme Carcopino.[198]

✻ 3 ✻

Bodies and Minds

THE WORK OF ARCHAEOLOGY

A motif running through the abundant correspondence between Antonin Morlet and Salomon Reinach recalls Morlet's day job as a doctor whose practice moved between two spa towns, Nice in the winter and Vichy in the summer and fall. In March 1927, Morlet expressed his concern that Reinach's brother and fellow academician Théodore (1860–1928) was experiencing arthritis pain; he went on to propose a little known but very effective treatment.[1] At other times he suggested medication for Reinach's wife's rheumatism and offered reassurances about the accessibility of the Glozel site even to those, like Reinach himself, who had difficulty walking.[2] Yet the image of Reinach's descent in a cart drawn by cows (fig. 2.2), widely publicized just a few weeks later, belied Morlet's optimism, and offered a stark reminder of the bodily infirmities afflicting many of the key players in the Glozel controversy. Later in the spring of 1927, a distressed Morlet reported that Charles Depéret, a geologist and staunch Glozelian, was in the hospital after suffering a serious stroke; he died the following year, aged seventy-four.[3] In 1932 another Glozelian, Emile Espérandieu (1857–1939), sought a medical excuse to avoid testifying in the Fradin family's defamation suit against René Dussaud. He was, he wrote Reinach, "fatigué," that all-purpose French term for feeling unwell, and feared the mockery to which his deafness would inevitably give rise. In his next letter he commiserated with Reinach on the latter's own court appearance: "I fully expected that the Glozel trial would exhaust you," he wrote.[4] Eight months later, Reinach was dead at seventy-four; Camille Jullian died the following year at the same age.

It would be easy to dismiss these exchanges as the physician-fed anxieties of the elderly. Although the scientific study of aging dates to the mid-nineteenth century, scientists have long disagreed on whether old age can be considered a pathology, and phenomenology has been notably reticent about the subjective experience of aging. Yet, as Elizabeth Barry has observed, "foregrounding the body as locus of selfhood makes sense" in the context of aging; aging parallels illness and disability as the source of a perceived disjunction between body and mind.[5] Moreover, matters corporeal recur with great frequency in the archaeological archive, from passing annoyances like boils to family illnesses to chronic diseases that threaten to derail both scholarly work and the *besogne* (drudgery) that sustains it. More than epiphenomenal, questions of embodiment—of how archaeologists used, talked about, and even deprecated their bodies—cast light on their attitudes toward the knowledge they produced, as well as toward the wider world in which they lived.

The anthropologist Thomas Csordas defines embodiment as at once "our fundamental existential condition" and "an indeterminate methodological field defined by perceptual experience and mode of presence and engagement in the world." Csordas borrows the phrase "indeterminate methodological field" from Roland Barthes's definition of textuality, thus likening modes of embodiment to texts that "can interact with each other in endlessly proliferating ways."[6] The simile drives home the ways archaeologists had to negotiate the complex relationship between their corporeal and intellectual selves, and between their bodies and the scholarly field they also sought to embody. Such insights reflect the enduring influence of Michel Foucault, Norbert Elias, and others who explore the connection between technologies of bodily control and larger structures of power and knowledge.

Scholars in the burgeoning field of disability studies, however, have over the past two decades pushed back against understandings of embodiment that emphasize the production of larger social and cultural meanings over narratives of individual experience. Tobin Siebers, in limning "a theory of complex embodiment" that fuses critical theory and disability studies, has pointed out "that the body has its own forces and that we need to recognize them if we are to get a less one-sided picture of how bodies and their representations affect each other for good and for bad."[7] Chris Mounsey proposes that bodies be understood in terms of "variability," of individuals each with a unique set of capacities, rather than as fully capacitated or disabled.[8] Jason Farr's study of connections between disability and queer sexuality in eighteenth-century British fiction starts from the understanding that embodiment has both physical and social dimensions.[9] Such approaches offer valuable ways of probing individual accounts of bodily experience, enriching rather than contradicting the Foucauldian point that, in Siebers's words, "embodiment and social location are one and the same."[10]

My reading of the archaeological archive in terms of embodiment owes much to two pioneering works published in 1998. Equal parts learned, thoughtful, and amusing, the essay collection *Science Incarnate: Historical Embodiments of Natural Knowledge* calls into question the age-old distinction between the bodily and the ideal. Contributors explore, in the words of the editors, Steven Shapin and Christopher Lawrence, "how one might write about the history of scientific (and related) ideas through a history of their embodied forms and vicissitudes," seeking ways of understanding the relationship between knowledge and corporeality that go beyond the old trope of disembodied knowledge. For Shapin and Lawrence, "bodily practices that visibly and publicly portray the status, identity, and worth of knowledge help *create* the notion of what knowledge is."[11] In England beginning in the late eighteenth century, for example, first doctors and then surgeons, the latter a group long lumped together with knife-wielding tradesmen like barbers and butchers, went to great pains to create a more genteel image of themselves. That effort demanded not only suitably grave clothing but the creation, in the nineteenth century, of distinct disciplinary codes.[12] Of particular pertinence to the depiction of archaeologists are the multiple ways in which scientists may or may not be shown with their professional tools.

For those seeking to understand the study of the past in terms of both scientific aspirations and bodily engagement, Bonnie Smith's work remains essential. In *The Gender of History*, Smith engages in close analysis of historical writing in the heroic age of discipline-formation, from the mid-nineteenth century to the early twentieth. In their private correspondence and, occasionally, their published work, historians like Leopold von Ranke, Augustin Thierry, and Gabriel Hanotaux vividly depicted both the materiality of the archives, full of dust, dirt, and traces of decay, and the bodily discomforts they endured in consulting them. They also used bodily imagery, usually fetishistic and often explicitly sexual, to describe the work they undertook in libraries and archives, where, predictably but dramatically, they overcame material obstacles and realized "the possibilities for transcendence" that science offered them. These accounts formed part of a larger project to wrest historical writing from the enlightened amateurs, many of them women, who had hitherto made significant contributions to the field.[13] Although the forms of archaeologists' bodily engagement typically differ somewhat from most historians', Smith's analysis offers a productive framework for considering how archaeologists in the late nineteenth century were also attempting to define their field as scientific, using similar metaphors, dynamics, and exclusions to do so.

This chapter explores the multiple components of the concept and practice of archaeological work in the early twentieth century. Differences between archaeology and that other science of the past, history, though superficially striking, turn out to be largely matters of visibility and articulation. Archaeology is, for the most part, and more explicitly than history, a collective activity, and at the time under study, from the 1890s to the 1930s, a hierarchized one. If the dirty, physically taxing work of excavation typically involved workers with expertise in the mechanics of the local terrain (gravediggers, for example), archaeologists usually supervised that physical labor without fully engaging in it. Indeed, as Natalie Richard has noted, the eschewal of manual labor played an important role in associating archaeology with elite learning.[14] Archaeology as "the discipline of things" must always attend to the materiality of its evidence, whereas historians have only recently begun to do so.[15] In both fields, the work of research, at once physical and intellectual, ideally (though, practically speaking, not always) culminates in a "work," typically a published text, identified with an author or authors who thus claim credit for new finds and findings. Through publication, scholars seek affirmation in the larger body of scholars within and outside the field, reminding even historians who cast their work in individualistic terms that they exist within a collectivity, one that the work, in its multiple senses, serves to reinscribe.[16]

From the standpoint of the history of scholarly work, archaeology does have one advantage over history, however: it tends to produce a more comprehensive visual, and especially photographic, record of its own activity. Photographs have long been part of the work of archaeology, serving as notes, records, publicity, and in some instances sources of revenue.[17] They also, as Christina Riggs has noted, "make visible an embodied experience of labour that sometimes distinguishes foreign archaeologists from indigenous workers."[18] Further, the tendency of photographs to raise questions about their own making, about the position (physical and social) of the photographer, about incidental details that run counter to the central message of the photograph, and about what the photograph leaves out, can both recall the asymmetries built into any collective enterprise and provide a way of incorporating them into a historical account.[19] Photographs and other visual images of archaeologists and archaeological work, in other words, reveal how archaeological embodiment functions as at once flawed process and always incomplete project.

The chapter comprises three parts. In a prelude, the second director of antiquities in French Tunisia, Paul Gauckler, fails to reconcile his bodily drives with his archaeological endeavors, and must resign his post in consequence. At one level a sad story of closeted same-sex desire in a colonial setting, the Gauckler affair, as carefully archived by one of his successors, came to stand as a cautionary tale about the difficulties of channeling bodily effort into archaeological work. The second section probes the way the Frenchmen responsible for archaeology in Tunisia represented both the corporeal and the intellectual aspects of their activity. Photographs of excavations at once complement and contrast with correspondence in which Alfred Merlin, Louis Poinssot, and colleagues from across North Africa commiserate about fatigue, illness, and other obstacles to carrying out their work. In the final section, pictorial evidence offers clues as to how the Glozel affair simultaneously challenged and reinforced the image of archaeologists and archaeological work in the wider culture.

The Archaeologist's Police File

"What a pitiable end!" Merlin wrote Poinssot in a brief note concerning Gauckler's suicide in Rome in December 1911 at the age of forty-five.[20] The entry on Gauckler in Ève Gran-Aymerich's comprehensive and reputable *Dictionnaire biographique d'archéologie* attributes the suicide to an unspecified "illness" (*maladie*) and proffers a somewhat murky version of Gauckler's departure from Tunis six years earlier: "In difficulties in Tunis, where he confronted the hostility of part of the local government and of a few private individuals, he resigned in 1905, returned to France and was entrusted by the Ministry of Education with a stipend for research in Rome."[21] This story conforms in many respects to Gauckler's own running account of his life, which portrays him as the victim equally of personal vendettas and bodily infirmity. Myriam Bacha tells us somewhat more, noting that the director's homosexuality had made him the object of a slanderous campaign in the Tunisian press, rendering his position untenable.[22]

Police reports indicate that although never himself the target of an investigation, on two occasions Gauckler attracted the attention of police working on other cases: a domestic dispute in 1899 and a case of procurement in 1905. The latter investigation produced the most abundant documentation. In their statements to the police, young men describe Gauckler's approaching them in public places, staring at them suggestively, asking if they needed money, then giving them instructions for a later rendezvous. Typically, the assignations began in the colonial city the French were constructing just outside the Tunis medina, often in front of the Grand Hotel; in one instance, Gauckler instructed his trick to look for him to doff his hat and then follow him onto a streetcar. At the director's residence—one of the statements noted that it was full of books—the two men would undress, and, after a period of licking the visitor's body, Gauckler would typically fellate him, though one of the witnesses stated, "at his request I fucked him, then he in turn fucked me."[23] At the end of the encounter Gauckler would pay the young man and send him on his way. The depositions do not report much conversation or negotiation, and they suggest that Gauckler would rarely choose the same partner more than a few times or for longer than a few months. That these acts form part of a pattern emerges in several ways: most of the witnesses state that they had seen other young men, whom they knew only by name or sight, entering Gauckler's house; several report that they were approached by an intermediary rather than Gauckler himself.[24] According to the witness who reported anal intercourse, sixteen-year-old T, "M. Gauckler told me that the police knew what he was doing with young men, that there was nothing to fear since these obscene acts took place in his home."[25]

In an overview of the affair prepared in July 1905, two months after Gauckler's hurried departure for France, a police official called Léal observed that "M. Gauckler has no problem admitting that he is a pederast, but he attributes this vice to an infirmity resulting from his poor health, and the fact that in having unnatural relations with boys older than thirteen he commits no crime leaves his conscience clear."[26] Along similar lines, when C, the only one of Gauckler's partners who acknowledged reaching a climax, asked Gauckler why he swallowed the sperm, he replied, "because the doctors ordered me to."[27] Correspondence between the highest-ranking French official in Tunisia, the resident general, Stephen Pichon (1857–1933), and the director of higher education in Paris, Charles Bayet (1849–1918), that same July reinforced this picture of a tendency rooted in a physical ailment. Bayet, whom Gauckler visited numerous times in an effort to work out a satisfactory exit, found him "very ill and over-excited by sickness and exhaustion; he is convinced that being in Tunisia at this time would endanger his health, and in addition believes himself the object of personal enmities."[28] Pichon agreed that Gauckler was both ill and overexcited, and noted that these conditions were not new, claiming that as resident general he had made special allowances for the director's "unhealthy susceptibilities." Although Joann Freed has established that Gauckler was the object of considerable criticism, including at the Residence, Pichon attributed Gauckler's sense of victimization to his overall state: "M. Gauckler," he declared firmly, "is not—and has never been—the victim of any enmities. He is simply possessed by a vice, which I attribute to his illness and which makes him incapable of filling any position like the one he still holds in Tunis."[29]

A vice attributable to an illness—this concept, and the lack of moral judgment attached to it, in its very nebulousness reflects the shifts and uncertainties in Belle Epoque understandings of sexuality. As Jason Farr has noted for the eighteenth century, "Representations of physical and sensory impairment often anchor heterosexuality to emergent notions of health and ability, relegating its doubles, homosexuality and disability, to the role of unnatural supplement."[30] But the late nineteenth and early twentieth centuries mark a significant moment in the medicalization of same-sex desire, with the invention of the modern concepts of "homosexual" and "heterosexual."[31] Gauckler knew the law: same-sex acts in themselves had not been crimes in France since the Revolution; they could be punished only if they occurred in a public place or involved corrupting minors, and the age of consent, raised from eleven in 1863, was indeed thirteen.[32] Gauckler's and Pichon's sense of the former's sexuality as connected to illness corresponded to the medical consensus of the previous decades, notably the work of Jean-Baptiste Charcot and Valentin Mangan, which continued to treat same-sex eroticism as evidence of a serious disorder. Unless he followed the medical literature closely, Gauckler is unlikely to have been aware of the first scientifically credible assertion of the normality of same-sex desire, Marc-André Raffalovich's monograph on what he called *unanisme*, published only in 1896 and for decades to come a minority opinion. Indeed, descriptions of the director's "unhealthy susceptibilities" echo many of the characteristics ascribed to the pathological type of the hysterical male, who, according to Vernon Rosario, was "regularly accused of effeminacy and of possessing an excessively impressionable imagination, supposedly like that of women and children."[33] Yet the medicalization of homosexuality had not entirely displaced the legal peril: in 1904 the trial of the socially prominent Jacques d'Adelswärd-Fersen for indecent assault and incitement of minors attracted much media attention and likely frightened closeted men like Gauckler.[34]

In some respects, Gauckler's object choices too resonate with what we know about same-sex desire among elite European men of the time. Although the documents provide no details of their appearance, his partners all gave their ages as between fifteen and eighteen at the time of their initial encounters with Gauckler. The director may well have had a taste for the slim, boyish type known as the ephebe, and in his liaisons could have cast himself in the senior role in the tradition of ancient Greek pederasty, although his alternation between top and bottom roles diverges from that model.[35] But some aspects of his encounters differ from the Orientalist sexual tourism briefly but influentially evoked by Edward Said.[36] All the witnesses identified themselves as Italians, indicating that Gauckler did not seek out Arab or Berber partners. He may have preferred ephebes of the types staged, photographed, and widely if discreetly distributed by the gay photographer Wilhelm von Gloeden, who was based in Sicily, which had longstanding ties to Tunisia.[37] André Gide, in rapturous accounts of his travels through North Africa around the same time, pioneered a still durable trope of same-sex identification with, as Jonathan Dollimore puts it, the "cultural and racial other."[38] In Gide's travel diaries published under the title *Amyntas* in 1906, the sinuous byways of the casbah and medina prefigure and intensify the thrill of discoveries at once erotic and exotic.[39] Gauckler, in contrast, preferred to cruise the modern city; one of the witnesses, C, said he first encountered the director "under the arcades of the Avenue de France," a prominent thoroughfare in the colonial city where another trick, M, had been working at a construction site. M noted that his first encounter had lasted about an hour, and ended with payment and a summons to return the following Tuesday at the same time, an unsentimental efficiency that smacks more of the department store than the *souk*.[40]

A specialist in Roman archaeology, and a resident of Tunis for over a decade, Gauckler was, strictly speaking, no more an Orientalist than he was a tourist. Yet, as Jarrod Hayes has usefully observed, sexual tourism, once it is commodified and committed to print, no longer necessarily entails travel.[41] Gauckler was clearly making use of the benefits his position in a colonial power structure afforded, notably money and the privacy of his own home, to seek sexual pleasure. "In the escape from sexual oppression and sexual repression," Dollimore writes, "individuals have often, like Gide, crossed divisions of class and race."[42] Class certainly—Gauckler's sexual partners included, besides the construction worker, a waiter, a chauffeur, and a tailor. Race perhaps, given that southern Italians were often classified as "dark." And, no doubt, cultural difference as well. Eve Kosofsky Sedgwick remarks on the irony of the gender of object choice becoming the key determinant of "sexual orientation," a concept anything but self-evident in Gauckler's day. But the presence of this evidence in a police file drives home her observation that "the now chronic modern crisis of homo/heterosexual definition has affected our culture in its ineffaceable marking of the categories secrecy/disclosure, knowledge/ignorance, private/public . . ." and more—she lists a total of twenty-two binaries.[43] It is this "marking," even more than Gauckler's obvious internalization of homophobia, or of pseudoscientific diagnoses of his sexuality as a disorder, that brought his private life into the archaeological archive.

Gauckler hurriedly left Tunis for France in the third week of May 1905. Over the next several months, he worked his networks in Paris and Tunis, pleading with Bayet for interim arrangements that would save face, such as a medical leave, while desperately seeking news as to whether his police file would result in charges. At one point he suggested that his contact in the Residence send a coded telegram, using archaeology as the code, to tell him what was happening: "'Excavation credits awarded'" if the prosecutor had decided to take no action, "'suspended' if there is still danger. 'Denied' if the matter is still to be decided." In the latter case, Gauckler declared his intention to return to Tunis to defend himself, saying he had already written the necessary statement.[44] In his letter to Bayet in July, Pichon, expressing real regret, intimated that Gauckler's misjudgments rather than his sexual preference made his departure unavoidable. Léal's report on Gauckler noted that in 1903 the director had resorted to intimidation to try to keep a potential source from creating a scandal, while more recently the contact known as T was preparing a knife attack on one of his rivals for Gauckler's attention. "A scandal was sure to erupt one day," the report concluded.[45] His mood oscillating between panic and depression, Gauckler vainly sought a meeting with Pichon during the summer holidays. When he finally submitted his resignation in September, he declared to a confidant, "Now I have nothing left but to die, and I am preparing to!"[46]

As the endnotes testify, the building blocks of this narrative come not from a police or diplomatic archive but from the Poinssot papers, where the archival moments devoted to a file blandly and misleadingly entitled "Procès Paul Gauckler (vers 1899–1905)" were among the most memorable of my career. How the Poinssots acquired materials from outside the family is not always clear; in most cases it was probably by gift. But this file almost certainly did not come from Gauckler's heirs, as it contains materials, notably the Bayet–Pichon correspondence, to which Gauckler himself would not have been privy. Close scrutiny of the documents reveals that both Bayet's and Pichon's letters are copies in the hand of Bernard Roy (1845–1919), who was Gauckler's primary contact in the Residence. As secretary-general of the Tunisian government for thirty years (1889–1919), Roy, a distinguished Arabist, was the highest-ranking French civil servant in Tunis, the chief intermediary between the French and the nominal Tunisian state. He was also an old friend of the Poinssot family—Poinssot's father Julien (1844–1899), also an archaeologist, had met Roy while the latter was a telegraph operator and spy in western Tunisia—and Louis Poinssot's protector from the time of his first trips to the Protectorate at the turn of the century. Only someone as highly placed and discreet as Roy could have copied out letters to and from Pichon, the man to whom he reported, while also obtaining official copies of the police reports. Roy had no family of his own, and Poinssot seems to have acquired all his scholarly papers not long after his death; his wife Paule Poinssot completed and brought to press Roy's edition of the Arab inscriptions of Kairouan.

The question remains, however, of why Roy and Poinssot would have considered this material worth preserving—and, indeed, of curating as its own virtual archive, separate from the rest of Roy's correspondence. A clue emerges in one of Gauckler's preoccupations in the months between his departure and his resignation: preventing Louis Carton from succeeding him as director. "I know," he wrote Roy in June 1905, "that Dr. Carton is already announcing *urbi et orbi* that he is going to replace me as head of the Antiquities Service."[47] Gauckler and Carton had had a close relationship over the years: the director had supported Carton's applications for excavation subventions from Paris, which he had coordinated with Gauckler's own program. In 1904, however, growing mistrust led Carton to begin a letter to the director, "I really don't believe I can any longer call you a friend, after all the malice toward you for which you hold me responsible." A conciliatory reply from Gauckler led to a temporary warming, but Carton was soon again denying that the Sousse Archaeological Society, which he had founded, had any role in newspaper articles hostile to the director. (Freed has concluded that they were "almost certainly planted" by Carton.[48]) Roy quickly reassured Gauckler that his preferred successor, Merlin, had a lock on the directorship, noting that it was not even clear that Carton would become a candidate.[49] But the previous year's exchange suggests that Carton loomed large among the *inimitiés* of which Gauckler was complaining at the time of his departure, the warmth of *amitié* having stoked the fire of its opposite.

So does this carefully assembled archive amount to nothing more than another artifact of Poinssot's mortal quarrel with Carton? Gauckler supported Poinssot's work in Tunisia at the very beginning of the twentieth century, and his letters to the young archaeologist reflect a strong mentor-mentee relationship, occasionally stern but usually warm.[50] Soon after leaving Tunis for the last time, Gauckler anxiously asked Roy whether Poinssot had any inkling of what was happening; Roy replied that he did not.[51] More generally, Gauckler's fretting over press coverage, both in his correspondence with Carton and after his departure, could have shaped Poinssot's own suspicion of the media.[52] Yet the significance of the Gauckler archive arguably goes beyond the personal. In putting together this dossier and entrusting it to Poinssot, Roy likely intended it as a cautionary tale about the proper relationship between bodies and minds. In early sexology, Heike Bauer observes, the body "came to be understood as a product of the past and a marker of the future, a measure by which individuals and larger groups of people could be identified and classified in relation to normative ideals about civilization and progress."[53] Virginia Woolf made a similar point in a different way, describing the activities expected of the healthy, "to communicate, to civilize, to share, to cultivate the desert, educate the native," and saying that in illness, "we cease to be soldiers in the army of the upright; we become deserters."[54] However they understood the goals of their own scholarly enterprise, the positions Roy, Gauckler,

Merlin, and Poinssot occupied as colonial officials placed them within this same normative framework, one actively engaged in identification and classification. Further developing—even, perhaps, subtly challenging—such "normative ideals" through conscientious scholarship required at a bare minimum that they control their own bodies. Gauckler's story offered a chilling example of what might happen if they could not. Its constitution as an archive was clearly intended as a warning; everything we know about Poinssot, in both his public and his private lives, suggests he heeded it.[55]

The history of archaeology does not lack for colorful characters, mortal feuds, and bad behavior of many kinds, some of them familiar in all walks of life.[56] Two decades after Gauckler's forced exit, for example, the alcoholism of one of the American archaeologists working at Carthage aroused considerable concern among the team's leaders.[57] Some of these stories, like that of Gauckler, have a poignancy that illuminates their connection to histories outside of archaeology. But the very human interest of these stories also raises important questions about the practice of archaeology in the era of discipline formation. It was not simply a matter of the representatives of official archaeology enforcing standards or setting a good example. Whether in a domestic or colonial setting, archaeologists in the early twentieth century operated as agents of various kinds of authority. Within the dig site, however eagerly or reluctantly, they embodied the power over society assumed by educated white European (or Euro-American) males. At the border of and beyond excavations, archaeologists constructed, even as they sought to wield, the authority of scholarship, of *science* in the broad French sense. Ideally, these forms of authority would be mutually reinforcing, but a flaw or weakness in one could easily undermine the other, and with it the whole edifice that made archaeological work possible.

Bodies at/of Work

Jason Farr has elucidated how, from the Enlightenment forward, the normative linking of reproductive health, heterosexual desire, and primogeniture created a parallel link between illness and queer desire: in a key example, a character in Horace Walpole's *Castle of Otranto* (1764) "stands outside of the normative frameworks of heterosexual desire and patrilineal succession due to his 'sickly' constitution."[58] This complex of illness and abnormality loomed large in Gauckler's communications with Roy and Bayet in the anxious summer of 1905, as he strove desperately to attribute his departure from Tunis to poor health. Certainly no one who knew him would have had reason to doubt this explanation. Gauckler's earlier letters to Poinssot continually invoke various maladies, from the flu to bronchial laryngitis to "a cholera-like epidemic caused by the deplorable sewer system with which the city of Tunis has been poisoned."[59] While Gauckler may have invented or exaggerated his "sickly constitution" to mask his sexual desire, in so doing he reinforced the conceptual and discursive linkage that he actively sought to embody (in the most literal sense of the term). In keeping with the insights of disability studies, this is in no way to minimize the physical symptoms Gauckler described. But Gauckler's case history and its archival traces remind us that archaeologists' bodies carried multiple charges and associations, including but not limited to the qualities they sought to embody, and not all of which they could control or even monitor. The complex relationship between individual bodies and the body of archaeological work can be understood only by keeping in mind that these are always mutually constitutive, even if stories of individual embodiment rarely come down to us in the kind of detail the Gauckler archive affords.

Gauckler also complained of being overworked (*surmené*), leading to a physical fatigue so overwhelming that on one occasion he had to repair to his bed. Such symptoms would likely not have raised an eyebrow at the time, however much mention of his bed may do in retrospect.[60] For similar complaints crop up frequently in Poinssot's correspondence, many echoing Poinssot's own (absent) letters. At the very end of 1912, Merlin sent Poinssot, by then his deputy, sympathy for the painful subcutaneous infections—boils, carbuncles, and abscesses—then afflicting him: "they must," he wrote, "be very painful and bring with them a whole parade of suffering and remedies that can only be very disagreeable."[61] A few months later, Merlin reported that he himself had been consigned to bed with compresses over his eyes.[62] As Poinssot entered middle age, with a number of young children at home, his correspondents regularly expressed concern about his health. In early 1927, for example, Eugène Albertini (1880–1941), Poinssot's counterpart in Algeria, wrote of his hope that Poinssot had recovered from the flu of which he had been complaining.[63] In all these cases, as Janet Browne has persuasively demonstrated with respect to Charles Darwin, ill health formed part of a repertoire of resources that shielded scholars from unwanted social and professional obligations.[64]

In a world before antibiotics, in a climate that northern Europeans understood as dangerous to their health, concerns about illness and wishes for recovery stemmed from more than routine politeness. Alain Corbin has shown that despite general awareness, beginning in the 1880s, of Louis Pasteur's influential discoveries, the search for microbes did not immediately displace previous ways of thinking about the body as subject to general environmental influences as much as to specific sanitary conditions.

The concept of coenesthesis, a consciousness of one's health rooted in the totality of bodily sensations, coexisted with medical advances focusing on specific symptoms.[65] Thus Georges Marçais (1876–1962), another Algeria-based scholar and museum director, could write Poinssot as late as 1936 that a severe E. coli infection had kept him in bed for two weeks, and that although he was again sitting up, "the collibacilli have not yet abandoned my kidneys and other battlefields." The stakes of this battle come in the following sentence, when Marçais says he hopes to resume normal activity the following week. "It's more than time," he concludes, "because work is pressing."[66] A few years earlier, when Poinssot was recovering from eye problems, Marçais wrote, "I'm happy that you are rid of these miseries and the reduced activity to which they must have condemned you."[67]

But how did archaeologists understand the work that caused them such fatigue? Typically, *surmenage* came from tasks archaeologists regarded as *besogne*, a term introduced in chapter 1 and most simply translated as drudgery: paperwork, routine personnel matters, budgets, and endless meetings. Such humdrum tasks took scholars away from actual work, or *travail*, a term that usually referred to intellectual labor. The assignment of particular tasks to one category or another varied among individuals, and the borders between categories could be blurry: in the fall of 1923, Poinssot's deputy Raymond Lantier sent him wishes for a year "even more full of scholarly chores [*besognes scientifiques*]," but he set this in direct contrast with the truly disagreeable work of dealing with Dr. Carton and his acolytes.[68] Lantier's correspondence with Poinssot during the summer holidays conveys a good sense of what constituted the work he valued but found too little time to do at other times during the year: installing new acquisitions in the Bardo Museum, sending periodicals to a binder, and, above all, writing for publication, an enduring preoccupation for archaeologists and often a collaborative effort.[69] Not simply the expected result or goal of excavation, publication created its own dynamic, as when Carton wrote, after a period at home in Lille, that he was running out of material to publish and would soon need to return to Tunisia to dig for more.[70]

Fieldwork, which many archaeologists have long regarded as, for good or ill, the core of their disciplinary practice, occupies an odd median position in the spectrum running from *besogne* to *travail*. Despite the inevitable physical discomfort, few complained about conditions in the field. When Poinssot began work at Dougga in the spring of 1903, Gauckler wrote that he was glad "that you are happy camping [at Dougga]. It must certainly be lacking in comfort, but it is much the best solution for your projects [*travaux*], and it spares you the waste of time and fatigue of a daily commute from Teboursouk," the nearest town. Although he would lose

FIG. 3.1. "Carthage—Nouvelles fouilles au Théâtre." Postcard, ca. March 1905. Photographer unknown. Bibliothèque de l'Institut national d'histoire de l'art, Collections Jacques Doucet, archives 106, box 48, dossier 4. Published with permission.

out on conversations with the locals, Gauckler continued humorously, "I know you are enough of a stoic to sacrifice yourself without protest on the altar of science!"[71] It took unusually extreme conditions for an archaeologist to complain. When Merlin began a letter to Poinssot, "So many challenges and worries! So many setbacks and so much lost time!" it was 5:30 a.m. and he was sitting in a boat monitoring one of the first significant underwater excavations, of a Greek vessel that sank off the Tunisian coast in the first century BCE.[72] Known by the name of the coastal town closest to the site, Mahdia, the excavation, which ran for several seasons, made Merlin's reputation and arguably his career.

The frustrations of excavation could take a number of forms. At the end of June 1923, Lantier wrote that exploration of a hoped-for Punic site at Carthage had thus far yielded nothing, but that he was pursuing it in the interests of site integrity. In the same letter Lantier described answering a summons to see Henri Doliveux, the director of public instruction and Poinssot's direct superior, with whom Poinssot was on bad terms; he arrived, he wrote, "covered in sweat and dust to show him that he was disturbing me on my return from Carthage."[73] Lantier's body indexes the specificity of the work of archaeology, one linked to his subject position by the first-person pronouns with which he describes work on the site: "I nonetheless continued the trench from east to west"; "There the trench gave me nothing."[74] Lantier probably had the time to change clothes before reporting to Doliveux; his work that day had consisted of showing the site to an incognito visitor, the Czechoslovak president Tomás Masaryk. The choice not to do so gave his appearance a performative function in the sense to be discussed in chapter 4. Yet those first-person pronouns also perpetrate a deception, albeit one so conventional that Poinssot, the recipient of the letter, would have perfectly understood their sense.

For the physical work of excavation, the literal heavy lifting of dirt and rock to clear the spaces where archaeologists deployed their trowels, was almost invariably carried out by local laborers. Lantier's "I," in other words, incorporates hidden others, whose names, as Nick Shepherd observed in a pioneering article on archaeological labor, only rarely find their way into the historical record.[75] Site photographs provide evidence of their existence, and of the physical effort their work entailed: in a view of the 1905 campaign at Carthage, considered suitable for use in a postcard (fig. 3.1),

FIG. 3.2. Photograph from the sequence "Premiers sondages; premières tranchées," Roman theater, Carthage, Summer 1904. Photographer unknown. Bibliothèque de l'Institut national d'histoire de l'art, Collections Jacques Doucet, archives 106, box 48, dossier 1. Published with permission.

FIG. 3.3. Photograph from the sequence "27 novembre–1er décembre 1904," Roman theater, Carthage. Photographer unknown. Bibliothèque de l'Institut national d'histoire de l'art, Collections Jacques Doucet, archives 106, box 48, dossier 2. Published with permission.

the use of carts on rails does not diminish the effort demanded of the figure pushing one. Other photographs, as Riggs notes, offer "glimpses of personal interactions, physical contacts and haptic gestures."[76] In an image from the Roman theater of Carthage in summer 1904—one of six pasted to an album sheet labeled, in Gauckler's hand, "First Surveys, First Trenches" (fig. 3.2)—the wielding of shovels has an almost rhythmic quality, the slight blurring evidence that the photograph was not posed. To the extent that, as Riggs argues, photographs of archaeological workers reflect their relation to each other as well as to the camera, the evidence is ambiguous. In one photograph (fig. 3.3) of the Carthage dig in fall 1904, workers in the foreground seem to be gesturing toward the camera; one figure points out another with a smile suggesting a friendly acquaintance. In a second (fig. 3.4), labeled with the same dates, a pensive European man almost literally, in photographic terms, consigns the workers to the mid- or background, his stillness an index of cogitation, perhaps reflection on the next steps in the process.[77] In another photograph from the same series, the photographer's trace looms as the shadow of a man's hat in the foreground.

The physical effort evident in the photographs of archaeological labor strikes the viewer in the manner of a Barthesian punctum, an incidental detail that pricks the eye, rather than the primary meaning of the photograph.[78] Those primary meanings vary within a fairly narrow range: the sequential images record for the archeaeological archive the "irreversible actions" that durably alter the site.[79] A scene judged worthy of transfer to a postcard, such as figure 3.1, pictures both a process judged newsworthy—"New excavations at the Theatre," the caption states—and a glimpse of artifacts of which archaeology celebrates the discovery. But all these photos also image reassuringly well-organized, purposeful, and productive colonial labor. As the camera, and by implication the viewer, assumes the position of the watchful European supervisor, the gaze absorbs, essentially subsuming, the difference signified by the workers' dress. Of course, that otherness could be enlisted in other types of denotation, as in an image

FIG. 3.4. Photograph from the sequence "27 novembre–1er décembre 1904," Roman theater, Carthage. Photographer unknown. Bibliothèque de l'Institut national d'histoire de l'art, Collections Jacques Doucet, archives 106, box 48, dossier 2. Published with permission.

UN VANDALE MODERNE
(Le chercheur de pierres tient à la main le couffin à l'aide duquel il a remonté la terre de son trou, que l'on aperçoit à droite. A gauche, on voit un chapiteau ionique brisé. Derrière lui, le tas de pierres arrachées aux monuments qu'il exploite comme carrières.)

FIG. 3.5. *Un vandale moderne*, full-page photograph in Louis Carton, "Pour Carthage! Histoire d'une ruine," *Revue de Carthage* 13 (1906): 369. Photographer unknown. Public domain. Image courtesy Bibliothèque Gernet–Glotz, Paris.

published as the frontispiece to an article by Louis Carton two years later (fig. 3.5). The figure characterized as a "modern vandal," the highly didactic caption tells us, is exploiting an archaeological site as a personal quarry, using something resembling a baby carriage to remove stones for new construction. Having himself called for and supervised the demolition of houses viewed as obstacles to excavation, Carton well understood the privations archaeology could inflict on locals. On at least one such occasion, he consulted with Gauckler about the possibility of making usable rubble available so that those whose homes were expropriated could rebuild nearby.[80] Yet this photograph directs the viewer's compassion not to the glowering "vandal" but to the broken Ionic capital to the left.

European archaeologists rarely expressed much sympathy for those doing the physical labor of excavation. Supervising or managing these workers, indeed, emerges in many letters as another form of "work" that archaeologists found frustrating but inescapable. The behavior of the Greek divers working on the wreck loomed large among the annoyances of which Merlin complained at Mahdia; in 1910 he had to make a court appearance after some of them got into (unspecified) trouble with the law.[81] More frequently, the wage demands of local workers presented a real challenge to ambitious excavation programs: in late 1919, Merlin wrote that "at

Sbeitla, the Arabs will no longer work for less than 3fr per day; at Carthage, it's 4.50." A few years later, Lantier lodged similar complaints about the eighteen "indigènes" he had hired at Dougga, but commented, "I assure you that I am making them earn their money."[82] Typically only workers filling managerial roles receive mention by name: "Ahmed Cherif," Lantier wrote a few weeks later, "has acquitted himself very well of his responsibilities as foreman; he is very serious and well understands how to carry out the work." He also managed to impose a slightly lower salary scale than the workers were demanding.[83] The memoirs of the American archaeologist known as Byron Khun de Prorok contain so many disparaging references to "Arab" workers as lazy, greedy, infantile, and excitable that these slurs become part of his underlying message. For Prorok, who, unusually, describes his own occasional physical exertions on the dig, the seriousness of Euro-American archaeology emerges in part through contrast with the childishness of the racial others who performed its physical labor.[84]

Stephen Quirke and Bonnie Effros, among others, have observed that workers who lived in and around the area under excavation often brought significant local knowledge to the archaeological site.[85] Gauckler may have had this in mind when he commented that the efficiency of making camp at Dougga came at the cost of potentially valuable exchanges with the locals (he may also have made the comment facetiously). Even Prorok acknowledged that some Tunisian workers had an innate sense of the territory where they were digging, even if he attributes this to self-interest.[86] Among the scholars working in Tunisia in the first decades of the twentieth century, none had a clearer or more generous sense of the contribution of native workers than Georges Marçais, whose field of interest, medieval Islamic art, and knowledge of Arabic probably had something to do with this attitude. Although based in Algiers, where he was a professor and curator of the museum of ancient and Islamic art, in the 1920s Marçais carried out several excavation campaigns in Kairouan, Tunisia's medieval holy city. In his first letter to Poinssot, Marçais wrote that, "while not a total novice in archaeology, I still have much to learn," and hoped the director would be able to advise him.[87] The following year, describing the challenges he faced in excavating a residential structure built not of stone, as he had expected, but of crushed brick, Marçais acknowledged that this confusion had cost him a good five days of work. The materials made it difficult to distinguish between the walls and rubble, necessitating great care: "The workers and their boss must learn from experience." He went out of his way to credit his "chief laborer," whom he did not name but whom he had already promoted, for discovering some intact glass carafes inside an earthen vessel.[88] Marçais also trusted his workers to act as local informants and guardians; in 1925 he mentioned giving a bonus to one Ali, who had gathered chance finds of pot shards in expectation of his boss's return.[89]

One type of work, scholarly publication, emerges in archaeologists' correspondence as the most valued, no doubt because of its close connection to credentials and careers. Soon after Poinssot assumed the directorship in 1920, Cagnat put him on notice: the funding his department received from Paris would dry up without regular bulletins about finds. "But let's be clear: don't think we are expecting a scholarly treatise from you every time"; a photograph or a squeeze (*estampage*) of new discoveries would suffice. This was, Cagnat made clear, a "tradition religiously followed by all your predecessors."[90] In a letter to Poinssot during the war, Merlin had called this type of reporting "feeding the North Africa Committee," and grumbled that even in normal circumstances, "I scarcely have any free time, as you know, and it is a big effort."[91] The Ministry of Education's North Africa Committee met monthly in Paris and published its proceedings in the *Bulletin archéologique du Comité des travaux historiques et scientifiques*.[92] A single report from Merlin in April 1908 comprised no fewer than eight separate discoveries, mostly of inscriptions, brought to his attention by his contacts all over Tunisia; in 1908 alone, his third year as director, he had seven such reports in the *Bulletin*.[93] Directors not only brought together reports from those they had commissioned, subsidized, or simply authorized to conduct excavations, they often revised them to play down difficulties and setbacks. One can practically see Marçais winking as he sent Poinssot a report on his disappointing preliminary dig at Kairouan in December 1921, with the hope that, in Poinssot's hands, "my half-failure will take on the gleam of a genuine success."[94] Like most things, scientific truth could benefit from some light editing.

The published work sometimes represented a kind of academic surplus value: it could make careers while not actually advancing knowledge. In the fall of 1909, Merlin was asked to present a report to the AIBL on that year's campaign at Mahdia; although he begrudged the time this would entail, he told Poinssot that he had no choice, since the Academy had helped finance the very costly expedition. A year later, he noted that his report to the AIBL, as covered in the press, had gone very well, with Salomon Reinach, among others, calling the discovery the greatest find of bronze statuary since Herculaneum and Pompeii. With a trace of exasperation, Merlin predicted that the praise would not lead to an increase in support for the excavation. Moreover, "as a consequence I could have done without," several academicians were insisting that the finds be published not only in the Academy's proceedings but in its most prestigious and lavish publication, known by the name of its benefactor as the *Monuments Piot*.[95]

The first of several articles to appear in this journal focuses on just two bronzes, one of which Merlin attributes to Praxiteles. It begins with the sponge fishermen's discovery of the wreck in June 1907, but its second footnote refers to nine previous reports already published by Merlin and Poinssot.[96] Minus the photos, the attentive reader could have gleaned all the pertinent information from the daily *Journal des débats* two days after Merlin's initial presentation.[97]

At once burden and reward, publication also presented a complex web of rivalry and collaboration, as scholars rushed to get into print first.[98] Throughout the archaeological archive, scholars worry about mail service as they send each other the raw materials of research—notes, drawings, photographs, sometimes even books—as well as markups of drafts and proofs.[99] In some instances the final publication literally becomes its own reward, as when Albertini tells Poinssot that he can pay him only in offprints.[100] Along with offprints, photographs constituted the most valued currency of scholarly collaboration and exchange, the most worried over when late or lost. Well after his return to the metropole, Lantier continued to coauthor articles with Poinssot based on research he had carried out in Tunisia. Such collaboration reflected deep mutual trust and esteem, but also the slowness of archaeological publication. It also enforced rules of priority and, occasionally, flew the national flag. During Lantier's term in Tunis a report that he was cowriting an article with an Italian scholar for an Italian journal raised eyebrows on the North Africa Committee because the Italian was listed first, "as if Tunisia were an Italian domain."[101] A few years later, when the Yale scholar Michael Rostovtzeff (1870–1952) asked Lantier if one of his students could write a thesis on Dougga, the latter told him firmly that he and Poinssot were preparing a major book on the subject, that it had occupied the better part of Poinssot's career, and that they regarded it as their territory. Rostovtzeff, Lantier reported to Poinssot, quickly backed off.[102] Around the same time, Poinssot told Rostovtzeff that he would of course authorize another American scholar to study the monuments of Dougga, but reminded him of the principle that French scholars had priority on publishing "discoveries made in the course of excavations for which France has unceasingly made the heaviest of sacrifices." The director blamed the delay in publication on "the misfortunes of time" but said that copiously illustrated books were forthcoming.[103]

Rivalries were not necessarily international. Carton had a strong competitive streak, and Gauckler accused him of taking credit for the discoveries of others—likely including Poinssot—in order to bolster his candidacy for correspondent membership in the AIBL.[104] Although Carton's later disparagement of Italian and German scholarship no doubt reflected his genuine feelings, particularly in the aftermath of World War I, it can also be regarded as something of a smokescreen.[105] On occasion, scholars would admit to other motives: in thanking Poinssot for contributing to a journal he had just revived, for example, Albertini wrote that he hoped it would become indispensable to scholars and thus a commercial success, because we are "motivated by such low thoughts of propaganda and lucre." Lantier, commenting on an article he had drafted for *L'Illustration*, went so far as to use the dreaded word for publicity, though his scare quotes—"in the current state of things, we should not neglect the question of '*réclame*'"—were perhaps intended to keep it at a safe distance.[106]

In a letter to Poinssot in 1935, Marçais admitted that he was enjoying working on a project aimed at the general public, a new edition of the volume on Tunis and Kairouan for a trade series called "Les Villes d'art" (art cities).[107] Given his temperament, Marçais perhaps enjoyed the opportunity to put more of himself into a work than scholarly conventions allowed, but no less a figure than Cagnat also published in the series. Ian Hodder, in a concise but suggestive study of British archaeological reports from the eighteenth to the twentieth centuries, argues that the emergent discipline asserted itself by evacuating embodied first-person voices and replacing them with dry, technical accounts, typically in the passive voice. "The authority of the text is no longer personal," Hodder writes, "but lies in adherence to abstract codes. The self, history and uncertainty must be denied."[108] A similar study of French reports might yield slightly different results, as French archaeologists never completely abandoned the literary flourishes they had learned at *lycée*: Merlin and Poinssot's first report on Mahdia, for example, quotes Callistratus's *Descriptions* to justify their attribution of a statue to Praxiteles.[109] But archaeological publications in France followed the same general trajectory, with individual agency giving way to the larger authority of the discipline, the collective body on which archaeologists depended for sustenance. In this way, they replicate the subordination of the physical act of excavating—and of the workers who carry it out—to the scholarship it makes possible. Certainly the Mahdia reports do not mention the Greek divers or the rocking boat from which Merlin monitored them.

In their scholarly publications, the "works" archaeologists most valued, the printed page smooths out the personal quirks, infirmities, emotions, and desires that threaten to derail scientific achievement, but that smoothing is always provisional and never complete. The strain posed by this subordination of body to mind emerges in some genial correspondence between Marçais and Poinssot about a perpetually delayed publication on medieval Islamic bindings from Kairouan on which they were ostensibly collaborating. Marçais's letters betray a fondness for metaphors: in 1923 he compared the two reports he had to submit, one to Poinssot and one to another funder, to a hare to be served with two sauces.[110] Poinssot clearly kept him up to date on additions to his family and their health, and Marçais alternately congratulated and commiserated with him. This element of their correspondence makes it difficult to determine when Marçais is speaking metaphorically, but the image of the publication as newborn crops up with some regularity in his letters. Noting in 1925 that he had not yet received the proofs he was expecting, Marçais wrote, "It's not that I'm impatient to see my youngest appear. He will enter the world in due course."[111] A decade later, referring to the book on bindings, Marçais conveyed both understanding for the director's other commitments and a trace of urgency: "I would, I admit, be happy to know that the child will soon be born." The same metaphor comes up in a letter sent, as it happens, nine months later: "Without being terribly impatient, I would like to see our great work completed. Do you think the baby could see the light of day in 1938?"[112]

The letters in which these metaphors occur invariably cover a number of professional and personal matters, and Marçais may have intended simply to inject a lighter tone, and perhaps slyly to evoke Poinssot's personal situation as the father of a growing brood and a man with many demands on his time. Nonetheless, allusions to an expected child suggest several things beyond well-meaning male cluelessness, some of them contradictory: that work follows natural rhythms, that its completion is not entirely in the hands of those carrying it out, that the outcome is uncertain. The trope may even be seen to invoke a kind of mystery in the process of producing work. Work seems to involve a passage through a strange body, the female as usual signifying both nature and the exotic, before coming to "the light of day." However we interpret the metaphor, its persistence places in a different register the body-mind conundrum posed by the Gauckler story. Marçais is effectively marking that relationship as a lifelong agon, the ultimate rewards of which—the durability of a name or reputation—scholars by definition do not live to see. Thus endowed with a life force, archaeology and science in general also seem fragile, their ultimate fate unknowable.

Looking Like an Archaeologist

Two images, created in different media several decades apart, offer rough visual approximations of the processes through which archaeologists identified themselves with their work. In a print by an unknown photographer from a cracked negative (fig. 3.6), a group of European men in dark suits and hats visits the Roman theater of Dougga around 1900. In the middle ground, a bewhiskered figure holding a walking stick appears to be

FIG. 3.6. Visitors in the Roman theater at Dougga, early twentieth century. Photographer unknown. Bibliothèque de l'Institut national d'histoire de l'art, Collections Jacques Doucet, archives 106, box 55, dossier 2. Published with permission.

addressing the men, suggesting that a tour is in progress. Most of the other figures are standing within earshot of the speaker, though a few are seated on the architectural fragments, and in the foreground one is leaning with his arms draped over a cornice, left boot crossed over right, his body almost one with the stone. The bodily contact with archaeological fragments carries over to a 1927 painting by Giorgio de Chirico (plate 1), one of the first in a series of works the artist executed over more than four decades, in media ranging from lithography to sculpture.[113] De Chirico used mannequins in a wide range of works from the mid-1920s onward. Abstracting facial features and other distinguishing marks allowed the artist to explore the ways people identify themselves and each other through costumes, poses, and appurtenances.[114] In *The Archaeologists*, however, he has gone even further, cloaking the mannequins in antique garb and literally constituting them of the stuff of their work, which, he suggests, may also be their passion or obsession. The different settings—an archaeological site in the photograph and a generic interior in the painting—makes the problematizing of the archaeological body in both images all the more striking.

Taken together, photograph and painting also suggest that, until the 1920s at least, no standard image of the archaeologist existed; this is unsurprising given that the visual representation of much more familiar scientific professionals, physicians and surgeons, had begun to crystallize only a generation or so before.[115] It is not even clear that the men in figure 3.6 are archaeologists; the group leader bears some resemblance to other depictions of Bernard Roy, who was primarily an administrator, in his spare time a scholar who compiled Arabic inscriptions, but not an archaeologist. Some of the figures were probably visitors to Tunisia, people Gauckler alternately boasted and complained about in correspondence for the time and attention they required: attendees (*congressistes*) at a conference of the Ligue de l'Enseignement in 1903, a parliamentary delegation in 1904. But only the most intrepid visitors made the trip to Dougga.[116] In parallel to the textual descriptions in the previous sections, archaeological photography typically shows only locals engaged in physical effort; even the most generous and considerate Europeans, like Marçais, distinguished these workers from archaeologists themselves. Exceptions, like highly posed photographs of Howard Carter at crucial stages of the Tutankhamun tomb opening, still show Europeans in leading or supervisory roles.[117] Photographs of Gauckler and Poinssot tend to picture them in their studies, in a jacket or shirtsleeves, surrounded by books and papers, like any other bourgeois professional with university training. That archaeologists would resemble other Europeans of the same class more than their local workforce makes sense in that, as Riggs observes, "archaeologists were, after all, tourists of a kind themselves."[118] But the resemblance also reinforces the convergence between sociopolitical and scientific authority that archaeology both depended on and actively constructed.

By the time of the Glozel affair, images of the archaeologist had stabilized around a set of standard features, as indexed in a cartoon (plate 2) by Henri-Gabriel Ibels (1867–1936). Published on the cover of the satirical weekly *Cyrano*, the drawing brings together two current news items, Glozel and a minor financial scandal involving postwar Hungarian bonds with faked seals that increased their value in certain countries, including France.[119] Alluding to the Glozelians' claims to have discovered the earliest alphabetic writing, Ibels has the archaeologists examine what appear to be pieces of paper, one exclaiming "Papyrus?" and then correcting himself, "no, Hungarian bonds!" to which the other replies, "Maybe these are authentic ones!" The archaeologists are all wearing greatcoats and bowler hats, and one of them is carrying an accessory long identified with bourgeois respectability, the umbrella.[120] Like Charles Darwin in widely circulated photographs half a century earlier, their clothing and lack of specialized tools signify, above all, respectability; Janet Browne's observation that Darwin "could as easily be a member of one of a number of solidly prosperous Victorian professions," from banking to university

FIG. 3.7. Antonin Morlet (far left) and members of the international commission in front of the Fradins' "Museé de Glozel," November 1927. Photographer unknown. Agence Meurisse photo. Source: gallica.bnf.fr/BnF.

teaching, applies to them as well.[121] The worker in the trench, his role signified by the shovel handle he holds, wears much lighter, more casual clothing, suitable for someone performing physical labor, though perhaps not for a December day in the mountains.

Two photographs, both widely reproduced at the time of the international commission's visit to Glozel in November 1927, reinforce the class assumptions inherent in this pictorial division of archaeological work. Both show groups in front of the one-room museum the Fradin family had opened in their farmhouse at Glozel. The visiting archaeologists are dressed for the most part as bourgeois professionals, as is Morlet in his role as host (fig. 3.7); they appear relaxed, jovial, self-confident, even if Morlet, gesticulating with his right hand, seems slightly overeager. In contrast, the Fradins—Emile third from right, his grandparents at far left (fig. 3.8)—look awkward and uncomfortable in front of what is, after all, their own home. The caption in the Paris daily *Le peuple* drives home this awkwardness, describing the Fradins as "quite surprised to find themselves before the photographer." Their sabots jut toward the viewer, accentuating their association with tradition as opposed to the modern technology of the camera, although, except for the grandmother, their clothing is not particularly dated. Lest the reader miss the point, the newspaper describes the building behind them as "the house of honest farmers."[122] Some of the caption's predictions would prove off the mark: the site would never be listed as a historic monument, and Glozel has not become a major tourist attraction for Americans. But the class implications of the photograph come across clearly as an element of its central message.

Yet in the visualization of archaeology, as in so many other realms, Glozel would prove destabilizing, in several distinct but interrelated ways. Consider this description in the Paris daily *Cri du peuple* of the international delegation at work:

> All these learned scholars had put on blue overalls of which the roughly finished fabric gave them the air of apprentice locksmiths or freshly decked-out electricians. Father Sabret [Favret] was wearing a khaki number worthy of the best airplane mechanic. Like her colleagues, Miss Garrod had put on a very becoming *béret basque* that barely covered her short hair; only M. Ferrer [Forrer], from Strasbourg, was working in a bowler.[123]

A photograph published in the British illustrated weekly *The Sphere* (fig. 3.9) provides a reasonable visual approximation of this depiction,

FIG. 3.8. The Fradin family in front of their museum at Glozel, November 1927. Photographer unknown. Agence Meurisse photo. Source: gallica.bnf.fr/BnF.

SEEKERS OF HISTORY AT GLOZEL: Members of the Archæological Commission who are trying to ascertain the period of the remarkable discoveries made in Central France

FIG. 3.9. "Seekers of History at Glozel," *The Sphere*, 19 November 1927, p. 13. Photographer unknown. Courtesy Mary Evans Picture Library.

even if more bourgeois hats than berets are visible, and the archaeologists seem to be searching with their eyes or with only the most delicate of tools. Dorothy Garrod later described the blue overalls that commission members wore as "virtually the uniform of French field archaeology."[124] But this image had yet to become standard in the press, and the comparison to mechanics or apprentice locksmiths, notwithstanding the humorous tone, conveys a certain anxiety, as though Glozel has made it impossible to tell what an archaeologist looks like, or who really is an archaeologist. The difficulty comes to a head in the depiction of Emile Fradin, who had been assisting Morlet with his excavations and who came under suspicion for forgery. Several widely circulated news photos show the young man wearing a beret and a respectable, if hardly couture, tweed jacket (fig. 3.10); only his sabots give him away as an outsider in the world of archaeology.[125]

Images of a smiling, craggy-faced Denis Peyrony, curator of the Prehistory Museum in Les Eyzies and a former schoolteacher, also depict him in a beret, implying that even those from a modest rural background could become recognized archaeologists.[126]

But if the media could cope with the fungibility of class at Glozel, the treatment of the sole woman member of the international commission, Dorothy Garrod (1892–1968), suggests that gender posed a more intractable problem. Readers of the *Paris Times* would have learned of Garrod's sterling credentials as a prehistorian: study at Cambridge, then with Henri Breuil in France; important excavations at Gibraltar; the recent award of a prestigious international prize.[127] Several papers recounted her willingness to scramble headfirst into cavelike openings at Glozel: a page-one spread in the mass-circulation *Petit parisien* included an inset of her on the

FIG. 3.10. Emile Fradin, in a photograph credited to Worldwide Photos and published in an unidentified publication. Source: Archives de l'Institut de France, K50.

ground with only legs and torso visible.[128] But most of the French press seemed more interested in her clothing and her chain-smoking, which some attempted to feminize, as in this description from the *Carnet de la semaine*: "miss Garrot [sic], wearing blue overalls, was smoking tobacco of the same color, which, on a gray November day, created a lovely tonal effect."[129] Others mentioned her torn stockings.[130] *Le Matin* even used Garrod's cigarette habit to signal a decisive moment in the inquiry, when the commission was absorbed in examining a find: "Shoulder to shoulder, the scholars looked [at the tablet], both attentive and delighted, and for the first time in three days miss Garrod, dressed in lilac suede overalls . . . let her cigarette go out."[131]

Garrod also became the object of controversy when, shortly after the commission's report appeared, Morlet accused her of having surreptitiously used a knife to carve holes in the soil, in an attempt to forge evidence of prior manipulation of the site.[132] Some reports claimed that Garrod had admitted to this act when Morlet confronted her. This charge went along with others Morlet brought forward in response to the commission report: that Bégoüen had manipulated the choice of investigators, so that the commission was biased against Glozel, and that Peyrony had scratched a stone with his penknife to create the impression of a forgery. As *Le Figaro* observed as a justification for reporting these charges, "It is no longer a question of prehistory nor of the authenticity of the finds," and Morlet was not responding directly to the "well-organized and thorough report of the commission: he is denouncing biased investigators and fake excavations."[133] In this sense, Morlet was following in the spirit of the declaration of the three academicians who compared him to Galileo. Well before the commission submitted its report, moreover, journalists sympathetic to the Glozelian side had accused Garrod and her Belgian commission colleague Joseph Hamal-Nandrin (1859–1958) of bias against Glozel.[134] Yet of all these charges, the ones against Garrod received the most media attention. The way she and her colleagues chose to respond yields valuable insight into the multiple bodies of archaeology during and after the Glozel affair.

For some more misogynist respondents, Garrod's alleged manipulation of the site conformed to existing stereotypes. "You don't know who miss Garrod is?" asked the ironically titled daily *Homme libre*. Its snide answer: "She is a woman, still young, who pursues old relics—Miss Garrod, also a member of the international commission, supposedly made some holes in the pit face. Holes to create the impression that new objects had been fraudulently placed."[135] Through his choice of language, the journalist fuses Morlet's charge with the image of the sort of young woman

FIG. 3.11. Paul Ferjac (pseudonym of Paul Fernand Levain), "Vieux Débris." *Le Rire*, 17 December 1927. All rights reserved.

who seduces old wrecks, a common image at the time. In a cartoon that appeared ten days earlier, for example (fig. 3.11), headlined with exactly the same phrase, "Vieux débris" (old wreck), a young man mockingly asks a well-dressed young woman if she found her boyfriend at Glozel. Less humorously, Jules Amar (1879–1935), a former professor at the Conservatoire des arts et métiers, gives a political cast to his misogyny in a column supportive of the Glozelian side. Observing that the international commission had lacked methodical and discerning scholars and "let itself be 'garrodized,'" presumably a pun on "garroted," he concludes with dire irony, "*Ah! Votes for women!*"[136] Amar does not mention Garrod by name, but he does not have to. Look, he seems to be saying, what happens when men give women a voice.

Although speaking in the name of science, and criticizing Dussaud for intervening in a field outside his own expertise, Amar was not an archaeologist but an exercise scientist specializing in the physiology of labor.[137] Archaeology surely contained its share of misogyny as well, but in this instance archaeologists in one of its emerging subfields, prehistory, rallied around one of their own. In mid-January 1928, the *Journal des débats* published a letter from Bosch-Gimpera, Garrod's colleague on the commission, recounting his version of the incident: Garrod, he made clear, denied any wrongdoing; she had, she said, simply been using a scraper to verify that signs in the plaster seal placed over the previous day's dig were intact. The Spaniard also implied that Morlet had broken a promise not to take the matter to the press, something Morlet denied in his inevitable reply.[138] Two weeks later, Garrod used her *droit de réponse* (right of reply) to publish a letter in her defense from nineteen scientists and members of learned societies, including Hamal-Nandrin, Breuil, Vayson de Pradenne, the anthropologist Paul Rivet, and the paleontologist Pierre Teilhard de Chardin. Based on Hamal-Nandrin's recollection, the letter described Garrod's work on the signs as "collation," repeated that her scraper might have caused some minor flaking, and asserted that "Miss Garrod could have left fifty similar flakes without having to justify, much less deny it." Indeed, the letter suggested that both the flaking and Garrod's subsequent acknowledgment resulted from the violence of Morlet's interpellation.[139] And it noted that Morlet had quickly taken his charges to a group of reporters standing to one side of the dig, notably including the *Journal des débats*' Henry de Varigny. The letter-writers concluded by insisting that Garrod had joined the commission "to lend her absolutely disinterested support to the pursuit of truth," and that the public criticism to which she was being subjected was "entirely contrary to the scientific spirit" as well as to French traditions of courtesy and hospitality.[140]

To an extent, Garrod had it both ways, avoiding the indignity of replying to Morlet's charges on her own by delegating her legal right of response to others. By associating themselves with members of other learned societies, the archaeologists who likely organized the letter were defending not just Garrod but the scientificity of archaeology itself. The letter portrayed Garrod not only as carrying out scientific activities (collation) in a professional manner but as imbued with the qualities that make for good science: disinterest, collegiality, and above all the pursuit of truth. The particularities of her own body, they were declaring, counted for nothing in comparison to her membership in the larger disciplinary body of archaeology. Garrod, it should be noted, possessed a sufficient number of credentials and bodily signifiers, from her upper-middle-class background and university degrees to her proper attire, to allow her professional affiliation to become her primary embodiment, one that overshadowed even her gender. In the formal photograph (fig. 3.7), like the gentleman next to her but even more carefully, she carries her hat, prepared to don it as soon as the posing is over.

Garrod's own account of Glozel, a five-page essay published in *Antiquity* four decades after the commission report, and just a few months before her death, makes no mention of her gender beyond her having been afforded the use of a spare bedroom in the Fradins' farmhouse to change into her overalls. Her detailed description of the dispute with Morlet differs little from her colleagues' response at the time; that she found herself alone to check the plaster seal she attributes to her youth, not her gender: "I suppose as the youngest I was the most nimble member of the Commission." After Morlet's half-hearted apology, she agreed, "most unwillingly," to shake hands with him, and she considers a press photograph making visible her dislike of Morlet an accurate rendering of her reluctance.[141] Something like this attitude—that she was a scholar first, but possessing the necessary proprieties—must also have been operative at the time

FIG. 3.12. "Ce que les savants ont trouvé dans les fouilles de Glozel," *Excelsior*, 8 November 1927. Photographer unknown. Source: gallica.bnf.fr/BnF.

of Garrod's election to the Disney chair in archaeology at Cambridge in 1939, which made her the first woman in any field to hold a Cambridge professorship.[142] Garrod seems not to have much enjoyed the experience, though less because of her gender than because she much preferred field archaeology to academic study, and small-group teaching to the lectures expected of a professor. She also, like other scholars considered in this book, disliked administrative work.[143]

To return to Garrod's *droit de réponse,* the letter contains one surprising note: after relating Morlet's movement toward the reporters, the scholars describe de Varigny, the *Journal des débats* reporter, as "the stabilizing element in this painful episode." It is not clear exactly what they mean, since after approaching the journalists Morlet repeated his accusations and his "coarse insults," which compelled Garrod, "infuriated and to put an end to this dispute of no consequence," to concede that she might have detached a flake.[144] The letter may have been alluding to the ostensible commitment, which Morlet later disputed, to keep the incident out of the press. Or the signatories may have sought to imply that even a journalist had behaved in a more equable manner—the word stabilizing (*pondérateur*) comes from the terminology of weights—than the intemperate doctor.[145] Whatever the case, the reference to journalism, in a letter defending scientific qualities *in the press,* conveys a realization, perhaps resigned, perhaps even mildly hopeful, of the imbrication of these two systems of representation, science and the media. And they make clear archaeologists' awareness that however disinterested their motives, however professional their conduct, they and their bodies could not escape media scrutiny.

Many press accounts comment on the presence of journalists at Glozel, and Garrod among others expressed annoyance at working under close media attention.[146] Photos like those in *Sphere* (fig. 3.9) and *Excelsior* (fig. 3.12) go beyond simply documenting presence. They cast archaeologists not only as consumers of and contributors to the press but as performers in the media ecosystem, a status of which they must have been well aware. As Elizabeth Edwards, among others, has observed, the photograph itself has a performative quality: it offers the viewer a framed and staged fragment of reality.[147] Whether understood in theoretical or in physical terms, and whether individual, disciplinary, or both, embodiment clearly involves a performance—codified, repetitive, occasionally innovative or subversive—of the self. The next chapter explores the performative dimensions of archaeology in these two controversies.

✳ 4 ✳
Reality Effects
STAGING ARCHAEOLOGY

The Poinssot papers contain relatively few documents in the hand or name of Louis Poinssot, and many can be found in one box, labeled with Poinssot's full name and title and the word "Contentieux" (Disputes). One folder in that box, with plentiful notes in Poinssot's hand and carbon copies of letters he signed, concerns two incidents that took place in Tunis on 10 March 1926.[1] First, after a heated exchange with his superior, the Protectorate director of public instruction and fine arts Henri Doliveux, Poinssot was ejected from a meeting of a special committee on ancient theaters. Later that afternoon, after being jostled by Louis Carton's widow Marie Thélu Carton on a sidewalk near the Cathedral, Poinssot, fearing that she was planning to take a revolver from her purse and shoot him, grabbed the widow's wrist and loudly called for police assistance.[2] This encounter, which Poinssot referred to in his notes as "the wrist incident," resulted in his suspension from the directorship for several months and a high-level administrative inquiry. After a brief legal proceeding, Poinssot was adjudged a small fine for minor assault and returned to work.[3] In subsequent correspondence Poinssot refused to back down from his remarks at the committee meeting, regarding a planned tour of Roman theaters by actors from the most prestigious company in France, the Comédie française.[4] Writing the resident general in June, after the performances had taken place, Poinssot, restored to his post, referred to one of the plays, a historical drama about St. Louis (King Louis IX) entitled *Le Croisé* (The Crusader), as "according to unanimous public opinion a scandal" and called for Doliveux to himself be investigated.[5]

Notwithstanding her venomous personal relations with Poinssot, in this instance Thélu Carton could not disagree with his verdict. In uncertain verse and at what many considered excessive length, the play depicts the sovereign most beloved of the church as a vacillating fop who has an affair with the Bey's wife as a way of uniting his European kingdom with the people of North Africa. Unsurprisingly, and as Poinssot noted by referencing the full political range of Tunis dailies, the play provoked widespread scorn, even indignation, in the face of which Thélu Carton reportedly extended personal apologies to the Catholic primate.[6] The imbroglio must have been all the more bitter for her given that the author of *Le Croisé* was none other than her late husband's protégé and propagandist, Jaubert de Bénac.

If the twin incidents of 10 March 1926 form part of the long history of Poinssot's conflict with the Cartons and their supporters, their significance goes beyond the personal. First, as noted in chapter 1, the idea of animating archaeological sites with theatrical and other events had long been a subject of contention between the two sides. For the Cartons, performances, in attracting residents, tourists, and publicity, proved the worth of archaeology not only to the economy but to the larger imperial project. For Poinssot and his predecessors, such events were at best a distraction, at worst an active threat to the scholarly work of archaeology, which depended on the integrity and careful preservation of sites. In justifying to his superiors his serious "reservations" about plans for the 1926 tour of the Comédie Française, Poinssot referred to unspecified problems that had occurred during performances at Carthage five years earlier, around the time of his appointment. But Poinssot had a long memory, and he was undoubtedly aware that Carton had organized theatrical events at Carthage as early as 1906, just a year after Gauckler's initial excavation of the theater. Consciously imitating performances in the Roman theaters of Orange and Nîmes in the metropole, the theatrical reuse of archaeological sites brought debates over the stakes and uses of archaeology literally and visibly into the public arena.

If much of this chapter looks directly at performances that can be understood as the staging of archaeology itself, it also engages with the related but distinct notion of performativity, which Mieke Bal has succinctly defined as "an aspect of a word that *does* what it says."[7] Here a second aspect of the "wrist incident," Poinssot's turn to popular media to understand Thélu Carton's motives, offers an important clue. One of the handwritten notes in which he seems to be working out his own motives has the header, "The wrist—seized because I was afraid of 'the pulp fiction gesture' [*le geste feuilletonesque*]."[8] Just over a decade earlier, Henriette Caillaux, aggrieved at what she considered the besmirching of the reputation of her husband, a prominent politician, had gained entrance to a newspaper editor's office in Paris and shot him with a revolver. Poinssot might well have had that event and image in mind—Caillaux's trial for murder was a media sensation in the summer of 1914—but he chooses to describe the act he feared in more generic, but nonetheless quite evocative, terms.[9] Less than two years later, a newspaper used the same term to describe Glozel, asking rhetorically if, "having started as a passing curiosity, is it not now turning into a serial novel [*roman-feuilleton*]?"[10]

The word *feuilleton* technically refers to any kind of serialization of a narrative, appearing in a daily newspaper or (more recently) on radio or television; most of the novels of Charles Dickens and Emile Zola, among many others, were first published in that form. Poinssot seems to have been using the term in its vernacular sense, to indicate a story filled with heightened emotion and punctuated with cliffhangers that leave readers eager for more. Having been warned by Merlin, Cagnat, and others about Thélu Carton's mental instability, Poinssot wrote that for months he had worried about finding himself on the (literal) same boat as she, since he thought her capable "in a moment of madness" of throwing one of his children overboard.[11] In another note, headed, "The wrist incident, why I might consider myself under threat," the director wrote, "Knowing how Mad. Carton took pleasure in transforming those she did not like into the Atreides [the family of Agamemnon, known for mortal feuding], I could not be surprised to see her transform herself into the heroine of a *feuilleton*."[12] Only as the incident was unspooling, Poinssot wrote later, and Thélu Carton began "loudly" and repeatedly describing his eviction from the committee meeting to the gathering crowd, "did I understand that my initial hypothesis was mistaken. It was not in order more easily to fell me with a melodramatic gesture that Mme Carton had jostled me, but only to create an incident" that enabled her to publicize his earlier humiliation.[13]

With veiled or direct references to two kinds of theater, classical (the Atreides) and melodrama, as well as to Thélu Carton's "relations with the theatre world," Poinssot's imaginings fairly bristle with drama.[14] James Lehning has pointed to melodrama's central role in French political culture in the nineteenth and twentieth centuries, its tropes of heroes and villains helping to shape public understanding of key events and ideological divisions.[15] Even more important in the context of this book, the term *feuilletonesque* retains the newspaper press, one of Poinssot's longtime obsessions, as frame of reference; on this occasion he complained that some papers gave free rein to the exaggerations and distortions of members of the Carton faction.[16] Poinssot's revised assessment of the wrist incident amounts to an acknowledgment that he had misjudged, not Thélu Carton's character, but the nature and intended audience of her performance. In contrast to performance, a unique (if repeatable) event, performativity, whether conceived as a speech act or the enactment of socially conditioned identities such as gender, typically involves citation, repetition, or both. Jonathan Culler observes that for Judith Butler gender constitutes "an obligatory practice, an assignment, say," and even if performative statements such as "I pronounce you married" or "We find the defendant guilty" work differently, their utterance invokes, renews, and very occasionally modifies powerful discourses that are the operative modes of culture.[17] Poinssot quotes Thélu Carton as concluding the stream of imprecations directed against him with the statement "I am Madame Carton, president of the French Ladies [Association]."[18] This utterance, performing both class and gender identities, conveys to him her true intentions.

If in Carthage actual performances offered a proving ground for archaeology's divergent and sometimes clashing claims, the widespread popular interest in Glozel manifested itself in a number of performative acts. In venues as familiar as the vaudeville stage and the courtroom, and as unusual as the Glozel hamlet itself, these acts or events were performative in that they sought, not always successfully, to constitute the legitimacy or illegitimacy of the site, the various actors, and the larger field of archaeology. Like performances, performativity works via the repetition of familiar "*fabrications* manufactured through corporeal signs and other discursive means."[19] Some of these events, even those occurring well away from theaters, can certainly be considered performances in their own right. But what links actual performances with the performative is the centrality of staging in the production of subjects. When Mieke Bal writes that "theatricality is the subject's production, its *staging*," she is referring to the production of human subjectivities, but the observation works for disciplinary "subjects" like archaeology as well. As a "travelling concept" staging or *mise-en-scène* migrates between practice and theory, originating in the theater, where it denotes "the overall artistic activity whose results will shelter and foster the performance." Ultimately, Bal argues, *mise-en-scène* serves as a kind of mediator between the performative, with its repetitive and potential qualities, and the performance.[20] The passions aroused by archaeology at sites such as Carthage and Glozel call attention to the performative dimension of the field and to the performances—or, in Culler's terms, the assignments—their practitioners regularly seek to carry out.

Performance and the writing of history have a complex relationship. As the theater scholar Freddie Rokem has observed, the phrase "performing history," the title of one of his books, can refer both to the original enactment of historical events and to their later recreation.[21] Florence Fix makes a similar point, observing that in French "the term 'drama' ... applies as much to a theatrical work as to a historical event"; she argues that this proximity merits further exploration.[22] In Rokem's analysis, whereas displays of archaeological objects tend to efface the temporal distance between the moment of their making and the viewer's present, performance foregrounds the artifice of the present moment—costumes, makeup, props—thus highlighting the multiple temporalities at work in all performances.[23] Fix observes that even when unfaithful to documented facts, history on stage "offers a guarantee of authenticity in that it installs a tangible relationship to the real."[24] This is a theatrically specific way of describing what Roland Barthes calls the "reality effect," the phenomenon by which the incidental details of a narrative or other set of representations lend credibility, even authority, to its underlying message.[25] Barthes describes the reality effect as emanating from a moment of narrative excess: an unnecessary detail, an aside that does not contribute to the narrative drive. This "vertigo of notation," contained only by an aesthetic purpose, points to the real world beyond narrative.[26]

The performative joins reality effects in straddling the unstable, highly political border Judith Butler has identified between the phantasmatic and the real, the place where "a phantasmatic construction receives a certain legitimation after which it is called the real."[27] A standard feature of historical writing since antiquity, the reality effect appears much later in fiction, most notably in nineteenth-century realism. For Barthes, this moment coincides with the "development of techniques, of works, and institutions based on the incessant need to authenticate the 'real'"; these include photography, "exhibitions of ancient objects ... [and] the tourism of monuments and sites."[28] Archaeology's contribution to this development comes from the plethora of objects it discovers, details that function to materialize narratives of hitherto unsuspected pasts or mythical places. The performativity of archaeology consists in actuating fantasies to build a convincing vision of the past through and around the remains it unearths. Actual performances offer a fitting locus for putting these versions to the test, but archaeologists have always been conscious of the need to maintain a balance between the vision and the artifacts it mobilizes.

Performing Carthage

If the Carthage pageants of 1906 and 1907 celebrated the maturation and potential of archaeology in Tunisia, they also functioned as a warning about the threats the development of Carthage as a suburb posed to archaeological work. Through the media coverage they carefully cultivated, the pageants spotlit what archaeology could offer empire: at once precedents, in the imperial cultures unearthed; an example of technical achievement; and a road map both literal and figurative. In 1889 Carton delivered a paper at a geographical congress entitled "On the utility of archaeological study from the standpoint of the colonization of North Africa." The paper celebrated Roman achievements in agriculture, construction, road-building, hydraulic engineering, and sanitation. "There is no vestige of Roman colonization," he declared, "that does not repay study with a lesson."[29]

What Stefan Altekamp and Mona Khetchen refer to as "the ideological appropriation" of archaeology certainly depends in part on the types of objects found, the extent of excavation, and the state of knowledge about them at any given time.[30] But the stakes of archaeological staging extend beyond the meanings that could be attached to particular artifacts or sites, as the pageants gesture nostalgically toward both the longed-for homeland and an imperial past. The sense of innocence in imperial nostalgia is, as Sarah Bracke has argued, "never far from guilt": nostalgia effectively transfers agency from people to the impersonal passage of time, thus obscuring the violence of imperial conquest.[31] Reality effects are momentary, and colonial rule is never stable or monolithic. Indeed, the year or so during which the two pageants took place was marked by significant upheaval in Tunisia, including the first sustained violence since the establishment of the Protectorate a quarter century before. In April 1906, a month before the first pageant, riots in two small towns in western Tunisia, Kasserine and Thala, resulted in the deaths of a French settler farmer, two members of his family, and an Italian who worked for him. Bedouins angered by land restrictions took other settlers hostage before the brief revolt was brutally suppressed.[32] That same year also saw the first stirrings of nationalism among the educated Tunisian élite, as a new group, the Jeunes Tunisiens, called for representation of Tunisians in consultative bodies, equity in salaries and working conditions, and greater educational opportunities. A newspaper espousing these views, *Le tunisien*, was founded in February 1907.[33] For all these reasons, scholars have identified 1906–1907 as crucial in the history of Tunisian resistance to colonialism.[34]

It is at once necessary and insufficient to position these events as a mostly unacknowledged background to the Carthage pageants of the same years. Necessary, because their obscurity is in part a function of the pageants' archive, which consists largely of published accounts and visual documents. But also insufficient, because it is possible to read public performances in a way that insists on their implication in the larger colonial enterprise. For performances not only occlude violence but mobilize what Ann Laura Stoler has called "the imperial sensibilities of destruction and the redemptive satisfaction of chronicling loss."[35] The novelist Myriam Harry's 1907 account in a Paris daily of the trial of the young marabout

Amor Ben Othman, accused of instigating the Kasserine-Thala revolt, exemplifies this process. Although Harry's sympathy for the marabout reportedly shocked the French settler community at the time, her article is replete with passages describing her state of mind as she wandered through "crude" (*fruste*), "rough" (*revêche*), melancholy spaces, reflecting on the timelessness of the landscape and the pettiness of the French as its inheritors.[36] Rather than in any way challenging the centrality of the Western self, Harry's sympathy for the unfortunate Muslim reinscribes it, in a sense offering her own performance of compassion, notably in a train ride after the trial with those found not guilty, as a form of Stoler's "redemptive satisfaction." Reading the pageants critically, keeping in mind their expressed purpose of countering loss, makes it possible to see them in the same light.

In March 1906, Carton, as president of the Institut de Carthage, acted as both guide and chronicler of an excursion to the site for members of the French community in Tunis. A published account of the excursion includes a reprint of an article written by "one of the charming ladies" who made the trip. It begins in a lyrical vein: "O the sad charm of traipsing through fields sown with asphodels, the flower that so loves ruins—and to hurt oneself on the broken stones of vanished cities . . ." Carton is praised for his learning and his ability to bring a "corner of Roman Africa" to life. As a guide, his imagination, "not the imagination that has been called 'the mistress of error and falsehood,' but that creative imagination that is the first element of genius," reconstructs amphitheaters, restores missing walls, and reconstitutes the plan of the Roman city.[37] The distinction recalls as it inverts the contrast Gustave Flaubert drew, in an account of a visit to the Neolithic site of Carnac in Brittany, between reverie and an imagination chained to a sterile "science."[38] For Flaubert reverie could accede to the infinite, whereas scientific efforts to reconstruct lost societies were doomed to failure. In the account of Carton's excursion, however, archaeology as emergent science has taken charge of guiding the imagination in safely practical directions. If the theater could, as in Orange and Béziers, be used for regular performances, "what an attraction for foreigners, what a resurrection for Carthage!"[39]

The reference to Béziers and Orange clearly places the inspiration for the Carthage festivals at the intersection of several recent trends. The quarter century from 1890 to the outbreak of World War I saw significant efforts to renew theatrical performance in France. From different directions, naturalist and symbolist movements attacked traditional dramatic artifice and sought to create a truly popular theater.[40] Extended into the 1920s, the same period marked the high point of open-air theater in Europe, as many saw in outdoor theatricals a way of finding communion between art and nature, breaking down class barriers through common seating, and returning to the primal force of theater without elaborate sets and distracting special effects.[41] Although some outdoor theaters were specially constructed for the purpose, the staging of classical drama and opera at the restored Roman theater of Orange, where regular performances began in the late 1880s, and the arenas of Béziers and Nîmes, also in Provence, attracted special attention and large audiences. As the theater scholar Sylvie Humbert-Mougin has pointed out, however, the outdoor theater movement was not free of contradiction. Claims to present historically accurate productions of Greek drama ran up against the desire for spectacular effects, and lyrical evocations of starlit nights gave way to the early adoption of artificial lighting.[42] In a similar paradox, although the Institut de Carthage excursion involved modern transportation, logistics, and conveniences—the visitors came by train from Tunis and lunched in a hotel banquet room—the published report about it was shot through with anxiety about further damage to be inflicted by larger, presumably less discriminating, crowds of visitors. Yet preserving the site would entail bringing still more people to it: Carton describes the desired performance as "an imposing artistic event, which would attract hundreds of spectators and aid the local economy."[43]

Billed as an entertainment or *divertissement* added to the annual meeting of the Institut de Carthage, the first *fête de Carthage* took place on 27 May 1906. After opening speeches, the program of the festival consisted of orchestral, vocal, and choral selections, theatrical performances, and the reading of original poems about Carthage and its history.[44] Highlights included a soprano singing two arias from Ernest Reyer's operatic adaptation of Flaubert's *Salammbô*, a performance of the final act of Corneille's verse drama *Polyeucte*, a chorus from Saint-Saens's *Samson et Dalila*, and "evocations," in which actresses allegorized Punic and Roman Carthage and modern France. The reasons for some of these choices are clear: *Salammbô* is set in ancient Carthage, and Delilah is a Philistine, a member of the group known in Latin as the Punic people and in modern languages as Phoenicians; the chorus presented had the informal moniker "Les Philistines."[45] The selection of *Polyeucte*, which is set in second-century Armenia and tells the story of a martyr who refuses to abjure his Christian faith, probably was intended to gesture toward parallel histories at Carthage, but the choice seems to have irritated many. Although the December 1905 law on the separation of church and state in France did not apply to Tunisia, the Protectorate had seen its own controversies over schools run by Catholic priests and nuns, and a number had been shut down under a 1903 decree.[46] Without referring to this situation specifically, Carton

LES EVOCATIONS   (Cliché Soler.)

FIG. 4.1. "Les Evocations" (performance at the 1906 Fête de Carthage), published in that year's *Revue tunisienne*. Photographer unknown. Public domain.

acknowledged the general problem in his postpageant report, saying that a performance more directly connected to the history of the site would have been preferable.[47] The following year would see this wish fulfilled.

One of the most striking elements of the 1906 pageant was "Les Evocations" (fig. 4.1), of which the institute's journal, the *Revue tunisienne*, published the script. A young Bedouin girl (*la Béduine*) enters carrying a sickle and a sheaf of wheat; she notes the omnipresence of ruins on the land she is supposed to cultivate and complains that when she asks about them, her brothers respond only, "Mektoub," it is written. Her words awaken the Punic Carthaginian, who, according to the stage directions, rises from her tomb to tell the Bedouin girl of the greatness of Carthage, the rival of Rome. The mention of her name rouses the Roman Carthaginian, who speaks of the glories of Roman dominion of this fertile terrain and concludes, "Nothing after her . . . nothing!" At this the figure representing France appears, gently reproaches Rome for her pride, then summons the young girl, saying, "It is a mother who reaches out to you." She addresses the two allegories of Carthage as her sisters, declaring that "France believes in vanished things, and the remote, glorious memories that you were evoking are still present in her recollection!"[48] To conclude the piece, the *Béduine* offers her wheat sheaf to the figure of France, a gesture that, as Clémentine Gutron puts it, "acts as the moral of the story."[49]

As a performance the "evocation" materializes the different eras that nostalgia attempts to knit together while simultaneously emphasizing their separation. The use of familial tropes to portray colonialism in affective terms—what Françoise Vergès has called "the colonial family romance"—was common at the time, as was the blurring of maternal and sisterly relations.[50] Charlotte Legg has pointed to a nearly contemporaneous image of a maternal Marianne, personification of the Republic, comforting an Algerian girl, and Myriam Harry referred to Tunisia as France's "adoptive daughter" and "younger colony," using a term (*puînée*) typically applied to siblings.[51] Ideologically, as Carton made clear in his speech, the Institut de Carthage conceived of its activity in terms of the larger educational mission of colonialism: "France, which came to this country as an educator; France, the heir to ancient Rome and which for centuries has fought for every noble and generous idea, cannot allow the work of destruction to continue!"[52]

In his account of the pageant, Carton expressed a few regrets, one of them that "we were not able to prevent spectators in jackets and vests from standing behind or next to the stage" (see fig. 4.2).[53] The following year Carton thanked a last-minute volunteer for policing the stage to prevent the intrusion of people in modern dress, but audience members still commented on "those by the wall, smoking their pipes."[54] Carton's concern went deeper than discomfort with anachronism. Although the size of the crowd was a point of pride, these members of the twentieth-century audience not only diminished the visual pleasure of the event, "they are all too visible in the photographs taken of the performance, and the reader . . . will certainly be shocked by the promiscuity that largely contradicts all the accounts of the pageant."[55] The concern, in other words, has to do less with

FIG. 4.2. Postcard showing the performance of an excerpt from *Salammbô* at the 1906 Fête de Carthage. Photographer unknown. Bibliothèque de l'Institut national d'histoire de l'art, Collections Jacques Doucet, archives 106, box 48. Published with permission.

spoiled illusion than with a representational record that fails to convey the reality effects of the performance. Carton wants the photographic archive to testify to and to model the successful use of archaeological knowledge to reconstruct the past in situ, but without revealing the mechanics of its own production. Here one of the meta moments in the "evocation" is especially revealing: France, after asserting her love of antiquities, urges her sisters to "look at this ancient theatre in front of you, which is awakening from nearly twenty centuries of sleep: our scholars [*savants*] have discovered these ruins, and our voices and music are here today to awaken their echoes."[56] Carton clearly hoped that the pageant would lodge in viewers' memories as a seamless picture, but he could not entirely control its particulars. The 1906 pageant thus illuminates the complexities of policing the line between reality and fantasy, suggesting that reality effects, as part and parcel of modern technology and knowledge production, do not always land where they are intended.

Planning for the second pageant, scheduled for the Tuesday after Easter, 2 April 1907, began the previous November, giving the Institut de Carthage nearly five months to prepare, twice as long as for the first pageant. This time Carton wanted to make sure the event was perfectly adapted to its setting; accordingly, the Institut commissioned two original verse plays. One came from the respected playwright and librettist Charles Grandmougin (1850–1930), the second from Lucie Delarue-Mardrus (1874–1945), a French poet traveling in North Africa with her husband, Joseph-Charles Mardrus (1868–1949), the author of a French translation of the *Arabian Nights* published between 1899 and 1903. Grandmougin penned a three-act drama, *La Mort de Carthage* (The Death of Carthage), Delarue-Mardrus a shorter play, *La Prêtresse de Tanit* (The Priestess of Tanit). Paris-based actresses were engaged to play the leads in both plays, with secondary and supernumerary roles, over two hundred of them, going to locals. A special committee of the Institut de Carthage raised nearly 16,500 francs for the event, of which over three-quarters came from government bodies, the rest from private contributions.[57] Carton's report emphasized the logistical complexity the pageant entailed; indeed, the cost of bringing all the participants (nearly three hundred people, including the orchestra and stage crew) by train from Tunis to Carthage made it possible to hold only one on-site rehearsal. Ticketing was managed by volunteers working out of a local travel agency; the Tunis hotel syndicate paid for a press run of thirty thousand publicity brochures, which were sent to learned societies, hotels, and travel agencies throughout Europe.[58] The publicity worked: estimates of attendance ranged between four and ten thousand.[59]

Set at the end of the Third Punic War, during the siege of Carthage by Scipio in 146 BCE, *La Mort de Carthage* features, in the first three scenes alone, an onstage throat-cutting of Roman prisoners and a haruspex (a type of divining priest) examining their entrails to predict future doom.[60] The violence continues with the blinding of another group of Roman prisoners, Rome's retribution by blinding Carthaginian captives, and the execution of a Carthaginian general, Abdozir, who has challenged the judgment of the vainglorious leader Asdrubal. At the end of the play Asdrubal's nameless wife denounces her husband, kills their children in front of him, and throws herself into the fire set by the victorious Scipio to destroy the city. In the avant-garde literary journal *Mercure de France*, Alexandra David, best known to posterity as the explorer-writer Alexandra David-Néel (1868–1969), dismissed *La Mort de Carthage* as "a crude and very ordinary play sprinkled with enough violent atrocities" to impress a popular audience.[61] Carton, in response to such criticism, observed that Grandmougin "was following the classical rules of our plays; his verses contained nothing strange or decadent." For the organizer, indeed, *La Mort de Carthage* constituted a model for the niche genre of plays suited to performance in ancient theaters, where "the enormous scale of the stage requires the movement of large and picturesque groups, and where scenes laden with dramatic effect—whatever the critics may say—are the ones that work best."[62] In order to work in the visual sphere as a form of descriptive excess, the reality effect must ensure that every member of the audience has something to look at.

When Carton wrote that the Grandmougin play had "nothing... decadent" about it, he was using one of the keywords of Symbolism, a movement with which Delarue-Mardrus was intimately familiar.[63] Her play, *La Prêtresse de Tanit*, unsurprisingly, engages with a number of Symbolist themes and staging practices. In the opening scene, set at an archaeological site in present-day Tunisia, a poet comes across a group of bored archaeological workers and their supervisor. As the workers toil, the supervisor greets the poet and expresses surprise at the latter's knowledge of the ancient Phoenician or Punic language.[64] It is not long before the workers dig up the sarcophagus of the title figure, a reference to a tomb discovered by Father Alfred-Louis Delattre at the necropolis of Bord-el-Djedid in November 1902.[65] When the priestess mysteriously comes to life, only the poet can speak her language, and the bulk of the play consists of a dialogue between ancient priestess and modern poet, who exemplifies the ability of contemporary France to understand ancient cultures through modern scientific knowledge. At first the priestess, Arisatbaâl, is bewildered—"where is my city," she asks, recognizing the landscape but nothing else, and the poet tells her, "It is gone." Eventually the priestess becomes angry and summons a group of Carthaginian dead.[66] The ghosts are replaced by a group of young Bedouin girls wandering in, like the young girl in the previous year's pageant. As though a reality effect provided by the surrounding pastures, they represent, like the priestess, a timeless culture for which one might feel a nostalgic longing but which must give way to the progress of civilization. Recognizing the girls as compatriots (she has, understandably, no real sense of time), Arisatbaâl addresses them and, despite their incomprehension, manages to direct them to her side before the supervisor chases them off as "intruders." Eventually the poet coaxes the priestess back into her tomb, assuring her that ancient Carthage has achieved immortality.

To judge from an article synthesizing critical reaction via three fictionalized composite voices—an ordinary spectator, a devotee, and a cynic—the overall public response to *La Prêtresse de Tanit* was unfavorable.[67] Whereas it would not have been difficult to cast the play as an allegory of the lyric imagination, and therefore indirect praise of his own efforts, Carton simply noted in response that it had been written exclusively for its lead actress, Jeanne Delvair of the Comédie française.[68] Some critics were more enthusiastic: Myriam Harry enthused in *Le Temps* about the play's "remarkable formal beauty" and the "breath of poetry and beauty" that

Delvair brought to the role.[69] The fictive spectator in the composite account hit upon the consensus view: "The idea was very original, it was perhaps a bold attempt, but it was ill served by the details of its excecution."[70] This is perhaps a way of deploring the *lack* of superfluous, reality-effecting detail in *La Prêtresse*; the overladen stage of *La Mort de Carthage* seems to have elicited a more positive audience response. But Delarue-Mardrus was contributing to the larger reality effect of the pageants by underlining archaeology's role as a new device for conjuring the past.

If Carton's seventy-page account of the 1907 pageant included compliments for all the actors and other participants, he devoted special attention to its visual dimensions. His narrative stresses the ways the pageant draws on recent archaeological finds as both inspiration and legitimation, but it also draws on a wider visual culture saturated with tropes of the distant past. The search for an archaeological authenticity that could be represented in easily accessible visual form began with the poster for the 1906 pageant (fig. 4.3), by the Orientalist painter Louis Flot (1865–1942). The general description of the central figure as "a virgin symbolizing Carthage, rising after a long sleep" is perhaps intended to justify the lack of precise detail,[71] although at least one of the theatrical masks to the lower left is reminiscent of a Punic mask found at Carthage and probably on view at the time in the Bardo Museum in Tunis (other examples could be found around the Mediterranean).[72] Most significant, however, is the rendering of the site itself, shown in its ruined state and with natural landmarks in the distance that clearly identify the site as both ancient and modern Carthage. The words "Théâtre Romain" thus serve a double purpose, identifying both the scene depicted and the site where the pageant will take place.

For the 1907 festival, the Institut de Carthage organized a competition for the poster, with an exhibit of the competition entries. The winner (plate 3), by the painter Alexis de Broca (1868–1948), is described as

> The Carthaginian priestess, of whom an admirable statue is on view in the Carthage Museum [plate 4], rising slowly, in her strange costume, and in the hieratic position of prayer, against an intensely luminous sky. A young French woman, laying a floral wreath at her feet, symbolizes French civilization awakening the soul of Carthage. In the background we see an excavation. Watched over by a cleric, workers remove objects from a deep ditch. Farther along, the wonderful site of the ports of Carthage, the gulf of Tunis, and the imposing sihouette of Mount Bou-Korneïn.[73]

The poster provides a prelude to *La Prêtresse de Tanit*, which takes as its starting point a scene of excavation, and pictures the very priestess statue

PLATE 1. Giorgio de Chirico, *The Archaeologists*, 1927. Oil on canvas. Rome, Galleria Nazionale d'Arte Moderna. © 2025 Artists Rights Society (ARS), New York/SIAE, Rome.

PLATE 2. Henri-Gabriel Ibels, "The Glozel Discoveries," *Cyrano*, 11 December 1927. Source: gallica.bnf.fr/BnF; color image, Archives de l'Institut de France, K51.

PLATE 3. Alexis de Broca, poster for the 1907 Fête de Carthage. Public domain. Source: Centre des Archives diplomatiques de Nantes.

ANTHROPOID SARCOPHAGUS OF A PHŒNICIAN PRIESTESS.

PLATE 4. "Anthropoid Sarcophagus of a Carthaginian Priestess," found at Carthage, 1902. Frontispiece from Mabel Moore, *Carthage of the Phoenicians in the Light of Modern Excavation* (London: W. Heinemann, 1905). Public domain.

PLATE 5. Three Carthaginian cinerary urns in the Middle East Study Room, British Museum, March 2022. Author's photograph.

PLATE 6. Mark Dion, *Concerning the Dig*, 2013, mixed-media installation, dimensions variable. Commissioned by the Museum of Contemporary Art, Chicago. MCA Chicago Collection. Photo courtesy the artist and Tonya Bonakdar Gallery, New York/Los Angeles. © MCA Chicago.

PLATE 7. Michael Rakowitz, *May the Arrogant Not Prevail*, 2010. Arabic–English newspapers and food packaging, glue, and cardboard on wooden structure, 597.5 × 493.4 × 95.3 cm. Courtesy the artist and Rhona Hoffman Gallery, Chicago. Photo: Nathan Kaey, © MCA Chicago.

PLATE 8. Jean-Luc Moulène, *Le Monde, le Louvre*, exhibition view, Le Louvre, Paris, 2005–2006. Courtesy the artist and Galerie Chantal Crousel, Paris. Photo: Florian Kleinefenn, © 2025 Jean-Luc Moulène/Artists' Rights Society (ARS), New York/ADAGP, Paris.

FIG. 4.3. Louis Flot, poster for the 1906 Fête de Carthage as reprinted in the *Revue tunisienne*, 1906. Public domain.

brought to life in the play. Chosen in preference to two designs showing modern visitors to the site, it combines the major elements of the 1906 pageant's pictorialization, with the key addition of a scene of archaeological work.[74] The poster also emphasizes the "strange" or novel features of the pre-Roman world.

Without the multiple voices of the Delarue-Mardrus piece, which lend it a certain complexity, the poster constructs archaeology as a kind of sacred inheritance of primitive ritual. The priest in his cassock, clearly a White Father like Delattre, echoes the priestess with her offering to the goddess Tanit, but not exactly, as he is *receiving* an offering from a worker standing in what most viewers would likely read as a tomb.[75] Casting the priest pictured as excavation leader as Delarue-Mardrus's "surveillant" situates the poster's viewer firmly in the action of her play. Along with its versification and evocative imagery, the play offers a strikingly unvarnished view of archaeology. The opening "chorus" of the archaeological workers is something like a lament: "We work for pay, / But it matters little why . . . / If we struggle, if we suffer, / It's because we must render to the soil / The eternal sweat of our brows. / We work for pay, / But what care we why?" The workers are also conscious of their own ignorance; questioned by the supervisor, one worker notes that the poet has come to the site before, and "he knows things better than we do."[76] Delarue-Mardrus seems to have grasped well the hierarchy of archaeological work: the very framing of the play, as of the poster, stages archaeology's subordination to a higher art, poetry, and in his limited role the *surveillant*, whom I have called the "supervisor" but who could also be a humble guard or watchman, embodies that auxiliary status.[77] The playwright recognizes as well the duality at the heart of archaeology, which disturbs a literally settled past to increase knowledge in the present. At one point the awakened priestess tells the poet, "But these tombs, they were the last city, / and you are destroying it!" He retorts, "For posterity!"[78] Archaeology's justification lies in the present-day knowledge and creation it enables and for which it serves as the ultimate reality effect.

As the publicity brochure had promised costumes and "scènes de moeurs" (lifestyle scenes) based on the latest discoveries, Carton was at pains to demonstrate the archaeological basis of his costume designs.[79] In the 1906 pageant, the actress playing Pauline, Polyeucte's reluctant fiancée and the daugher of his tormentor, is described as "wearing a garment of which the draping, as well as the diadem that rings her forehead, is, it is said, the copy of one of the most delicious statues in the Bardo Museum," found just a short time before at Carthage.[80] But Carton acknowledged that his own designs also represented an interpretation, one that emphasized the evolution of Punic fashion from a purely Egyptian style to one showing clear Greek influence.[81] Carton allowed others to sketch out the costumes worn by the mercenaries accompanying Asdrubal, including "superb black Sudanese wearing red hats" and groups of Gauls, Sardinians, Egyptians, and otherwise unspecified "Orientals, clothed in long, elaborate tunics."[82] The exactitude or consistency of the details matters less than their profusion; that is the nature of the reality effect.

At another level of visual apprehension, the large numbers of performers filling the stage evoked the logistical and organizational efforts of which the pageant formed part—those of empire as well as of archaeology. At moments in the pageants, archaeological exactitude joins with fantasy to provoke emotion, as in this passage from Carton's report:

> Who, dreaming of an evening in the ruins of ancient Africa has not, in the fever of its images, had a vision of ancient inhabitants wandering through the abandoned city? This time the illusion has almost become reality, through this apparition, which at first allows us to make out, in the distance, the swaying of the white fabrics caressed by the gold borders, then the undulation of the luxuriant hair tied—on such lovely foreheads!—with gilded bands, the brilliant metal of the bracelets and the fibulas on marmoreal bosoms.[83]

In Carton's musings, the site and the performance merge to offer a particular way of understanding the relation between past and present. Imbued with nostalgia, this understanding also firmly embraces the historical knowledge that archaeology offers. Historical understanding finds its most fitting representation in the distancing effect of performance, based in the audience's awareness of the elapsed time between the depicted moment and their present. Carton's emphasis on glints and fragmentary impressions, as well as his clear preference for a crepuscular light that creates atmosphere, artfully conveys the particular conditions in which archaeology can best be staged while at the same time filtering out unwelcome memories or inconvenient knowledge. If, for example, Grandmougin clearly relied heavily on Polybius's *Histories* as a source for his play, he notably omits Scipio's famous comments on the transience of empire.[84]

Press accounts from the metropole suggest that Carton largely succeeded in his stated aim of calling attention to the importance of preserving and further excavating Carthage. Maurice Pottecher, whose presence as an official delegate of the minister of education was noted in several accounts, devoted the last long paragraph of his article in *Le Figaro* to this worthy goal. In this chorus of praise, Alexandra David-Néel's dissenting voice stands out for its prescience. As a local effort, she wrote, the pageant could be counted successful, but the publicity aimed at foreigners she considered false advertising: "To promise a unique event, when after all is said and done all that was on offer was a miserable imitation of the shows presented in Orange, seemed to many like an excessive bluff."[85] Even more revealing, however, is David-Néel's judgment of the archaeological enterprise. "At Carthage, where there are many pebbles and no ruins," her article begins, "there once existed a sort of circus built up against an earthen hill." Archaeologists, she continues, "saw there the site of a Roman theater," taking as evidence marble bench seating and a historiated fragment showing a stage. The decision to clear the site as a revived theater was "the beginning of the end": for the 1907 pageant, the stones that littered the theater floor had to be cleared and new wood seating installed. "Apart from a few capitals given the honor of decorating the stage, all the bothersome product of the excavations was ejected from the enclosure."[86] In conclusion, David-Néel returns to the asphodels: "To transform into an empty hole, similar to an abandoned quarry, this last corner of a field where, among tufts of asphodels and golden chrysanthemums, the mutilated marble offered a melancholy evocation of the memory of Carthages past and gone, borders on the criminal."[87]

Subsequent scholarship has not borne out David-Néel's assertion that the theater of Carthage was not really a theater at all. Frank Sear's definitive study of theaters in the Roman world dates the Carthage edifice to the mid-second century CE, details its structural elements, and notes that it pioneered an architectural feature—semicircular niches enclosing the main doors in the stage wall—that became fashionable across North Africa.[88] David-Néel's clear preference for ruins and disorder over the tidied results of archaeological excavation has a long pedigree, yet few thoughtful archaeologists today would discount the larger point she makes. Archaeology, David-Néel posits, is about desire, but it seeks to control the desire it mobilizes. Like the *surveillant* shooing the Bedouin women off the stage, archaeology functions as a kind of border police between the phantasmatic and the real. The legitimation it performs as it chooses which elements of fantasy can be accepted as "real" comes at the price of certain kinds of willed blindness. David-Néel's doubts implicate archaeology's core practices as well as its ideological entanglements, yet neither she nor any of those more conscious of cycles of empire is entirely free of colonialist assumptions. Harry, for example, who in her account of the Kasserine-Thala trial had overtly expressed her sympathy and respect for the Muslim population of Tunisia, lightly refers to her barefoot guide as "the little savage."[89] Delarue-Mardrus reproduces this unthinking colonial racism in her play: a colonist calling women identified as Bedouin "intruders" is simply a plot device, a way of literally clearing the stage. In the end, the priestess returns to her grave, and archaeology has secured, provisionally, the boundary between past and present. To that extent, the Carthage performances confirmed the performative value of archaeology as disciplined knowledge. The performativity of Glozel would, in contrast, fall victim to a *mise-en-scène* that seemed to privilege spectacle as it relegated science to a distinctly secondary position.

Glozel and the Performative

Although Glozel produced nothing directly comparable to the Carthage pageants, it nonetheless offered many outlets for the performativity of its participants. From the academy to the masonic lodge, from the vaudeville stage to the lecture hall of the Collège de France and finally to the courtroom, multiple and diverse publics had the opportunity to consume and appraise different versions of Glozel. Significant distinctions could and should be made among these various examples: those restricted to small groups, for example, versus those open to the public; free versus paid events; live performances as opposed to textual "events" in which performativity remains mostly virtual. Yet the voracious appetite of the press for Glozel-related stories, particularly at the height of the controversy from late 1926 to mid-1928, tends to lessen those distinctions. At the very least this press coverage reminds us that *media* comes from the same Latin root as the verb "to mediate," placing written accounts inescapably in the middle of efforts to access and comprehend performances and the performative.

Much of the performative activity around Glozel points to the ways a discipline functions, like the Carthage pageants, as a kind of border police to allow and disallow claims not just about the discipline's operations but about its quintessence. To construe disciplines as themselves performative is to say that their essential nature emerges from the practices in which they engage, that, in Judith Butler's terms, "their very interiority is an effect and function of a decidedly public and social discourse."[90] Not all the performances spurred by Glozel took a position on the site's authenticity, but in summoning the codes of their own discursive fields, such as satire and farce, they suggest that it could be understood in familiar cultural terms other than those of archaeology. At the same time, many were making a judgment of Glozel's persuasiveness as an archaeological performance, with the ultimate failure of the Glozelians' performance marking a new boundary for disciplinary practice.

To understand the role and significance of performance and performativity in the Glozel controversy, it makes sense to begin with their pervasiveness as metaphors. Take, for example, a cartoon that appeared in *L'Oeuvre* at the end of February 1928 (fig. 4.4). Many newspapers had described the search and seizure carried out at the Fradins' farm as a *coup de théâtre*, a mostly dead metaphor for a sudden change in fortune.[91] But these two gentleman, well on in years, have clearly been discussing Glozel for some time: "What had I told you?" the gesticulating figure says to the other. Glozel was "une comédie." The word can obviously be translated literally, but in French, as a metonym for the theater, it has a more general sense than the English "comedy." Yet the *coup de théâtre* ending conveys the impression that this is not great drama. Several observers thought that only a great (comic) playwright like Molière could do justice to Glozel, but one was a lawyer for the Fradins, and a second, citing the notorious hypocrite *Tartuffe*, was effectively indicting them.[92] As early as September 1927, and continuing into 1928, a number of commentators said that Glozel reminded them of the farcical comedies of the prolific playwright Eugène Labiche (1815–1888). Several singled out Labiche's *La Grammaire* (1867), in which a wily servant, seeking to conceal his clumsiness, buries crockery he has broken in his employer's garden; the employer, an amateur archaeologist, then unearths some of the fragments and joyously declares them Roman.[93] Another writer noted that "Human vanity also offers fine subjects for vaudeville." The critic Jacques Coutant used vaudeville to dismiss a particularly discreditable journalistic gambit, interviewing nonexperts, such as gravediggers at Père Lachaise or a Vichy postman, for their take on Glozel. "Once more," he declared "they approached the question as though considering a scene in an end-of-year revue."[94]

In fact, by one critic's count, Glozel inspired no fewer than seven sketches in the variety shows put on in Paris cabarets and music halls in the winter of 1927–1928.[95] Some of these revues even boasted names that made sly references to the Glozel affair, such as "Suivez la fouille!" (literally "follow the excavation," a pun on the familiar expression "Suivez la foule" or follow the crowd) or "Gloz . . . ons," a wink at the verb *gloser*, to comment on or annotate, as in the English to gloss. The latter offered a sketch satirizing only the divisiveness of the controversy, and thus its evocation of the Dreyfus affair: a group of friends of varying political stripes gather in the salon of a dowager, improbably agreeing on topics ranging from disarmament to the country's financial situation, but when someone mentions Glozel, a brawl breaks out between Glozelians and anti-Glozelians.[96] Other revues put characters from the affair on stage, notably the year-end

FIG. 4.4. R. Guérin, cartoon from *L'Oeuvre*, 28 February 1928. Source: gallica.bnf.fr/BnF.

review at Les Deux Ânes, a recently founded cabaret (still extant today) at the foot of Montmartre, known for its clever spoofs of current events. In one scene, according to a reviewer, "Salomon Reinach is closely examined by the pharmacist and schoolteacher of Glozel as a prehistoric figure, and Salomon Reinach engages in stage business that proves he is indeed very old."[97] Much of the criticism of this piece focused on the performance in the role of Reinach (and many others) of a young actor, Marcel Dalio, who would go on to become a star of the French screen.[98] On the first day of 1928, the daily *Le Journal* offered its own textual "Revue de fin d'année 1927," in which characters from the news follow each other in disconnected chatter. Reinach enters last, withdraws a brick from a large sack and says that its inscription, "in the purest magic Phoenician," predicts "the complete and detailed history of 1928."[99]

Given the nature of the form, dominated by what one critic called "clownish tomfoolery," cabaret performances, which assume their audience's familiarity with the news of the day, do not bring much subtlety to the historical record.[100] Imagine a hypothetical audience member, Francophone but from a distant country and not up on the news: what might she glean from these shows, or from the brief gloss of a friend in the know? Not much more than that Glozel involves aging scholars arguing about old things or, to be more specific, old bricks with strange signs on them. The concept of excavation does occasionally come up, as when a clown at the Cirque d'Hiver, mocked for a particularly dirty hat he is wearing, declares that he unearthed it at Glozel.[101] All of this is fairly predictable, as is the choice of Reinach as emblematic figure. Alone among the major protagonists, he bore a name, celebrated since the Dreyfus affair, that resonated with the public at large, and both his clearly aging body and his statements lent themselves to caricature, tapping into traditional images of the doddering old man. The Fradins, especially Emile, would come to embody a parallel, indeed complementary, stereotype as wily peasants, but their exposure in the national media lagged significantly behind Reinach's, dating only from the early days of 1928, too late for inclusion in year-end revues.

Another elderly performer who attracted the attention of the media was Joseph Loth (1847–1934), a specialist in Celtic languages and one of the three academicians who likened the report of the verification commission to the persecution of Galileo. In early January 1928 Loth began a series of lectures on Glozel at the Collège de France, where he had held a professorship since 1910. As courses at the Collège were traditionally open to the public, they had long attracted large, curious audiences and strong media interest: a photo spread in *Excelsior* just before World War I shows a crush of people outside trying to listen to Henri Bergson's lectures through open windows.[102] Accounts of Loth's lectures devote as much attention to the restive and unruly listeners they attracted as to their content, which was largely a rehash of the Glozelian doxa with the addition of disparaging personal remarks. The accounts thus add the public to the picture, for it is the public whose presence activates and registers the event as a performance. One report describes the audience on 10 January as "a crowd of Glozelians, with some rare adversaries mixed in." A claque of anti-Glozelians disrupted the proceedings with coughing and sneezing powders, provoking a group of Loth's supporters to kick them out.[103] Later in the month, *Le Journal* began its report on the latest lecture with close-up description of a few audience members, whose loyalty ostensibly reassured the professor as he began to speak: "the little *rentier* in a muffler, the old lady who had dreamed of being a schoolteacher, the skinny Czech student ceaselessly taking notes."[104] Shortly thereafter, however, a group of hostile students invaded the lecture hall and, after being repelled by guards, began throwing stink bombs through the windows, some of which were broken in the process.[105]

Loth was certainly well aware of the media interest in his lectures: *Le Journal* reported that, running a gauntlet of photographers on his way into his first lecture, he asked them to wait to photograph him as he removed his overcoat to reveal his service medal from the Franco-Prussian War.[106] He was able to use the Collège's summary *mise-en-scène* to his advantage: one newspaper photo shows Loth sitting, as is the custom in France, at a desk on a raised stage, reading from notes, a sea of white heads below him.[107] According to the staunchly anti-Glozelian *Action française*, whose young partisans had considerable experience in disrupting lectures and other public events, Loth did nothing to discourage the tumult, and in fact seemed to revel in it.[108] A report on the 10 January lecture describes the professor continuing to speak over the din, "unperturbed," and according to *Le Petit journal* he began the lecture on 21 January by declaring, "They will not interrupt the course, whatever they do!"[109]

A few days later, an article in *Paris-Midi* jocularly suggests that Loth give his course in the Olympic stadium in the suburb of Colombes, or perhaps in the circus; an accompanying cartoon (fig. 4.5) depicts him standing between two lions, cheerfully unperturbed.[110] Such a depiction likely pleased the professor. Yet in press accounts of this performance, the reality effects come from the stink-bomb throwers. Untroubled by the claque, *Le Petit journal* reported, Loth "commented eloquently on the written articles and gave his reasons for believing in the authenticity of the finds." This was a milder version of *Action française*'s acid observation after the first lecture that "up till now, at least, he has not touched on

FIG. 4.5. Cartoon by an unidentified artist showing Joseph Loth as a lion tamer, *Paris-Midi*, 26 January 1928. Public domain. Source: Archives de l'Institut de France, K51.

the scientific aspect of the question but was content to polemicize about newspaper and magazine articles."[111] The key word is the *Petit journal*'s "*croire*," to believe, a word that effectively reduces the lecture to one side in an ongoing controversy. The partisanship of the audience accentuates this discredit: *La Rumeur* describes "a crowd of Glozelians . . . come to hear the Gospel."[112] The accounts suggest that Loth is playing the part of a learned professor, in a play his listeners take as a window onto reality. His contradictors, relayed by the press, insist it is only a *comédie*.

Loth was also treading a fine line: in a long letter to Reinach after his first lecture, he describes a contentious meeting with the administrator of the Collège de France, Maurice Croiset, who had summoned him to complain of his ad hominem attacks on Bégoüen and reproach him for causing a "scandal," which Loth vigorously denied. But, Loth admitted, "what unfortunately gives him a slight advantage over me is that I had chosen *The Romances of the Round Table*" as the subject of his winter course, and though he had announced the change, he had not, per Collège regulations, submitted it to the whole faculty for approval.[113] Based on this technicality, rather than any declared or admitted breach of norms on Loth's part, the course was suspended after the second lecture, though Loth resumed it later.[114] This legalism served as an apt prelude to the judicial actions that began around the same time and brought the Glozel affair public attention for years after it had otherwise ceased to occupy the headlines.

Ranging over several years, the legal proceedings over Glozel encompassed several distinct cases, of which two attracted the most attention. The first comprised a complaint by the Société préhistorique de France claiming fraud, which led to the police raid at the end of February 1928 and eventually to Emile Fradin's indictment. The second involved a defamation suit the Fradins had brought slightly earlier against Dussaud; Morlet also sued several individuals and newspapers for defamation. Although brought in different jurisdictions—the Fradins' suit in Paris, the fraud case in Moulins—and turning on distinct legal questions, the actions yielded verdicts that typically favored the Glozelians. In dismissing the fraud case against Emile Fradin in 1931, the investigating magistrate observed that the continuing disagreement of scholarly experts ruled out any definitive evidence of fraud. *Le Moniteur*, a Glozelian newspaper in Clermont-Ferrand, headlined its story on this decision "Triumph at Glozel" and crowed that the decision "closes the judicial proceedings."[115] A year later, Georges Claretie, writing in *Le Figaro*, cited the magistrate's decision at length to make the point that judges cannot resolve essentially scientific disputes.[116] But after the fraud case ended at the point known in French law as the *instruction*, the largely closed-door investigation preceding indictment, the libel trials gave the attorneys on both sides the chance to show off their rhetorical skills.

The short-lived American daily *Paris Times*, reporting on the Fradins' visit to Paris in early 1928 to launch their suit, predicted that "oratory, invective, entertainment, ill-feeling, and perhaps even some science are expected to be brought forward at the trial."[117] More succinctly, a regional newspaper described as "an oratorical joust" the courtroom appearances of Maurice Garçon (1889–1967; fig. 4.6) for the anti-Glozelians and Dominique Audollent (1897–1972), son of Auguste Audollent, for the Fradins and Morlet.[118] Garçon in particular, whom Benjamin Martin describes as one of the two most celebrated members of the interwar Paris bar, was well known for his elegant, well-constructed briefs and rhetorical mastery.[119] But what does it mean to treat judicial proceedings as entertainment? Dianne Dutton has traced the conceptual relationship between trials and theatrical performances to Aristotle and such notable Roman rhetoricians as Quintilian and Cicero. The Roman writers, who had considerable influence on early modern French legal theorists, stress the artistic element of rhetoric and explicitly connect lawyers and actors.[120] Focusing on much more recent literature, Yasco Horsman grounds his

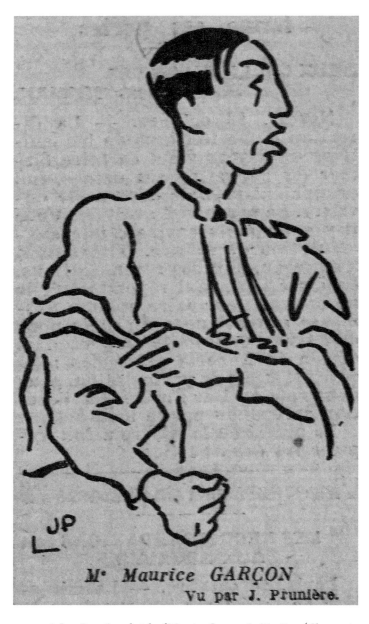

FIG. 4.6. Jean Prunière, sketch of Maurice Garçon, *Le Moniteur* (Clermont–Ferrand), 19 October 1929. Public domain. Source: Direction des affaires culturelles, Région Auvergne–Rhône Alpes, Service régional d'archéologie, Site de Clermont–Ferrand.

analysis of post-Holocaust texts about trials in a conception of the trial as "specific theatrical structure."[121]

In the French case, Yann Robert has shown that until the Revolution a professional code of ethics had maintained a strict separation between lawyers and actors; attorneys were supposed to take on as clients only those they believed to be innocent and to represent them with their own reputations for probity rather than eloquence. In the eighteenth century, reform-minded members of the bar called for the removal of these restrictions, which among other things banned actors from even attending court, much less testifying. The reformers argued that rhetorical eloquence imitating the stage could promote fair verdicts by conveying defendants' state of mind to judges and members of the public. With the Revolution came acceptance in France of the universal right to legal counsel, which meant that the Ordre des Avocats, an organization akin to the American Bar Association, no longer insisted that lawyers believe in their clients' innocence. Throughout the nineteenth century, however, the Ordre's acceptance of rhetorical eloquence, its recognition that lawyers "must play a role [*jouer la comédie*]," remained grudging.[122] The overlap between legal and theatrical terminology, which includes the metaphorical use of the term *coup de théâtre*, in French extends, as Dutton has observed, to the shared use of the word *représentation*.[123] Metaphors of performance therefore suffuse many accounts of the Glozel trials, especially those at which the site's authenticity was not at issue. These narratives of the "theater of justice" both reinforce and renew some of the tropes used to characterize other stagings of Glozel, while driving home its performative aspects.

As with Loth's lectures, the audience often claimed journalists' attention. Describing an appellate hearing of Morlet's case against the Société préhistorique and the *Journal des débats* in February 1930, the regional newspaper *La Montagne* noted that the courtroom in Riom contained "not an empty seat" and that, in the "large and select public [one saw] a few serious cassocks, the severe faces of some very Glozelian types, and many women, some of them young and pretty." The report proceeds as a kind of hybrid—part review, part comedy of manners. Garçon, a "young glory of the French bar," receives praise for his "elegant, alert argument, sprinkled with sallies and witticisms as well as adroitly controlled bits of nastiness." A riled Morlet, on the other hand, could barely contain himself, "feverishly ... taking notes he passed continuously to his lawyer, Mᵉ Audollent, who seemed quite bothered by his client's nervousness." Faint praise for Audollent leads into the apparent satisfaction of the audience, and the story concludes with a sly reference to Reinach, saying that the future career of this "young scholar"—Reinach was seventy-one at the time—would depend on the verdict.[124]

Journalists were in part relaying the tendency of attorneys to cast the parties as recognizable types: *Le Matin* quoted its own attorney, Henry Torrès, contrasting Emile Fradin, "a salt-of-the earth little guy we know," with Dussaud, a man of learning led astray by his "scholarly fanaticism."[125] The attorneys' notes and drafts show them building character portraits of both their clients and those on the opposing side: Audollent depicted Regnault, the president of the Société préhistorique, as a mildly dishonest intriguer in his professional life, whereas in a brief in the fraud case, Garçon cast Emile as a skillful forger.[126] But newspapers went beyond the lawyers' briefs to mold characters in their own stories. *Le Figaro* referred to the Fradins as "ploughmen" (*laboureurs*), noting in a mocking tone that they would soon be giving a lecture at an élite Paris club and likening the entire hearing to Gulliver's visit to the Lilliputians.[127] As usual, Reinach, when he testified in an appellate hearing on the libel case against Dussaud in 1932, offered a tempting target for caricature. The Glozelian *Le Matin* limited itself to quoting Reinach's own invocation of salt-of-the-earth imagery: "to attribute all these objects to a modest peasant like Emile Fradin is to defy common sense."[128] The anti-Glozelian *L'Oeuvre* went considerably farther, beginning with a portrait of Dussaud as stuffy bourgeois: "M. Dussaud, in pince-nez and black dicky, has an oily solemnity that contrasts with the thinness of his attorney, Me Garçon." But it reserved its most mocking description for Reinach, with "an Assyrian beard and a clearly Neolithic overcoat" and uttering a mild oath, "saperlipopette," that likely already seemed dated in 1932. Loth gets less attention, but his age (he was eighty-six) and "large white moustache" are mentioned, as well as his testimony in favor of Glozel's authenticity.[129]

Amusing as they are, these largely extraneous details function as reality effects in Barthes's sense of a "vertigo of notation" exceeding the demands of the narratives in which they appear. But what version of reality do they effect? What larger story do they help to legitimate? One clue lies in the self-consciousness with which reporters used stereotypes: describing Regnault at the 1932 appellate hearing, *L'Oeuvre*'s Pierre Bénard said his "*coup de vent* beard and floppy tie give him the conventional appearance of the 'old original.'"[130] As with Reinach's "Neolithic" overcoat and Loth's age, these punning descriptions have the effect of fusing the object of debate with the debaters, suggesting that half a decade after the beginning of the controversy, the Glozel artifacts, whatever their age, held interest only for the elderly. The Clermont-Ferrand daily *Le Moniteur* wrote in 1929 that, far from idealistic debates aiming for higher truth, the controversy involved "internal quarrels that, if the combattants' age did not forbid it, would often degenerate into hand-to-hand combat."[131] Jean Prunière (1901–1944), the sketch artist who produced figure 4.6, the image of Garçon that accompanied this article, seems to have picked up on this idea.

In a draft of his argument for an earlier case, Dominique Audollent says of the story in the *Journal des débats* that prompted Morlet's suit—written, as he knew, by Garçon but in the name of the Société préhistorique: "The history of Glozel can be found there, in broad strokes. We see there a procession of the main characters who there played a part."[132] The repetition of the pronominal adverb *y*, or "there," has the effect of emphasizing the power of the written account: *there*, in that document, in the world or reality it creates. But the recurrence also reflects back, probably unconsciously, on the ritual character of judicial proceedings, and on their repetitiveness.[133] Audollent won a symbolic victory in this case, but a court of law could not diminish the sense conveyed in the press, whatever its position, that Glozel was a kind of performance, with lawyers and witnesses playing predictable roles and following familiar scripts. Journalists' attention to the public, to surface details, and to the background, as well as their comments on lawyers' effectiveness, remind one of bored students tapping away on their smartphones. In these reports, clearly, legal jousting had prolonged the show well beyond its performative potential.

At this point, after this catalog of Glozel-related performances on stage, in the lecture hall, and in court, the reader might be wondering about my claim that Glozel tested the performativity of archaeology itself. The text in which Glozel emerges most clearly as a problematic archaeological performance actually predates the trials: *Glozel, vallon des morts et des savants* (Glozel, Valley of the Dead and the Learned), a novel by the right-wing writer René Benjamin (1885–1948) published in mid-1928, barely two months after the events it describes.[134] If a *roman à clef* refers to a work in which one can easily trace the real-life inspirations of pseudonymous characters, *Glozel* might be called an open-door novel, since it does not even bother with pseudonyms. Focused on the second verification dig, the one organized by Morlet in April 1928 and comprising almost entirely Glozelians, the novel features characters named Fradin, Morlet, Reinach, Loth, Audollent, and even Prorok. (Prorok pronounced the novel amusing but told Reinach that in the United States the author would have faced a dozen lawsuits for unauthorized use of personal names.[135]) Beginning with the arrival of the invited archaeologists in Vichy, where a journalist who has been covering Glozel identifies them to a younger colleague, the novel's action extends over three days of digging and includes several fine set pieces, notably lunch at the Fradins' house and the arrival of both élite visitors, including Prorok and his wife, and a crowd of curious onlookers. Throughout, Benjamin strikes a tone at once knowing, cynical, and humorous: one reviewer refers to his "unpitying, almost fierce observation, which discerns the clownishness of the very serious" and says that the book "lets off an irresistible gaiety."[136]

Though best known for his novels, Benjamin also wrote plays, and throughout *Glozel* the stage recurs as both a formal and conceptual device. At several points, for example when the two reporters are discussing the prior history of Glozel, or when one of the high society visitors is trying to arrange a tryst with a would-be mistress, Benjamin reduces the dialogue to exchanges arranged as in a play script, in the latter case referring to the speakers only as "Lui" and "Elle" (18–22, 135–40). At other moments, he draws back to describe the characters moving in a manner typical of boulevard farce, whether an agitated Morlet getting caught on a chain-link fence as he seeks out an intruder or a circle of people closing and reopening several times as they examine a find (176–77, 196). Throughout, media interest and the Glozelians' obvious desire to feed it add another layer to the narrative, from the journalists' opening description of the protagonists to the Gaumont camera operator recording one day's lack of finds to the photographs taken by a young girl among the tourists, just as the sun goes in (148, 144). Prorok's wife attracts attention not only because of her beauty but because she wields an elegant portable movie camera (194–95).

Beyond these obvious elements of *mise-en-scène*, Benjamin regularly introduces the allegorical figure of Comedy, *la Comédie*, as a framing device. "Comedy, behind her mask, is there in a corner, jubilant," he writes on the third page (11). "She couldn't care less about the truth: whether it's found or left alone, her only concern is to depict humans as they really are, and in the end she blesses journalists, who know how to inflame everything." After the first day, when to her amusement rain turns the dig site muddy, Comedy chooses to "renew her delight" by gilding the *savants* with "the sunshine that makes both roses and fools blossom" (109). Soon after, the only independent archaeologist in the group, Arcelin, begins to excavate in what he insists will be a truly scientific way. Then the easily bored Comedy, "because it is she, the charming girl, who has arranged this very pleasant décor," prompts the rest of the group, "grotesque and odd characters," to join the scene (113–14). The action thus "arranged" (*ordonné*) includes Morlet making a rapid find on the first day, Arcelin laboriously excavating without interference on the second and finding nothing, and the journalists' conversation with the passerby Morlet has accused of forgery, who tells them that an experiment he has conducted in a bowl has shown how objects can be inserted in the "archaeological layer" without arousing suspicion (184–85). The various figures from other stages, in both senses of the word, come together in this artfully disposed performance: Reinach, likened to an idol in the cart drawn by cows, who spends most of his time in a tent, seeing nothing (115, 154); Emile Fradin, who says very little but smiles a great deal (45–46); his stolid grandfather, several times likened to a log (50). The story even includes the Marquis del Alcantor's attempt to seduce an attractive young woman in his party via an explanation of an inscription he had found in the press. "On the Fradinian bricks," he tells her, "there are at once ancient Greek, second epoch Latin, letters from the Rhône and kabbalistic signs." "All that!" the woman says, "her eyes widening" (133–34).

The juxtaposition of a suspect find on the first day, after Morlet has pointed to supposedly virgin ground, irreproachable but unproductive digging on the second, and a genuine experiment with manipulating soil clearly puts *Glozel*, the novel, in the anti-Glozelian camp. This is unsurprising coming from Benjamin, a writer close to the Action française who would go on to serve as a high-level advisor to the Vichy regime. His hostility forms part of a broader antirationalism, as the critic in *Journal de l'Est* discerned: "he deplores scientific and demagogic obstruction, which detracts from the harmony of life and the flowering of hearts."[137] At the beginning of the dig a wild boar begins sniffing around the site, something most of the party misses; "What power! What truth!" Benjamin exclaims. "He is the only one to be where he should be" (63). The novelist dismisses the Glozel finds as "fragments of things without form, bones that might be teeth, stones that might be bones. Horrible!" (62). If the humorous tone of the novel would likely keep most readers from taking it too seriously, it nonetheless presents Glozel as an unconvincing performance of archaeology, and archaeology as a performance more of pedantry and human vanity than of anything worthwhile. Certainly Benjamin is condemning the pretensions and sterility of science as much as its presence in the media, but he also sees the particular manifestation of science that prompted this condemnation as indissociable from its spectacularization. In its multiple, varied, but ultimately repetitive performances over the course of several years, Glozel had hijacked the performativity of archaeology and consigned it to the domain of stale jokes, stereotypes, and dead metaphors. The final chapter will explore how, given these handicaps and in comparison with archaeology in Tunisia, the Glozel finds worked—or did not work—to nourish the archaeological imaginary.

✳ 5 ✳
Picturing Things
ARCHAEOLOGY AND THE IMAGINED PAST

The faces stare out at us, their asymmetry and quizzical expressions attenuated by the neat order of their arrangement on a plain black background (fig. 5.1). Each mask, made, according to the spare caption, of "glazed earthenware," retains an individuality to which the Arabic numerals below attest, while also forming part of the whole signified by the designation "Plate xxxv." Flipping back to the text of the volume in which these photographs appear, a 1900 catalog of the Carthage archaeological museum, the reader learns that these masks, "of bizarre appearance," comprise a necklace recently discovered in a Punic cemetery. The catalog entry divides the beads, as we may now think of them, into three typological categories, which it enumerates; it does not discuss each individually.[1]

Consider, in contrast, figure 5.2, facing pages of a much smaller-format book. Here two pieces of decorated pottery compete with surrounding text for the reader's attention. The work on the left is numbered "30 *bis*"—an indication (akin to "30a"), reinforced by the slight angling of the image on the page, that it was inserted at a late stage, when the rest of the figure numbers had already been set. The bold type just above figure 30 *bis* highlights a theory of the fusion of formal and practical elements, eyebrows becoming handles, in another work that, confusingly, is not illustrated at all. The indications of haste in this publication, the third of Antonin Morlet's "fascicules" (essentially pamphlets) on the Glozel discoveries, convey a sense not only of urgency but of insecurity, in the technical sense of an attribution or interpretation that lacks the authority of scholarly consensus. The contrast reflects, in part, the vast gap between the prestigious Paris publisher Leroux, one of the leading French scholarly presses of the day, and the Vichy printers who brought out Morlet's pamphlet.[2] But all of these text-image relationships raise larger questions about the nature of archaeology and its construction on the printed page.

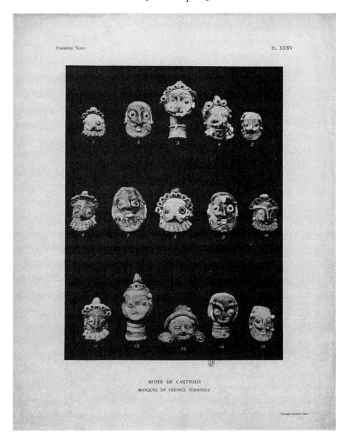

FIG. 5.1. "Masks in glazed earthenware," collotype by Berthaud, Paris. Plate XXXV in *Catalogue du Musée Lavigerie*, part 1 (Paris: Ernest Leroux, 1900). Source: gallica.bnf.fr/BnF.

This chapter explores the image bank of archaeology, both its visual modalities and manifestations and the verbal images that helped archaeologists make sense of their finds. The images in figures 5.1 and 5.2 pertain to one of the fundamental questions of archaeology: how—through what technical means, thought processes, and disciplinary protocols—can the objects unearthed by archaeologists create pictures of past societies? This large question of course involves many issues besides the visual, but recent scholarship has argued for the vital place of visual representation, and more generally of visualization, in archaeology's knowledge claims, in both scholarly texts and media intended for a general audience. In a series of articles going back to the early 1990s, Stephanie Moser has traced the importance of visual media in archaeology from the early modern period to the present, observing that "the creation of visual methods for recording data and communicating ideas has been critical to the establishment of archaeology as an independent discipline." Visual images, Moser emphasizes, have enabled archaeologists not only to depict the past but to construct and legitimate hypotheses about it.[3] More broadly, Frederick Bohrer asserts that "images, perhaps even more than artefacts, held (and perhaps still hold) the archaeological enterprise together, knitting together not only its practitioners themselves but also its sponsors and audience."[4] These images, it should be stressed, include not only visual images but verbal tropes.

Much of the recent scholarly attention to visual representation in archaeology has focused on photography, which emerged at virtually the same time as modern archaeology and played a key role in its development

FIG. 5.2. Antonin Morlet and Emile Fradin, *Nouvelle station néolithique* (Vichy: Imprimerie Wallon, 1926), 3:28–29. Public domain.

both as a scholarly field and as the object of widespread public attention.[5] Michael Shanks has explored the ways photographs act as guarantors of archaeology's objectivity, noting that "a photograph always seems to attest to some truth; it draws on scientific conceptions of value-free or objective truth, but more importantly upon an ethics of truth-telling rooted in a valuation of sight over words."[6] Yet this aura of truth, fundamentally a product of discourse, depends on a host of conventions, assumptions, and relationships that tie photographs to conceptual bundles known by such rubrics as realism and naturalism. In exploring the prephotographic history of archaeological imagery, Moser and others have shown how visual modes drawn from antiquarianism, botany, history painting, and biblical illustration helped shape the conventions of archaeological photography from the mid-nineteenth century until well into the twentieth.[7] In her influential work on photography's multiple histories, Elizabeth Edwards calls attention to the materiality of photographs, including their "presentational forms" and "surface interventions."[8] Such interventions testify to persistent doubt about photography's efficacy in conveying information, and indeed about its very truth value.[9] Thus what Stefanie Klamm and others have called "intermedial practices"—the use of multiple media such as drawing and cartography to supplement photographs—became standard in both scholarly and popular media reporting on archaeology.[10]

Of particular interest are what Edwards calls "confluences of scientific visibility and aesthetic arrangement."[11] Over the course of the nineteenth century, archaeological photography increasingly shied away from "aesthetic or experiential effect," instead opting for a flatly documentary approach following certain conventions that, Bohrer has argued, supported the field's scientific claims. This shift paralleled what Moser sees as the emergence of distinctive archaeological protocols in museum displays.[12] The site plans, maps, stratigraphic diagrams, and detailed drawings that often complement photographs in archaeological publications and museum labels seek to plug the inevitable holes in the photographic record. The photographic binary of presence (what the camera captures) and absence (everything outside the frame) corresponds to one of the fundamental challenges of archaeology, how to move from the visible—objects, structures, sites—to the invisible.[13]

The binary of visible and invisible, presence and absence, also sequences two of the key activities of the field, which could be abbreviated as search and find(ings). Archaeologists survey sites, map structures, unearth objects, and then offer, as an interpretive proposition, the possibility of envisioning or picturing the vanished worlds that produced those artifacts. They do this by various means, including the conjectural reconstruction so disparaged by Flaubert.[14] This chapter asks where those propositions come from, exploring the complex intersection, a veritable cloverleaf of mental traffic, of discipline, knowledge, and imagination. Just as vision, as a cognitive act, extends well beyond the processing of visual sensations, the imagination entails not just the retrieval and rearrangement of images but multiple acts of recognition, filtering, and categorizing, involving words as well as pictures.[15] A range of filters and categories shaped the archaeological imaginary, its component images, and the knowledge production it enabled. Even casual perusal of early twentieth-century French archaeological scholarship reveals the centrality of assumptions about ethnic and racial hierarchies and the so-called progress of civilization. These assumptions pervaded both scientific fields such as biology and what we would now consider pseudosciences, like eugenics.[16] Many archaeologists sought insights from ethnography, an evolving method of anthropological observation that in the early twentieth century often still had recourse to older practices, notably the measurement of skulls and bodily features in extant peoples regarded as primitive. These features were in turn thought to index intelligence and culture, the latter often in the specific sense of artistic achievement.[17] Such practices imprinted in the archaeological imaginary a host of images, many far removed from the material evidence of artifacts, and archaeology in turn lent many pseudoscientific theories of racial hierarchy a factitious veneer of antiquity. Even today, Matthew McCarty has observed, "these colonial, ethnographic models themselves survive and shape explicitly post- and anticolonial accounts of indigenous North Africans and their religion."[18]

The two controversies considered in this book posed, and pose, different problems of visibility, which makes their study from the standpoint of the archaeological imaginary at once challenging and rewarding. The temporal remoteness of prehistory confronted the first generations of prehistorians with what Rémi Labrusse has aptly called "a real void in the representation of humanity," a void largely filled by anachronistic fantasies that even the widely reproduced cave paintings discovered from the late nineteenth century on would never entirely displace.[19] Carthage, in contrast, contained vestiges of a number of cultures or "civilizations," to use fin de siècle terminology, that fall within the domain of classical and postclassical archaeology.[20] But the culture that most intrigued archaeologists there, and that fired the imagination of writers like Gustave Flaubert, proved the most difficult to visualize. The people variously known as Punic, Western Phoenician, or Carthaginian were famously defeated by the Romans in 146 BCE, their capital, Carthage, razed to the ground. They left few monuments and no indigenous literary or historical tradition.[21] With only slight exaggeration, Camille Jullian's description of the remoteness of prehistoric man could be applied to the Phoenicians as well: "We can reach them only by means of half mutilated monuments, nameless, mute, and abandoned helter skelter on the land, preserved by nature's caprice."[22]

Bill Brown has theorized things as comprised of "what is excessive in objects, as what exceeds their mere materialization as objects or their utilization as objects—their force as a sensuous presence or as a metaphysical presence, the magic by which objects become values, fetishes, idols, and totems."[23] Between search and findings—between, that is, excavation and publication—came significant intermediate steps, none more important than the sorting and classifying of excavated objects.[24] In Brown's framework, sorting and classifying, as well as other activities discussed in previous chapters—consultation, oral and written reporting, fundraising and planning—serve as agents in transforming objects into things. The various stages of archaeology endow its objects with, among other attributes, evidentiary value: they make visible the larger worlds whence they came. The taking of photographs and their circulation, as well as other forms of image-making, played a key role in all of these activities. Exploring the passage from finds to findings thus entails examining text-image relationships, the elaboration of categories and hierarchies, and, with respect to the two structuring examples, mutual interferences. For the study of the ancient Mediterranean, and in particular both what was known and what remained mysterious about the Phoenicians, haunted the Glozel affair in ways that help explain the enormous interest it aroused.

A comprehensive survey of the construction of archaeological knowledge about Carthage and its connections to Glozel in the early twentieth century would greatly exceed the scope of this book. Instead I offer a kind of textual microhistory, a sampling of a few volumes that typify the ways archaeologists transformed the objects they discovered into the signifiers of ancient societies. The sample is not random; it includes, besides Joseph Déchelette's *Manuel d'archéologie préhistorique, celtique et gallo-romaine* (1908–1914), a standard text much cited by Morlet, several works by scholars encountered in earlier chapters, notably Paul Gauckler, René Dussaud, and Stéphane Gsell, with occasional appearances by Camille Jullian and Salomon Reinach. Across these expansive works— Déchelette's "manual" comprises four volumes, Gsell's *Histoire ancienne de l'Afrique du Nord* (1913–1928) no fewer than eight—stretches a web of ideas and assumptions that undergirded contemporaneous knowledge of the prehistoric and ancient Mediterranean worlds. As is typically the case, many of the core assumptions are so deeply ingrained that they can be traced only at the level of language or, when they involve images, through decisions about layout for which no author can be clearly identified. Deciphering these "paper inscriptions," in David Jenkins's phrase, demands an approach fusing attention to archaeological norms and the material properties of images with sensitivity to the wider historical context.[25]

The book's structuring tension, between science and spectacle, in this chapter carries a somewhat different weight from the preceding ones. Across a range of media, archaeologists deployed visual images to support their claims to scientific status. Yet the science they invoked did not always offer plausible explanations for the objects they found. In this respect, Carthage and Glozel proved similarly troubling in the 1920s. Whereas at Carthage disconcerting evidence of ritual child sacrifice was just coming into scholarly focus, Glozel promised a radically new starting point for alphabetic writing. Faced with such instability, visual representations, particularly those like museum catalogs that followed well-worn conventions, could paradoxically endow the picture they painted with greater security, and scientificity, than the underlying findings. The chapter traces this kind of uneasy oscillation between science and spectacle. It looks first at museum catalogs from early twentieth-century Tunisia, arguing that their layouts and varying text-image relations correspond to and construct a range of understandings of the pictured objects. The second section homes in on the unflattering view catalogs and related publications construct of the pre-Roman inhabitants of Carthage, whom for convenience, and following recent usage, I generally call Phoenicians.[26] In the final section, the negative picture of the Phoenicians functions as a backdrop for Morlet's portrait of the Glozelians, which employs many of the images and tropes that dominate the archaeology of North Africa. Picturing the Glozelians as an isolated but culturally advanced community, Morlet's visual and interpretive program taps into growing anti-immigrant discourse in the 1920s and, more broadly, an ideological defense of Western civilization. Yet whatever its contemporary political resonance, in the end the portrayal of "Glozelian" culture could not overcome its failure to live up to the emerging standards of archaeological practice.

What Are These Objects?

The Carthage catalog with which this chapter opened forms part of the *Description de l'Afrique du Nord*, a series of museum catalogs published beginning in 1890. The title pages add, in smaller type, the words "Carried out by Order of the Minister of Education and Fine Arts."[27] In evoking state authority and its aspirations to comprehensiveness, the series calls to mind that pioneering work of scientific imperialism, the *Description de l'Égypte*, produced between 1809 and 1829 as a record of the scholarship carried out on Napoleon's expedition to Egypt in 1798–1799.[28] Adding to this emphasis, many of the catalog volumes bear another series title, in very large type—"Museums and Archaeological Collections of Algeria and Tunisia"—before identifying the museum covered in the particular book.[29] As with most aspects of French archaeology in North Africa in the late nineteenth and early twentieth centuries, it does not take long to find René Cagnat roaming the ministerial corridors behind the scenes of these publications. As discreet as he was energetic, Cagnat in 1886 succeeded Salomon Reinach as secretary of the recently created Committee for the Publication of North African Archaeological Documents, a position he would hold until his death in 1937.[30] The remit of the committee, itself part of an older committee within the ministry, the Committee for Historical and Scholarly Works (CTHS), extended from instructions for would-be archaeologists in the Maghreb, published in 1892, to regular reports on excavations and finds published in the committee's *Bulletin*, to the *Description* project.[31] For the latter, Cagnat enlisted practicing archaeologists with impeccable scholarly credentials.

It is important to understand that "documents," in the name of the committee, refers not only, or even primarily, to publications, but also to the objects, the finds, those publications put into print. As an example of this sense of the word, the *Petit Robert* dictionary offers an excerpt from the writer Emile Henriot's *Les romantiques*: "Portraits, statues, allegories, autographs, medals, frontispieces—all these documents speak to the eyes."[32] But accepting the idea of artifacts or antiquities as documents, the question then becomes, what do they document? Of what do they serve as evidence? And through what codes, forms, and methods of visualization do their spokespersons "speak to the eyes" to make their case?

An initial comparison of two volumes in the series, those devoted to the Bardo and the Carthage museum, makes clear that these questions call forth multiple responses.[33] The Carthage volume consists of three large folios, part 1 devoted to Punic antiquities, part 2 to Roman antiquities, and part 3 to the early Christian epoch. The Bardo catalog, in contrast, is organized by materials, dividing its contents among seven main sections: mosaics, architecture, sculpture, epigraphy, metals, ceramics, and "various." The metals and ceramic sections are further subdivided, for a total of fourteen numbered sections. Although the organizing principle of each volume effectively serves as a set of secondary rubrics for the other, the Carthage catalog offers only very summary media rubrics. Both catalogs group the images—engravings for the Bardo volume, mostly photographs for Carthage—at the back, but the Carthage editors seem to envisage a reader actively flipping back and forth between text and image. This approach seems to have caught on, and the supplements to the Bardo catalog, beginning in 1910, include plate callouts in the margins of the text.

Yet certain commonalities run through the Bardo and Carthage volumes as well. For each object, the catalog lists, at a minimum, the material; a descriptor (such as "head" or "bas-relief" for sculpture, "capital" or "column" for architecture, or the subject for ceramics—usually the name of a mythological figure or divinity); in most cases, a dimension, typically the height; and a find spot, though this sometimes precedes the listing of a group of objects found together. If the listed object has been previously published, the catalog entry provides the reference, usually in an abbreviated form for which the initial volumes do not always provide a key.[34] Dates of fabrication appear only rarely, and then approximately, for example, "Age of the Punic wars" for a lamp.[35] Dates of finds are given more frequently, sometimes with great precision: "19 August 1889" for an oinochoe, or wine jug, found in Carthage.[36]

The Bardo catalog contains much less descriptive prose than its Carthage counterpart, no doubt because it has far more objects to list. Yet both manifest a certain didactic ambition, at once buttressed by and transmitting to readers ways of looking at the objects in question (all those in the Bardo catalog were supposedly on view at the time of publication). The first part of section K of the Bardo catalog, for example, authored by Gauckler and devoted to lamps, bears the title "History of the Lamp," and begins by noting that numbers 1–55, arranged "as far as possible in chronological order," set out "the principal types used in Africa from the most distant past to the Arab invasion."[37] Occasional notes scattered throughout the list of these fifty-five objects point out changes in form, decoration, and regional origin. Images of the lamps appear, with a few omissions, over three separate plates (34–36), conveying the primarily visual nature of the lessons Gauckler is seeking to impart.

The catalogs offer other kinds of instruction to the eye, both explicit and implicit. In his preface to the Carthage catalog, Antoine Héron de Villefosse, curator of Greek and Roman antiquities at the Louvre, refers to the museum's collections as "treasures," and says that "some of them are as precious for the artist as for the archaeologist."[38] The author of this volume, Philippe Berger, describes that gilded bronze oinochoe found in August 1889 as "one of the most beautiful pieces produced by the Carthage excavations," its handle "describing an elegant curve."[39] But the catalogs direct the reader's and viewer's gaze in more subtle ways as well, through both text and image. The oinochoe is one of very few works to merit its own plate (fig. 5.3), with no other object to inform or distract the viewer (a

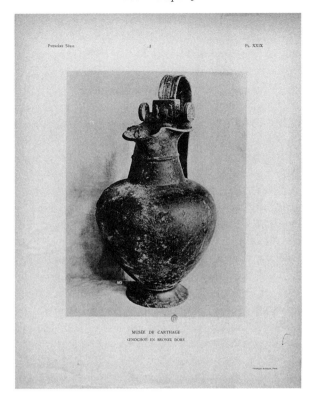

FIG. 5.3. "Oinochoe in gilded bronze," collotype by Berthaud, Paris. Plate XXIX in *Catalogue du Musée Lavigerie*, part 1 (Paris: Ernest Leroux, 1900). Source: gallica.bnf.fr/BnF.

few other plates show a single object from multiple perspectives, another form of distinction). More typically, plates present a group of objects, like the four neatly arranged mummy-shaped statuettes in the Punic section of the Carthage catalog (fig. 5.4) or, in the Bardo catalog, the five mosaics shown in plate 7, the ten stelae shown in plate 17 (fig. 5.5), or the twenty-eight pottery vessels in plate 42 (fig. 5.6). The Bardo plates feature engravings, many with tiny signatures of the museum staff members, employees of the DAA, responsible for them. As Geoffrey Belknap has found in his study of scientific and popular journals in Victorian Britain, the choice of media reflects a range of factors, including a widespread assumption that a skilled draftsman could reveal certain details better than the camera.[40]

Even a reader not following the text would likely notice that the stelae in figure 5.5 represent a selection from a much larger number, in fact over five hundred. The catalog text provides a single description for 538 stelae, whole or fragmentary, found at the site of Thignica in 1889, and already published in the committee's bulletin by Berger and Cagnat. The next entry covers five additional stelae fragments from a sanctuary atop Mount Bou Kornine, three of which are shown in the same plate.[41] Both entries describe the functions of these stelae in the cult of Saturn, syncretic with that of the Punic goddess Tanit; the latter notes that some better preserved fragments had gone to the Louvre, and that the Bardo display included "only those suitable to serve as types for figurative representations." The wary-looking bull that appears in several of the stelae, we are informed, is the usual "victim of the sacrifice." The entry for the Bou Kornine stelae describes them as "more skillful artistically" than those found at Thignica.[42] The text here reinforces what the viewer could infer from the images: that some objects, those accorded more prose and more space on the illustrated page, merit not only more attention than the others, but attention of a different kind. The oinochoe found on that precise day in August 1889 (fig. 5.3), the image proclaims, deserves a slow look, one attentive to its aesthetic qualities: the sweeping curve of the body set off by the compact foot, the combination of functionality and decoration in the tab connecting

FIG. 5.4. "Mummiform statuettes in terra-cotta," collotype by Berthaud, Paris. Plate XIV in *Catalogue du Musée Lavigerie*, part 1 (Paris: Ernest Leroux, 1900). Source: gallica.bnf.fr/BnF.

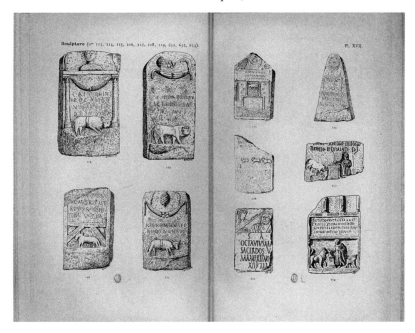

FIG. 5.5. Engravings of stelae. Plate XVII in *Catalogue du Musée Alaoui* (Paris: Ernest Leroux, 1897). Source: gallica.bnf.fr/BnF.

the handle to the spout. The pottery assembled in a gridlike formation (fig. 5.6), in contrast, lends itself to a quicker glance that takes in only its significant forms and variations. Such compositions correspond to and justify the amassing of related objects in a single museum vitrine.

In the catalogs, the language of aesthetic judgment, positive as well as negative, communicates to the reader the multiple signifying processes through which archaeology, in Bill Brown's terms, transforms objects into things. Such judgmental language recurs with considerable frequency, notably the word *grossier*, meaning rough or crude: "a crudely modeled face," runs the description of a pottery fragment in the Bardo; "Tanit with a roughly figured human head," reads that of a votive stela.[43] The Carthage catalog deems a group of mummiform statues (fig. 5.4) "rather crude imitations of Egyptian funeral statuettes," although the smallest one "is distinguished by a smiling mouth and eyes, and by a fine rendering of the features and expression of the face."[44] The reference to Egypt hints at a recurring preoccupation with cultural influence, borrowings, and hierarchies, tinged with what could well be called the crude theories of late nineteenth-century racial science.[45] The author of the Bardo catalog presents a group of eighty-five stelae just after those shown in figure 5.5 as "very roughly worked, but purely African: they are curious both for their neo-Punic inscriptions and for the figuration we see on them."[46]

Archaeology also mobilizes ethnographic references to convey meaning: a Roman pottery vessel earns a mention for its "exact resemblance to the type that the natives today use to make couscous."[47] On the pages of a catalog and in museum galleries, an object can become, literally, a thing of beauty, signaled as worthy of the viewer's artistic appreciation by the space and prose accorded it. Alternatively, the work's placement in an ensemble of similar objects summons an understanding, more intellectual in character, of the object as document: a jar, say, that provides evidence of the daily life, rituals, or burial practices of a defined group at a particular time. A single object can of course do both of these things, but the catalog and, usually, the museum layout guide the reader or visitor to a primary meaning. Beyond the aesthetic or the cultural, the properly archaeological signified appears when the catalog shows archaeologists at work, teasing out the meanings they are proposing.

The ethnographic parallel offers insight, in this regard, into some of the governing assumptions of modern archaeological inquiry. The three-ages

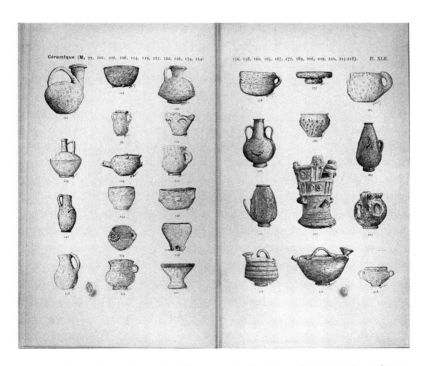

FIG. 5.6. Engravings of ceramics. Plate XLII in *Catalogue du Musée Alaoui* (Paris: Ernest Leroux, 1897). Source: gallica.bnf.fr/BnF.

theory—the sequence of stone, bronze, and iron ages—emerged in the 1830s and quickly became the foundation for archaeology's concern with tracing the evolution of human civilization. The brainchild of Christian Jürgensen Thomsen, founding director of what would become the Danish National Museum, the three-ages theory, supported by carefully organized stratigraphic evidence, became a beacon for the emerging field of prehistory.[48] But the tendency to measure human achievement in the making and manipulation of tools and the mastery of ever more complex material processes, from stone carving to metal forging,[49] extended to the more text-based domain of classical archaeology as well. The perfection of Athenian sculpture, to take one well-known example, reflected the technical mastery that undergirds cultural prowess. By the early twentieth century, Joseph Déchelette could assert in the introduction to his manual of prehistory that the methods of investigation and classification of prehistoric archaeology "differ in no way, despite what has sometimes been claimed, from those of classical archaeology."[50] To the extent it distinguishes between different degrees of skill and sophistication in craft production, the language of aesthetic judgment encountered in the catalogs indexes this overall framework. That framework also meshed with pseudoscientific notions of racial difference that built on, and ultimately perverted, theories of evolution, whether Lamarckian, the dominant strain in France throughout the nineteenth century, or Darwinian.[51]

For the most part discreet, the catalogs' invocation of archaeological modes of thinking and their connections to those of neighboring disciplines may have registered with only the most informed of readers. Other references, however, highlighted some of the basic methodology of the field, notably stratigraphy and contextual dating. At its simplest an equation of increasing depth with greater antiquity, stratigraphy occupies a central place in archaeology's self-definition as a science. Stratigraphy draws on geology, an unimpeachably scientific field, for the core idea of uniform strata or sedimentary layers.[52] If stratigraphy alone made possible only relative dating, contextual comparison offered the possibility of greater precision. The original Bardo catalog relates the discovery in 1895 of a group of vessels in a neo-Punic vault near the modern town of Tabursuq and adds, "They can be approximately dated to the middle of the first century BCE, by the observed presence in the tomb of a denaria [a coin] of the *gens Postumia*, from 64 BCE."[53]

"Observed presence": the phrasing uses the verb *constater*, from the same root as the noun *constatation*, at the heart of the international commission's report on Glozel. The catalog thus offers further evidence of the importance archaeologists attached to observation as an essential guide on the path from finds to findings. Contextual comparison also requires visual training that enables the identification of cultural influence. One plate in the Carthage catalog of Punic objects presents twenty-four pottery vessels, all in black lusterware but imported from different places. One, with incisions and impressions that recall objects found in Sicily, "seems to relate to Magna Graecia." Another plate identifies the headdress on a terra-cotta statuette of a seated goddess as an archaic Greek borrowing from "Eastern goddesses," and uses this relationship to propose a tentative dating.[54] Repeated references of this sort, though susceptible to racializing judgment, also point to the centrality of trade, exchange, and migration in the early Mediterranean. Understanding the objects one finds, placing them in a larger context, whether aesthetic or more broadly cultural, thus requires an eye open to movement and adaptation. Archaeology relies not only on profound local knowledge but on familiarity with broader, delocalized trends. This unspoken expectation would prove deeply problematic for Glozel.

The tentative language about influences and the approximate dating also testify to the uncertainty and doubt that hover over archaeological findings, particularly those at an early stage of research. The Bardo catalog describes a group of twelve limestone stelae as "VOTIVE STELAE still not fully explained": the divinity to whose cult they are dedicated remained unknown. An extensive discussion of the stelae gives only a likely find spot, the major site of Dougga, and notes that they came from a collection amassed, and subsequently dispersed, prior to the creation of the Protectorate, implying the superiority of European methods of preservation.[55] Text entries, notably in the more detailed Carthage catalogs, occasionally acknowledge disagreements among scholars, for example between Father Delattre, the dean of French archaeologists at Carthage, and Philippe Berger about the identification of a female deity.[56] Doubt could even find a visual register, as when the Carthage catalog provided line drawings of motifs on comparanda for a Corinthian-style vase (fig. 5.7). Visual information here accentuates but does not directly illustrate the "strange" foliate and foliform decoration of the vase: "These bizarre arrangements must reflect the traditional influence of some symbolic concerns the origin of which escapes us."[57] Here, to reverse the modernist paradigm, more (imagery) is less (certainty). Additional visual information serves most fundamentally to show something that photography cannot. In so doing, it takes the reader to the limits of scholarly knowledge, disrupting any simple equation of knowing with seeing.

One final aspect of the catalogs, their periodic multiplication, complements the list of find spots and dates and offers another kind of insight into

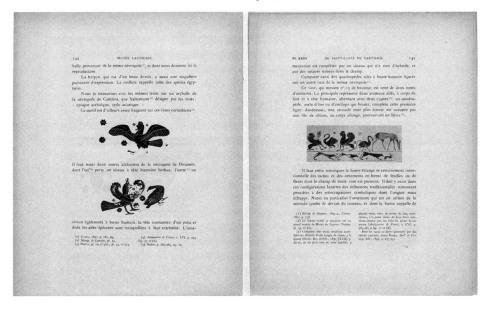

FIG. 5.7. Text and line drawings in the entry for a Corinthian-style skyphos or vase, plate XXIII in *Catalogue du Musée Lavigerie*, part 1 (Paris: Ernest Leroux, 1900). Source: gallica.bnf.fr/BnF.

the field. In his introduction to the Carthage catalog, Héron de Villefosse refers to the "archaeological ardor of the African missionaries," Delattre chief among them.[58] Fifteen years later, prefacing a supplement to this catalog, Cagnat writes that in publishing the original three volumes, "we were not unaware that Fr. Delattre would quickly take care of making them obsolete; that has not failed to happen, and at a scale well beyond what we imagined." Recent finds, Cagnat goes on, "have enriched the collections with numerous works, some of them of great importance: they have been admired by visitors and are of inestimable value to science."[59] His introduction to the supplementary Bardo volume, which appeared in 1910, confronted the plethora of new finds in somewhat blunter language. Cagnat praises a veritable who's who of colonial collectors, starting with Gauckler and including "civil servants, colonists, and officers of the occupying army," noting that their acquisitions had come about through both deliberate excavations and the accidents of agriculture and construction. Together they had made the Bardo "incontestably the most important" museum in Africa. But, Cagnat observes with the rhetorical equivalent of a smothered sigh, "the very abundance and speed of these discoveries were not without disadvantages; it became impossible to publish them as quickly as would have been desirable." Here Cagnat evokes the North Africa Committee's responsibility to publish archaeological finds as promptly as possible.[60]

A second Bardo supplement appeared in 1922. Both supplements followed the organizing principles of the original volume, even picking up the catalog numbering from the point where the 1895 volumes stopped, though they featured photographs rather than engravings. The Carthage supplement also followed that catalog's distinct organizing principles, the division by cultural epoch (Punic, Roman, Christian) rather than by material. Over and above these differences, the subtle forms of visual emphasis employed in the original catalogs persist. Of the eleven plates in the Punic section of the Carthage supplement, for example, only one presents a single object. This is the priestess sarcophagus (plate 4) used in the poster for the 1907 Carthage pageant, which the text describes as "for good reason the most famous of Fr. Delattre's finds."[61] As for the persistence of the different organizing principles, by cultural group or by material, they index both the challenges and the possibilities of archaeology in its formative period. They also cast archaeology as a variable archival practice, perpetually inventorying, classifying, and storing. In offering the new Bardo supplement to a broad public—"scholars and workers"—with a matter-of-fact comment about the continuity of rubrics and numbering, Cagnat seems to be acknowledging both the arbitrariness and the provisional character of the archival sorting these volumes recorded. Here are our things, he is effectively telling the reader. This is what we think they are. Make of them what you will.

Who Were These People?

Toward the end of volume 4 of his monumental history of ancient North Africa, Stéphane Gsell allows himsef a kind of metagrumble. After "once more" noting the limitations of textual sources about ancient Carthage, he writes baldly, "Archaeological discoveries do not make up for the poverty of the texts."[62] Since 1912 Gsell had occupied a chair in North African archaeology at the Collège de France, a recognition of his extensive and meticulous excavations, first of Etruscan tombs in Italy, then in Algeria, where in 1902 he became director of the museum of antiquities and Islamic art in Algiers. He also contributed a volume, on the museum in Philippeville (now Skikda), Algeria, to the *Description de l'Afrique du Nord*.[63] Subsequent generations of scholars have expressed something akin to awe with respect to Gsell's scholarly achievement. In a book published in 1962, the British archaeologist Donald Harden writes, "Our knowledge of the western Phoenicians must always be founded on the great work of Stéphane Gsell." In 1970, Pierre Cintas begins his *Manuel d'archéologie punique* by declaring Gsell's work "in all respects still fundamental," and justifies his own considerable work as an updating of Gsell that preserves the latter's structure.[64] Even a major history of Carthage published a century later cites Gsell often.[65] Yet the limitations of sources, both textual and archaeological, receive multiple mentions in the *Histoire ancienne*, a sort of leitmotif in a work that nonetheless paints a vivid and detailed picture of North Africa in the first millennium BCE.[66]

The books Gsell and other archaeologists published in the first quarter of the twentieth century fit comfortably in the scholarly world of the *Description de l'Afrique du Nord*. Even in works predominantly historical rather than archaeological in orientation, the language of aesthetic judgment, with that telltale word *grossier* (rough or crude), appears regularly, as a way of measuring what Dussaud, in his volume on the pre-Hellenic Aegean, refers to without elaboration as "the progress of art."[67] The historians all understand trade, migration, and cultural exchange as normal processes in the ancient Mediterranean world. At the same time, an interest in contemporary anthropological comparisons often bears the taint of dubious racial theories. Dussaud takes sufficient interest in the American archaeologist Charles Hawes's study of Minoan skull shapes to cite it prior to publication, in his final chapter on "Race and the Movement of Peoples."[68] Déchelette, in his *Manuel*, regularly compares his prehistoric subjects to "contemporary inferior peoples," but he also consistently puts forth evidence of material and cultural imports from the eastern and southern Mediterranean to what is now France.[69] All these authors discuss stratigraphic and comparative evidence as key to dating, and regularly cite important moments in the history of archaeology; Dussaud, for example, harshly criticizes Luigi Palma di Cesnola for his unsystematic excavations in Cyprus and his flouting of antiquities laws.[70] Dussaud and Déchelette also make extensive use of visual information. The former begins the preface to the second edition of *Les civilisations préhélleniques* by noting that the number of pages and illustrations has increased by half, and points with pride to the addition of fourteen plates, five in color, a way of signaling works of special importance.

This preoccupation with cultural advancement informs the early twentieth-century scholarly portrayal of the Phoenicians, especially the western branch typically referred to as the Carthaginians or Punic people. Today, a century or so later, scholarship on Carthage and the Phoenicians emphasizes two qualities: their elusiveness and their persistent neglect in scholarship and museum displays. For the first, Josephine Quinn devotes a well-documented volume to demolishing the notion of "Phoenician" as a form of self-identification in antiquity in any way akin to modern national or ethnic identities. Many of the essays in a volume Quinn coedited with Nicholas Vella echo this argument, as does Asher Kaufman in his study of the fabrication of Phoenician roots for modern Lebanon.[71] Similarly, Jonathan Prag's concise survey of linguistic usage concludes that "the ethnic label that we use so freely in modern discourse was used in a range of ways in antiquity, few of which map with any ease onto modern usage, and few of which, if any, equate to the normal range of ethnic labels in antiquity."[72]

Like many ethnonational constructions, "Phoenician," Vella shows, emerged as a category in the nineteenth century. One of the chief architects of this category, the pioneering French archaeologist and art historian Georges Perrot (1832–1914), regarded objects sharing some kind of formal or decorative unity as "ethnically diagnostic," wherever they were found. "The Phoenicians, however," Vella writes, "often subsumed within the generic rubric of 'Oriental' and 'Orientalizing,' were destined to remain ambivalent and stereotypical for much of the first half of the twentieth century."[73] In the 1960s the dynamic Italian scholar Sabatino Moscati began a campaign to recognize the Phoenician place in ancient civilization, culminating in the spectacular exhibition *I Fenici* in Venice in 1988. But this effort suffered from a tendency to homogenize a highly diverse culture.[74] More recently, Carolina López-Ruiz, while acknowledging the uncertainty about what Phoenicians called themselves, has made a strong case for moving beyond what she calls "Phoenicoskepticism." A combination of archaeological and epigraphic evidence, she writes, provides "enough internal evidence for an identifiable Phoenician culture." Further, "we can recognize the Phoenicians as a distinct collective by external indicia, and so could the Greeks, Romans, and others. It is difficult to imagine that they themselves did not."[75]

By the late fourth century BCE, Carthage, originally a colony established by Phoenicians most likely from Sidon and Tyre in what is now coastal Lebanon, had established a strong imperial identity distinct from its forebears. Carthage produced such objects as coinage with Punic inscriptions and symbols, invaluable for future archaeologists.[76] But this greater identifiability did not secure Carthage, in Gsell's time or in ours, a better reputation than its "Oriental" progenitors. Gsell describes the Carthaginians as "rather small people"—in this instance a moral rather than physical description—known around the Mediterranean as merchants whose cupidity made them untrustworthy.[77] Practically minded, one of their few virtues being their love of family, the Carthaginians, Gsell observes, contributed little to science and produced "neither scholars, nor poets, nor thinkers; or at least," he adds, acknowledging uncertainty in a way intended to minimize it, "history knows of none."[78] Arrogant, cruel, and insincere—they failed to seduce even those they flattered—"the antipathy they provoked was almost universal," and Gsell clearly believes they deserved it.[79] Gsell spends only a few pages on the most notorious aspect of Phoenician and Carthaginian ritual, child sacrifice; although it is a climactic moment in Flaubert's *Salammbô* (1862), solid archaeological evidence for the practice was only beginning to emerge in the early 1920s.[80] Nonetheless, on the basis of the evidence available, Gsell notes that child sacrifice lasted much longer in Carthage than in the Levant, and observes that "in time, natural feelings revolted against" this religious practice. But he quickly undercuts any impression that the Carthaginians may have been developing a greater sense of humanity: he cites a rumor, recorded by the Greek historian Diodorus, that wealthy families were secretly purchasing children to offer as victims while sending their own into hiding.[81]

A look back at the catalog of the Carthage museum suffices to situate this negative view of the Phoenicians within a clear scholarly consensus at the turn of the twentieth century. At the beginning of the Punic section of the original catalog, Philippe Berger presents borrowings from other cultures as the signal feature of Carthaginian art and concludes, "To put it plainly, Carthage could never create an independent, truly original art for itself; it was always the tributary of its neighbors, and it scarcely did more than imitate the models they offered."[82] Much of the language of "roughness" and crudity quoted in the previous section pertains to examples of Punic art, and the statuettes in figure 5.4, then deemed mostly inferior copies of Egyptian models, were discovered by Delattre in the course of his excavations of Punic cemeteries. Berger consistently attributes works that garner admiration, such as the oinochoe featured in figure 5.3, to Greek or Cypriot artists; he notes the Carthaginians' Hellenized tastes in his introduction as a factor making it difficult to determine "if we are in the presence of objects of Greek or Phoenician fabrication."[83] In the 1913 supplement, the young Hellenist André Boulanger follows this line: the first plate features a group of stelae in the form of statues, two of which "are the only sculptural monuments in the round found in Carthage that can be attributed to indigenous art of the Punic school." In the very next line, however, Boulanger says "they scarcely deserve to be called sculptures." Again, they are *grossier*.[84] Boulanger's discussion of the Punic anthropoid sarcophagi notes flaws in technique, and he attributes the most famous, the priestess (plate 4), a hybrid of Greek technique and eastern physiognomy, to "a Carthaginian working after Greek models. In the deeply

Hellenized Carthage of the third century [BCE], such models were not lacking, the city having been abundantly supplied with works of Greek art by trade and looting."[85] Hellenization wins Carthage no credit with Gsell, who observes that while the Carthaginians appreciated Hellenic culture, they also, through warfare, obstructed its reach and influence. Carthage, he concludes, "contributed very little to civilization in general. Its luxury scarcely benefited the arts."[86]

Only two aspects of Phoenician culture as it was understood in the early 1900s modified this negative picture: alphabetic writing, about which Gsell has little to say, and ethnicity, about which he says a great deal. The historian's lack of attention to the Phoenician alphabet has both geographical and scholarly explanations. First, the invention of Western alphabets, scholars then and now agree, most likely took place in the Levant and Asia Minor, not in North Africa, the subject of Gsell's opus. Second, like child sacrifice at Carthage, early alphabetic writing was a subject under active study, its contours not yet fixed, at the time Gsell's books appeared.[87] What scholars now regard as the oldest Phoenician inscription came to light only with Pierre Montet's discovery of the Ahiram sarcophagus at Byblos, in Lebanon, in 1923.[88] Nonetheless, even in the late nineteenth century, most scholars believed that the Phoenicians, in the territory of their initial habitation often called by the biblical name Canaan, played a crucial role in the transition from earlier forms of writing, both glyphic and cuneiform, to the modern alphabet. The leading authority at the time, none other than Philippe Berger, noted this consensus at the beginning of his *Histoire de l'écriture dans l'antiquité* (1891): "The Hebrew alphabet," he writes, long regarded as the oldest form of writing, "is nothing but a fairly recent transformation of Phoenician writing, from which came all the alphabets still in use on earth." But, he notes, the Phoenician alphabet itself was preceded by a number of other systems of writing.[89] After a discussion of pre-alphabetic forms of writing in part I, Berger begins part II of his book with the appearance of the alphabet on the Syrian coast at a date still uncertain, but around 1500 BCE, and writes, "The almost unanimous testimony of the ancients attributes its invention to the Phoenicians," though others credit it to a Syriac people.[90]

This painstaking inquiry into the beginnings of alphabetic writing (and its Egyptian and Assyrian antecedents) makes clear why the discovery of a supposed Neolithic alphabet at Glozel would provoke such controversy. Both Berger in 1895 and Dussaud twenty years later emphasize that the process of breaking down syllables into sounds and identifying them with letters must have taken some time, and acknowledge several candidates for the role of originary alphabet. Dussaud takes note of Reinach's hypothesis, based on pottery inscriptions found at the ancient Canaanite city of Lachisch, that the prototype characters came from the Aegean, but like Berger considers Phoenician a (slightly) more likely initial alphabet.[91] Whoever invented it, Berger concurs with Renan that the alphabet was "one of the greatest creations of human genius." He attributes this invention to the practical needs of trade, "yet another reason to attribute its paternity to the trading people par excellence, the Phoenicians."[92] Berger's volume is generously illustrated with drawings, photographs, and diagrams of inscriptions, including a full-page color plate showing an important stele from Byblos; Berger also makes liberal use of the comparative chart or table as a way of showing the relationship of Greek alphabets to their Phoenician cognates (fig. 5.8).

Although the beginnings of the Phoenician alphabet lie, as noted, largely outside Gsell's scope, he does acknowledge its persistence in the West, that is to say Carthage, well into the Roman period. Unmoved by the importance of the invention, however, he cannot resist observing that "Punic inscriptions, even those engraved with some concern for calligraphy, are of shabby appearance: small letters, composed in haste, crowded together on small stones; nothing that recalls the scale and majestic regularity of monumental Roman inscriptions."[93] This portrayal of the Carthaginians also participates in the Orientalizing to which Vella has pointed. Despite its location to the west of the Italian peninsula and its manifold borrowings from the Greeks, for Gsell Carthage remained "deeply oriental."[94] Yet the element of what Vella, citing Ann Gunter, calls the "ethnically diagnostic" earns Carthage a grudging credit, a civilizational left-handed compliment, from Gsell. More than the Greeks in the rival port colony of Marseille, the Carthaginians "educated the natives," and profoundly marked the Berber territory they colonized. "In a barbarous land," Gsell writes, "the Phoenicians had introduced an already advanced civilization; they had created urban centers and farms," and the Romans, after the defeat of Carthage, would take advantage of this social and economic organization.[95]

Gsell concludes his discussion of the Carthaginian legacy with a look at religion. By emphasizing "humble submission to the will of the Lord," the Punic religion, he argues, prepared North African subjects for Christianity and—via the proximity of the Punic language to Arabic—ultimately for Islam.[96] For Gsell, a scholar who had scaled the heights of French academic institutions, this judgment hardly amounts to a wholehearted commendation; the secular Republic wanted informed citizens, not humble servants. It does, however, remind us that the racial and ethnocultural hierarchies of the colonial era had many rungs and a complex set of positionalities.[97] If the French liked to think of themselves as heirs to the Romans, they could

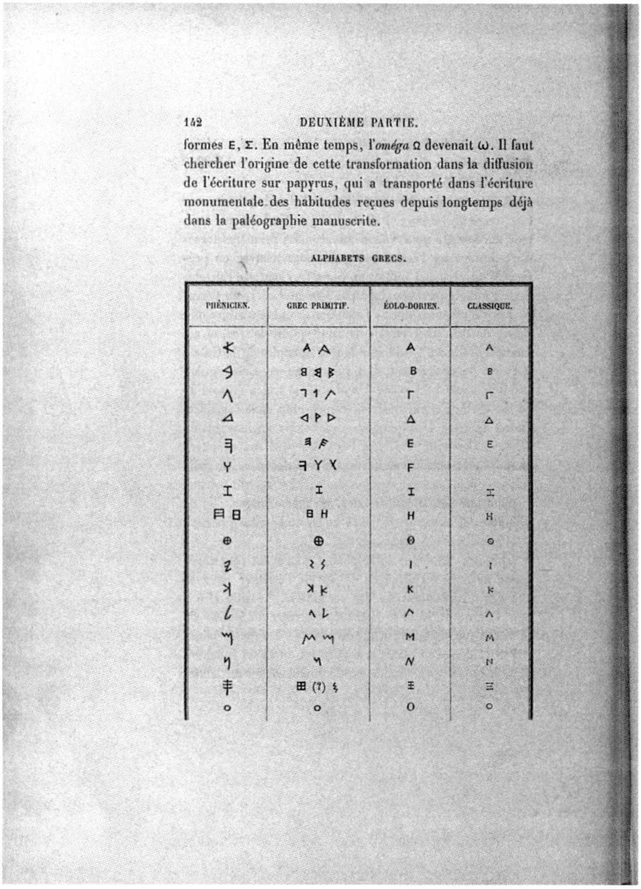
142 DEUXIÈME PARTIE.

formes E, Σ. En même temps, l'*oméga* Ω devenait ω. Il faut chercher l'origine de cette transformation dans la diffusion de l'écriture sur papyrus, qui a transporté dans l'écriture monumentale des habitudes reçues depuis longtemps déjà dans la paléographie manuscrite.

ALPHABETS GRECS.

PHÉNICIEN.	GREC PRIMITIF.	ÉOLO-DORIEN.	CLASSIQUE.
𐤀	A A	A	A
𐤁	B B	B	B
𐤂	Γ Λ	Γ	Γ
𐤃	Δ D Δ	Δ	Δ
𐤄	E E	E	E
𐤅	F Y X	F	
𐤆	I	I	Z
𐤇 B	B H	H	H
⊕	⊕	Θ	Θ
𐤉	S S	I	I
𐤊	K K	K	K
𐤋	Λ L	Λ	Λ
𐤌	M M	M	M
𐤍	N	N	N
𐤎	⊞ (?) Ξ	Ξ	Ξ
O	O	O	O

FIG. 5.8. Chart correlating alphabets, from Philippe Berger, *Histoire de l'écriture dans l'antiquité* (Paris: Imprimerie Nationale, 1891), 142. Source: gallica.bnf.fr/BnF.

regard the Carthaginians, literate colonists before them, as analogous to the Arabs: an inferior people but a literate one, closer to the French than the Berber "indigènes" to whom French colonizers were ostensibly bringing "civilization." That analogy had its own power, one that Morlet and Reinach would find difficult to contest.

Imagining the Glozelians

In the spring of 1926, Antonin Morlet, the spa doctor who was leading the excavations at Glozel, began publishing regular articles about the finds in the *Mercure de France*, a venerable bimonthly literary magazine revived in 1890.[98] Together with the many photographs in his self-published "fascicules," the *Mercure* articles, illustrated exclusively with engravings, offer a rich and varied assortment of images through which Morlet attempted to establish the authenticity of the Glozel finds and to make visible a "Glozelian" culture. Many instances of prose repeated verbatim in multiple texts track the urgency he felt in bringing Glozel to press first and exclusively, usually with Emile Fradin listed as coauthor. Through this mimesis of scholarly practice, Morlet was seeking to establish his own bona fides as an archaeologist.

Some of the most revealing moments in these publications arise when visualization and the norms of disciplined archaeology come into conflict. Toward the end of the third fascicule, intended both as an update on finds made in excavations since late 1925 and as a rebuttal to skeptics, Morlet presents a clay "human figurine," reproduced in a photograph occupying half a page. Although he has previously praised skillful animal engravings, in this work, the doctor declares, "The modeler demonstrates an extraordinary inability to represent the human figure, which here appears in a very pronounced state of barbarism. There is no verisimilitude and this work can in no way reveal to us the physical type of the neolithic tribes of Glozel." Yet if this figure "possesses no documentary value from the racial point of view," it does relate to a "known type of the neolithic idol."[99] These few sentences clearly place their author in the milieu of early twentieth-century archaeology, with its conception of objects as documents, its readiness to make aesthetic judgments, and its preoccupation with civilizational stages and racial classifications. Deliberately or not, Morlet evokes one of the basic dilemmas of archaeology, and especially of prehistory: how to visualize the invisible, specifically the belief system that produced the object.

Within the world of archaeological publishing, Morlet's articles and brochures (a term he occasionally used himself for the fascicules) more closely resemble initial archaeological reports than the highly ordered catalogs of the *Description de l'Afrique du Nord*. Such reports appeared in a variety of venues and illustrative modes. The account of the 1923 discovery of the Ahiram sarcophagus in Dussaud's journal *Syria*, for example, takes the form of a series of letters from the excavation leader, Pierre Montet, to Cagnat in his role as perpetual secretary of the Académie des inscriptions. The *Syria* article contains a number of maps and diagrams of the tomb being excavated, as well as one discreet line drawing; soon after, Montet also published the finds in the general-interest weekly *L'Illustration*, amply illustrated with photographs.[100] But Glozel posed several problems for the standard sequence of archaeological publishing. The seemingly unending productivity of the digs Morlet undertook, with only Emile Fradin to assist him, confined them to a perpetual reporting stage. More seriously, Morlet lacked the relays for definitive, scholarly analysis that professional archaeologists could expect. On the one hand, he isolated himself from local history and archaeology buffs: the *Mercure* published his letter of resignation from the Société d'émulation du Bourbonnais in 1926.[101] On the other, as Cagnat cut off debate about Glozel at the AIBL in the fall of 1926, the national arena closed for him as well. As a result, Morlet's texts often read like an uneasy combination of multiple genres: part news report, part textbook, part scholarly treatise.

The use of both photographs and engravings in Morlet's writings likely had practical explanations, including the house style of the *Mercure* and the greater ease of taking photographs than of commissioning a draftsman in provincial France in the 1920s. The multiple media also had a number of precedents. Déchelette's *Manual*, for example, is illustrated primarily with engravings, but it also contains a number of photographs, notably of a ceiling in the Altamira caves, the earliest site of authenticated cave paintings.[102] Frederick Bohrer has seen in the juxtaposition of photography and engraving "the traces of desire to find, or to construct, more than is in front of the eye," a mode of image-making that treats photography "as one tool among many, not a self-sufficient revelation of an unmediated reality."[103] For the Glozel finds, photographs serve a basic indexical function, recording the existence of various types of objects and, like the North Africa catalogs, indicating importance by the size of the reproductions and occasional group shots: one image in the third fascicule, for example, shows six flints in a more or less symmetrical arrangement.[104] Like many images in Déchelette's *Manual*, which groups like objects together efficiently and presents them clearly, sometimes with a centimeter rule at the bottom to show the scale, the prints in the *Mercure* articles convey detail more effectively than the fascicule photographs. An article appearing in October 1927, near the height of the controversy, for example, includes an engraving of a number of objects with roots growing out of them, a demonstration of their age but also of the malleability characteristic of what Morlet called "the first age of clay."[105]

Given the skepticism and accusations of forgery that surrounded Glozel, the questions any illustration of the finds might prompt take on a special edge. What does the side of the object not depicted look like? What aspects does the photograph or engraving highlight or omit? Some photographs, notably those of inscribed stones or tablets, show signs of manipulation, either in the lighting process or in printing, to enhance

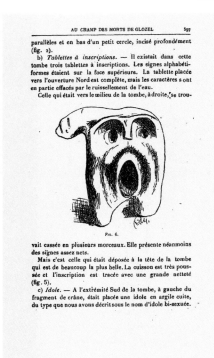 

FIG. 5.9. Antonin Morlet, "Au champ des morts de Glozel," *Mercure de France*, 1 August 1927, p. 597. Source: gallica.bnf.fr/BnF.

FIG. 5.10. Antonin Morlet, "La décoration céramique," *Mercure de France*, 26 October 1926, p. 261: figure 4 by F. Corre. Source: gallica.bnf.fr/BnF.

legibility; do they distort the object's appearance in other ways?[106] Did F. Corre, the artist who produced most of the *Mercure* images, see the actual objects, and if not, what kind and quality of documentation did he or she use? In some instances, the *Mercure* images seem to cast a veil of discretion over objects Morlet categorized as "phallic and Hermaphrodite idols," which in photographs—even small and dark ones—convey an impression of striking eroticism and vulgarity (fig. 5.9).[107] Yet some of Corre's drawings, notably of a set of ceramic vases with facial features (fig. 5.10), communicate the aesthetic appeal of the Glozel artifacts with such force as to make them seem too advanced and original to be plausibly Neolithic. Writing in *La Liberté* in late December 1927, shortly after the publication of the commission report on Glozel, Charles Omesssa describes the maker of the objects as an "incomparable modeler, sculptor, draftsman, and [theatrical] director"—praise that largely replicates Morlet's own, except that the journalist is applying the words not to some unknown Neolithic artist, but to an accused forger, Emile Fradin.[108]

Morlet deploys a number of rhetorical strategies to explicate and, in a sense, neutralize the images that might cast doubt on Glozel. In some instances, he uses the terminology of his medical training to describe objects in precise anatomical terms. An August 1927 article places a recently unearthed idol (fig. 5.9) in the Hermaphrodite category, taking the oval opening surrounded by a raised ridge as a representation of female genitalia.[109] He sees the phallic protuberance on the top as "the penile swelling prolonged at the front by the foreskin, pierced at the end. From this we can conclude that the neolithic people of Glozel did not practice circumcision."[110] His discussion of the figured vases like the one in figure 5.10 includes both specific physical description and a comparison to "'the human mask typical of the prehistoric idol' described by Déchelette."[111] Elsewhere Morlet notes the resemblance of the facial features to those on the so-called face pots unearthed by Heinrich Schliemann at Hissarlik (Troy) in the 1880s; in insisting on their human features he takes the opportunity to correct Schliemann's description of the pots as "owl-faced."[112] The invocation of a characteristic "neolithic mask" enables Morlet to link figured vases to idols, and to propose "a new theory of [their] origin and symbolic meaning."[113] These are, he claims, funerary vessels derived from the observation of skulls, in which the absence of the mouth symbolizes the "great silence" of death.[114]

Like the archaeologists discussed in the previous sections, Morlet connects individual artifacts to larger categories and groupings, and in so doing extrapolates from the visible object to the invisible world he is trying to construct. That world rests on a few key premises. The Glozel site he deems a tomb from the early Neolithic, understood as a period between 7000 and 6000 BCE; in the absence of any stratigraphy, all of the finds come from approximately the same time period.[115] The objects discovered provide clear evidence of a previously unknown culture that, Morlet claims, marked a transition between the Paleolithic and the Neolithic, and this in several respects. The inhabitants of Glozel, though becoming sufficiently sedentary to produce fine pottery (as well as rougher varieties), were still primarily hunters. Engraved and incised stones depicting reindeer suggest that the warming of the climate generally thought, by Déchelette among others, to have taken place at the end of the Paleolithic, sending these cold-weather animals far to the north, was still occurring at the time of the Glozel deposits.[116] These stones also testify to the artistic ability of the inhabitants, thus filling a puzzling gap previously noted by Reinach with respect to the cave art of the late Paleolithic, which he famously described as "a child born without a mother, a mother dead without children."[117] Most remarkably, a number of tablets and stones inscribed with lettering indicate that at least some portion of the population had command of an "already quite advanced" form of writing, with characters that also marked a transitional stage, from the ideographic to the syllabic.[118] The "Glozelian," then, like the Magdalenian before it, described a period of vital importance in the story of human evolution.[119]

Assumptions and methods from the wider field of archaeology undergird this picture and work to establish not only the authenticity of the Glozel artifacts but the legitimacy of Morlet's practice. Morlet regularly cites both standard texts, at one point referring to Jullian as "the illustrious historian of our national antiquities," and very recent scholarship, for example, that of the Norwegian prehistorian Haakon Shetelig (1877–1955), to make comparisons to analogous groups and artifacts.[120] He employs toolmaking as an index of civilizational progress, discussing the fabrication of cutting tools with a keen sense of the level of skill they demonstrate. Rémi Labrusse has observed that for the first generations of prehistorians, "the artistic gesture is not distinguished from the technical one," and Morlet follows this trend by evaluating both tools and design with the same vocabulary of aesthetic judgment seen elsewhere.[121] And whatever the artistic attainment of which he deems them capable, Morlet subjects the Glozelians to kinds of ethnographic comparison rooted in the racial "science" of the time. He regularly refers to the people of Glozel as "tribes" (*tribus*), which makes the comparison to Aboriginal peoples of the time he was writing seem all the more natural. "Back then as in our own time," Morlet declares, "the industries of primitive man varied by tribe."[122] At his cleverest, Morlet portrays himself as a well-read empiricist, immune to the lure of "categories and subdivisions"; he quotes one of the founding fathers of French prehistory, Emile Cartailhac (1845–1921), to the effect that in prehistory all theories must remain provisional.[123] As part of this somewhat disingenuous self-portrait, the doctor embraces the uncertainty that affirms archaeology's scientificity. But he does so selectively, about small matters such as terminology or the perpetual problem of dating, never about the authenticity of the finds.[124]

One of the most significant of Morlet's admissions of uncertainty comes in his extensive discussion of the ostensible Glozelian writing system: he acknowledges the impossibility of deciphering the inscribed tablets, and the unlikelihood of ever finding a Rosetta stone that would key the characters to a known language.[125] Morlet takes pains to connect Glozelian letterlike signs to the existing scholarly consensus about the beginnings of alphabetic writing. In the second fascicule, published in early 1926 and devoted entirely to "the Glozel alphabet," he makes clear his understanding that alphabets represent a fairly late stage in the development of writing, coming after glyphic or pictorial systems and syllabic ones. The principle of the alphabet is that of a one-to-one equivalence of sign (letter) and spoken sound, which permits great economy of signs. At one point Morlet corrects himself, describing the Glozelian alphabet as "currently comprising 81 alphabetiform symbols," which he illustrates in three figures, "or rather 81 ideograms, since we don't think that these primitive tribes had reached an intellectual level enabling them to transform their writing into the exclusive depiction of sounds."[126] Morlet places Glozelian at the syllabic stage, observing that a purely ideographic writing system would have had many more than eighty characters, but a purely alphabetic one would have had fewer: as a telling example he correctly cites the number of letters in the Phoenician alphabet, twenty-two.[127] The fascicule concludes with reproductions of a hand-drawn chart of the letters as well as one comparing "Glozelian" to two other lettering systems, Egyptian Hieratic and Phoenician. The charts, dated 18 February 1926, could be described as at best provisional, at worst amateurish.

Morlet makes a further concession to the consensus history of writing: he reaffirms the central role of the Phoenicians in the spread of the European alphabet. Given the religious significance of earlier, symbolic stages of writing, the revolutionary transformation to a purely phonetic alphabet, he asserts, "could only be done by a new people, having no reason to regard written characters with religious superstition. It was the Phoenicians who undertook [this revolution]."[128] The reduction of the number of letters came about as the Phoenicians chose letters common to the different languages of the peoples with whom they were in contact. Readers wondering how the "Glozelian" letters reached the Phoenicians would find a somewhat murky explanation involving "circum-Mediterranean tribes," an evocation of dolmens and primitive inscriptions in the Canary Islands, and citations of scholars like Gustave Glotz (1862–1935) and Léon Homo (1872–1957), who make vague allusions to Neolithic writing.[129]

This elaborate schema in the end boxed Morlet into a number of serious contradictions. The draw of the Glozelian writing system rested, he writes, on the spread of the "advanced civilization of the tribes established in central France from the beginning of neolithic times." But how can "primitive tribes" at the same time constitute an "advanced civilization"?[130] As though in response to this question, in the second fascicule Morlet imagines a stratified population at Glozel, with a "learned class" who, on certain stones, combined a simple inscription with a depiction of the animal it referred to. Such a fusion of alphabetic and iconic signs would have made the meaning accessible to "the common people" who had not yet learned the syllabic alphabet.[131] In an article a year later, Glozel has become "a religious burial center under the control of a learned and creative caste," spurring a burst of invention in both ceramics and writing.[132] The reader is asked to imagine a Neolithic culture with significant aspects of social organization, notably a scribal class or caste, that archaeological evidence has consistently associated with the foundation of cities, at least three millennia later. It is a dizzyingly improbable mix of worlds, in which reindeer continue to roam central France and a barely sedentary population, still predominantly hunters, supports a small group of literate creatives.

The stereotypical image of the Phoenicians discussed in the previous section looms over this portrayal of Glozel; it authorizes Morlet to imagine the Phoenicians as inveterate traders who "had opened great exchange routes between East and West, and their commercial activity reached into the most remote areas."[133] In the conclusion to the third fascicule, the Glozelians offer a stark contrast to this portrayal of the Phoenicians. Summarizing their accomplishments, including glass, ceramics, and writing, under the heading "Indigenous Culture," Morlet declares, in the bold type he favored, "None of the branches of Glozelian culture needs immigration or foreign contribution to be understood."[134] Could Phoenician merchants, among others, really have reached the "remote region" inhabited by this small band of culturally advanced Gauls without leaving any trace, material or otherwise? Morlet does not address this issue directly, but he has no need to: based on the stereotypical image of the Phoenicians as anti-intellectual and cultureless, he could confidently expect that none of his contemporaries would even raise the question. Under the final rubric, "Occidentalisme," Morlet allows that writing likely arose in more than one center. Reaffirming his allegiance to Salomon Reinach's view that civilization moved from west to east, however, he asserts that only in two places, Egypt and "among the Neolithics of central France," did writing really catch on, and that the two systems developed independently until the Phoenicians brought them together in simplified form.[135]

It would be difficult to overstate the extent to which this portrait of a thriving but autochthonous Neolithic group deviates from contemporaneous understandings of prehistoric Mediterranean societies. Dussaud, in his study of the ancient Aegean, disparages as a form of intellectual laziness the tendency to see the Phoenicians as the default intermediaries between "eastern" (a term encompassing ancient Mesopotamia and Persia) and Greek cultures. But his book seeks to show the importance of other groups and places, notably Cyprus, in a continuous, vibrant process of cultural transfer and exchange, and the Phoenicians, if far from the only intermediary, clearly remain in the mix.[136] Déchelette likewise distances himself from the idea that resemblance between objects from different places indicates some kind of filiation. At various points he expresses great respect for Salomon Reinach, including what he calls the "brilliant argumentation" of Reinach's celebrated article "The Oriental Mirage."[137] Yet in the same sentence Déchelette declares himself unconvinced by the argument that civilization moved from west to east; elsewhere he supports the consensus view of a movement from east to west, while stressing the need to look carefully at particular encounters to understand their complexity.[138]

Loyalty to Reinach's thesis from "The Oriental Mirage," the only allegiance that had given Glozel any kind of intellectual pedigree, certainly underlay Morlet's portrayal of a productive and self-contained Neolithic society. But Glozel as media spectacle was also playing to a wider audience. Consider the last few sentences of the third fascicule, a brief evocation of a later time:

> If the prototypes of copper and bronze tools certainly seem to have come from the southeast, at the same time as the Oriental hordes that would pour into our lands, the current of civilization had, in Neolithic times, a diametrically opposite direction.... If one day the lands of Asia would furnish people, it's in the West that art and writing were born: "Ex occidente ars et litterae."[139]

The mention of copper and bronze clearly alludes to later epochs of civilizational development as then understood; some archaeologists at the time placed an "age of copper" between the Neolithic and Bronze Ages. But the use of the future tense for the verb *déverser*, meaning to pour, to discharge, or to shed, though grammatically permissible, also creates a certain temporal ambiguity, as if the writer might be thinking about multiple time periods, including his own.

Recent scholarship has cast World War I and its aftermath as a watershed moment in immigration to France. Wartime labor shortages led the government to bring workers as well as soldiers from French colonies. These policies of course reflected the kinds of racial assumptions shared by the ethnography and archaeology of the time. Regarded as childlike and unassimilable, Black workers from sub-Saharan Africa were kept largely isolated, as the government assumed that they would be repatriated after the war.[140] The devastating impact of the war on the working-age population led postwar governments to continue actively to recruit foreign workers, from Europe (with preference to countries that had fought on the Allied side or remained neutral) as well as from overseas colonies. The 1920s thus mark the moment when France became one of the leading destinations of immigration—indeed, with restrictions imposed in the United States, over the decade ending in 1930 it took in more immigrants than any other country. In particular, as Ethan Katz has put it, "in the interwar period France became a major space of trans-Mediterranean migration, circulation, and cultural interaction."[141]

More than simply a practical measure, immigration had principled supporters, including Salomon Reinach's brother Joseph. The most politically active of the Reinach brothers, Joseph Reinach (1856–1921) had served as an aide to one of the founding fathers of the Third Republic, Léon Gambetta, and was the first member of Parliament to defend Alfred Dreyfus. In 1916 Reinach, again in Parliament and a prominent member of the Jewish community, published an article in the establishment daily *Le Figaro* calling for greater rights for Muslims, mostly Algerians, working in France. This position attracted favorable attention in Muslim circles and from the preeminent French human rights organization, the Ligue des droits de l'homme.[142] But the surge in immigration during and after the war also prompted criticism, anxiety, and resentment. As early as 1923, a murder committed by an Algerian in France received widespread press coverage and prompted an official inquiry that resulted in restrictions on the circulation of Algerians in France. In 1925, the Paris city council created a special surveillance service to monitor the activities of North African workers in the capital; similar bureaus were created a few years later in Marseille and Lyon.[143] "An ambient anti-black, anti-'oriental' racism pervaded the country," Clifford Rosenberg has observed, while noting that previous immigrant groups, including French peasants moving to cities, had also faced "hostility with racial overtones."[144] At the same time, the far-right *Action française*, a fringe political movement as well as a newspaper, continued to agitate against the "internal enemy," now, with the birth of the Soviet Union, associated with international communism. In 1928 the xenophobia of the far right found a new organ in the newspaper *L'Ami du peuple*, founded by the cosmetics magnate François Coty; by 1930 it had a press run of a million copies.[145]

If the media coverage of Glozel discussed in chapter 2 registers a scrambling of the normal political codes, it also reflected changes in the larger political and intellectual landscape of the 1920s. Morlet rarely addressed politics external to the affair even in his private correspondence with Salomon Reinach, and his own political views remain something of a mystery. Occasional remarks attributing anti-Glozelian views to Catholic obscurantism, however, make clear his awareness of the Reinach family's indelible commitment to Republican *laïcité*, or secularism, as well as to the Jewish community as an avatar of universal human rights, which would have been common knowledge at the time.[146] A more intriguing development came in the spring of 1928 with the publication in the *Mercure* of a letter from Victor Basch, president of the Ligue des droits de l'homme. Writing to the minister of justice, Basch protested irregularities in the police search of the Fradin farm and subsequent arrest of Emile Fradin; Morlet had served as one of the Ligue's sources.[147] Although Basch had impeccable leftist credentials, this gambit presents at least an outline of how Glozel, both the contemporary hamlet and its putative distant ancestors, could fit a conservative strand of French republicanism. Associated with the politician-ethnographer Louis Marin (1871–1960) and before him the fin-de-siècle nationalist Maurice Barrès, this strand celebrated the peasantry and traditional rural values, idealizing the self-sufficient community of small producers.[148] In the world of learning, however, such a portrayal risked dismissal as a form of what Dussaud, in one of his accounts of the affair, called "regional particularism."[149]

The ethnic slurs Morlet permitted himself in his private correspondence with Reinach are unexceptional for the time, and indeed can be regarded as logical extensions of the racialist terminology common in interwar archaeology. He compares one member of the Société préhistorique de France, whose legal proceeding led to the search at the Fradins' farm, to "the Hurons," and he tags Byron Khun de Prorok, whom he blamed for a negative 1928 article in *L'Illustration*, with the nasty racist epithet *métèque* (a bilingual dictionary offers "wog" as an approximate English equivalent).[150] But Morlet's personal attitudes and prejudices matter less for themselves than for their affinities with common tropes and discourses of the time. The portrayal of a Neolithic settlement in central France as one of the sources of Western civilization explicitly invoked the theory Reinach had put forward thirty years earlier in "The Oriental Mirage." But it also found an echo in a book published in 1927, the year the Glozel affair reached its height, Henri Massis's *Défense de l'Occident* (Defense of the West). Massis (1886–1970), a right-wing essayist close to the Action française, saw France, and indeed all of Europe, as threatened by the incursion of unreason and mysticism from Asia, with Russia and Germany serving as conduits. Sandrine Sanos has observed that Massis's "vision of Western civilization concerned at once race, culture, and aesthetics"—just like Morlet's portrayal of Glozel.[151]

Ultimately, however, Glozel had to meet standards not simply of plausibility but of rigor, the standards Dussaud associated with "the methods proper to archaeology."[152] Here a final glance at the illustrations of Morlet's publications offers some insight. The fifth and last of the fascicules,

— 22 —

Nous avons déjà vu deux animaux, vraisemblablement de jeunes cerfs vidés, gravés autour d'un galet. C'est un procédé qu'on retrouve fréquemment chez les Magdaléniens et qui s'est transmis à travers les siècles, aux artistes de l'époque néolithique.

La tête est courte, le mufle étroit et le front large et arrondi. La corne droite est courte et très large, alors que la gauche est étroite à la base et beaucoup plus longue.

Tête de bovidé sur anneau de schiste (fig. 24 bis). — Sur les faces d'un anneau de schiste se voient gravées au milieu de caractères alphabétiformes, trois têtes d'animaux dont une de bovidé. On y retrouve les cornes arquées et la tête longue de la plupart des représentations bovines de Glozel.

(fig. 24bis)

Canidés

Le gisement de Glozel nous a livré plusieurs espèces de canidés. Dans les deux tombes existaient des gravures de loups aux oreilles pointues et au long museau. Elles sont reproduites avec le mobilier funéraire de ces sépultures (1).

(fig. 25)

(1) "Au Champ des Morts de Glozel", *Mercure de France*, 1er et 15 Août 1927.

— 23 —

Une tête beaucoup plus fine, aux oreilles dressées, et portant à la partie supérieure du cou de légères rayures parallèles, destinées à représenter de longs poils, semble être plutôt la représentation d'un chien.

On voit en avant trois signes alphabétiformes (fig. 25).

Enfin, sur un galet plat, percé d'un trou de suspension comme une pendeloque, apparaît une tête d'un dessin tout à fait fruste et naïf. La mâchoire inférieure est trop volumineuse, mais le mouvement de la bouche entr'ouverte est bien caractéristique du chien ou du loup.

Animaux divers

Ours (fig. 26). — Un galet porte gravé, en traits nets et profonds,

(fig. 26)

le dessin d'un animal assez difficile à déterminer. Les pattes sont très courtes par rapport au corps qui est gros et arrondi. Trois pattes sont terminées par un trait horizontal, comme si on avait voulu représenter un plantigrade. Cependant la tête est loin d'être caractéristique.

Sur le cou, on distingue, en traits très légers, l'esquisse d'une autre tête.

Au-dessus de l'animal se voient nettement gravés trois signes alphabétiformes.

FIG. 5.11. Antonin Morlet and Emile Fradin, *Nouvelle station néolithique* (Vichy: Imprimerie Octave Belin, 1928), 5:22–23. Public domain.

entitled "The Animal Art of Glozel" and published after the pro-Glozelian "study committee" dig in April 1928, contains fifty-one images, most less than half a page in size, over thirty-six pages of text. Most of the figures (all of them photographs) reproduce images of animals, divided into engravings on stone, relief sculptures, and a few engravings on bone; stone objects, the first and largest group, are subdivided by type of animal. Unlike in the previous fascicules, where photographs often outran the text, here the text and the images closely correlate. Descriptions in short paragraphs precede the reproductions, so that generically the fascicule bears some resemblance to a catalog. Yet the images come in a wide range of sizes and formats (see fig. 5.11, for example, where the objects in Morlet's figures 24bis and 26 have been cut out, their shapes isolated against a white background, while most of the other photographs retain their rectangular format), and the layout of the publication never achieves any kind of consistency. The conclusion, moreover, largely repeats that of the third fascicule, including the emphasis on the alphabet and the phrase affirming that Glozelian culture needed no immigration or foreign contribution. Although technically justified by the presence of inscriptions on some of the engraved stones, this conclusion conveys an impression of stagnation and sameness, of a group of objects mobilized to signify one and only one thing, on a take-it-or-leave-it basis.

To compare the Glozel publications with the *Description de l'Afrique du Nord* or with Montet's volume on the Byblos excavations is to realize once again how much the archaeological imaginary depends on the standardized performance of those all-important intermediary activities between excavation and publication.[153] The conventions that structure the official catalogs result from various kinds of work, acknowledged and otherwise, including stratigraphic and contextual analysis, classification, scholarly debate, and careful editing. These processes radiate a power at once sociopolitical and intellectual: a power not entirely of archaeologists' own making, in other words, but one to which they indubitably contributed. The visual representation of Glozel offers the abundance of discovery and some attempt at classification, but it never progresses to the later stages of archaeological practice that the Tunisia catalogs make evident. Images of objects, many of them arresting, accompany Morlet's texts, in close association with the ideas intended to turn them into things. But the layouts, herky-jerky in the fascicules and whimsical in the *Mercure*, offer only random signs of scholarly seriousness, thin slivers of science sprinkled atop a large pudding of spectacle. In the end, the haphazard quality, in both literal and metaphorical senses, of the visual representation indexes Glozel's larger inadequacies. To paraphrase Reinach's much cited aphorism, the imaginary of Glozel, stunted by its failed attempt to meet the disciplinary norms of the time, was a child with multiple progenitors but no offspring.

Epilogue

After Bizerte: 1926–1933–1962

The Anglo-Irish archaeologist Donald Harden (1901–1994) entered my consciousness through a series of archival moments I can only partially reconstruct.[1] A search for Carthaginian materials on the British Museum collections website yielded a surprising number of funerary urns, with Harden listed as one of the donors. The identification of Harden as a former president of the London Society of Antiquaries led me to his papers in the Society's august reading room in Burlington House, and these yielded an intriguing picture. As a green but hardworking assistant on the American dig in Carthage in 1924 and 1925, Harden, then in his early twenties, caught the eye of Francis Kelsey, who offered him an opportunity to work at his home institution, the University of Michigan. The two men agreed that Harden should study the Carthaginian urns ceded to Michigan as part of the excavation agreement with the Tunisian Antiquities Service. Based on this work, Harden would produce a substantial publication, which would help launch his postbaccalaureate career. The question of whether Harden would actually pursue a PhD at Michigan was left open; Kelsey did not have a high opinion of that degree.[2]

Of course, things did not go according to plan. Harden arrived in Ann Arbor in the fall of 1926, well before the objects he was supposed to study. It soon became clear that the urns were casualties of the dispute between Kelsey, the abbé Chabot, and the French director of antiquities, Louis Poinssot discussed in chapter 1. Poinssot had absolute control over objects unearthed in (legal) excavations in Tunisia, and since he considered the Americans responsible for breaching their agreement, he refused to authorize the urns' dispatch to Michigan. Kelsey, furious but helpless—and also absent from Ann Arbor at the time of Harden's arrival—hit on a practical alternative: that Harden should study a collection of glass from excavations at Fayum, Egypt, also led by Kelsey. "It all belongs in your general field," Kelsey wrote, "and this would make a fine publication of itself."[3] Harden, nothing if not adaptable (though his accommodations in Ann Arbor apparently left much to be desired), agreed, and this became his project at Michigan. Thoughts of remaining there, however, were cut short by Kelsey's sudden passing in May 1927.

The story does not end there, however. In 1933, Harden, by now an assistant keeper (curator) of antiquities at the Ashmolean Museum in Oxford, initiated contact with Poinssot, ignoring the bitterness his name aroused among Kelsey's associates.[4] Poinssot replied cordially, explaining the "challenging" situation at Carthage, including the fact that since Kelsey's death taxes on the property the Americans had acquired there had gone unpaid. The director reminded Harden that all discoveries remained the property of "the Tunisian state," but also of his, Poinssot's, commitment to allow finders to display and publish them: "I have not forgotten that they [Kelsey and his colleagues] had put you in charge of publishing the pottery, and as far as I'm concerned, I see only benefits to your taking up the full publication of the ceramics." He even offered to help Harden publish a French translation of his eventual articles.[5] Thus encouraged, Harden obtained a small grant to make another trip to Tunisia that fall, and a few years later his lengthy survey of the Carthaginian urns appeared in the scholarly journal *Iraq*.[6]

Harden became an acknowledged expert on Roman glass and, after promotion to keeper at the Ashmolean, served as director of the Museum of London from 1956 to 1970. But he never lost interest in Carthage and the Phoenicians, and in 1962 he published a volume, *The Phoenicians*, in a Thames and Hudson series for the general public called Ancient Peoples and Places. In 218 pages, sprinkled with maps, diagrams, and line drawings, and followed by 115 well-captioned black-and-white plates, the book first offers an overview of Phoenician people, geography, and history, with a separate chapter devoted to Carthage. Harden then explores the Phoenicians' religion, language, urban settlement, industry and commerce, and art—the last and longest chapter. Harden wears his learning lightly, with endnotes rather than footnotes and a vigorous, conversational style. Describing continuing scholarly debate over the priestess sarcophagus (plate 4) discovered by Delattre six decades earlier, he writes, "If she was meant as a representation of the defunct within the coffin, it was wholly idealized, for the bones were those of a toothless old crone with a big broad nose and jutting jaws, of African even negroid blood, perhaps!"[7] The use of familiar racial terminology signals that, despite considerable advances in archaeological knowledge over the previous half century, Harden's view of the Phoenicians does not diverge greatly from those of the early twentieth-century scholars discussed in the last chapter. The book concludes with a mixed assessment of the ability and accomplishment of Phoenician artists, a renewed attribution of the most impressive works, including those anthropoid sarcophagi, to Greek artists, and, finally, this: "The Phoenician, though he possessed an artistic bent, was less interested in art for his own purposes than for the price he could get for it abroad."[8]

The most memorable moment in the Harden archive for me was the discovery of Poinssot's letter from 1933, typed but with his familiar crabbed signature at the bottom. Poinssot made (or kept) few copies of letters he wrote, so his positions, views, and general perspective usually have to be inferred from the letters of his correspondents. But the interest of that 1933 letter goes beyond its rarity: in its obvious respect for the recipient, its generosity of spirit, and its frank simplicity of tone, the letter manifests the characteristics recalled by one of Poinssot's admirers, praising, notwithstanding a sometimes difficult character, "an absolutely astonishing breadth of vision and boldness of conception."[9] In 1962, the year Harden's book on the Phoenicians was published, those qualities gained recognition as Poinssot, by then in his eighties, was finally elected to the Académie des inscriptions et belles-lettres. The honor emerged from a rare quiet campaign, of the sort reserved for war heroes and people considered above the fray, led by the archaeologist Jérôme Carcopino and carried out by Poinssot's youngest son, Claude, without his father's knowledge. To Poinssot's old friend Alfred Merlin, a member of the Academy since 1928 and since 1948 its perpetual secretary—a position once held by his father-in-law René Cagnat—Claude Poinssot wrote, "He has at times had the painful, if no doubt mistaken, impression that his effort and his work have been undervalued. This [election] would be a kind of reparation."[10]

Had he read this letter—the whole point of the procedure being that he would not—Poinssot might well have greeted it, and especially the word "reparation," with a sardonic smile or a raised eyebrow. The fact that Carcopino was directing this unusual election would have struck him as particularly ironic, given that Poinssot blamed Carcopino, as the Vichy minister of education, for forcing him to retire from the directorship in 1942.[11] Would he also have grasped the double edge of the word "reparation," now one of the keywords of postcolonial efforts to secure restorative justice? Poinssot and his wife had left Tunisia for France only after the Bizerte crisis of 1961, a reassertion, five years after Tunisian independence, of French authority over a military enclave on the coast. French forces repulsed the Tunisian incursion in 1961 but surrendered the base two years later, after the end of the Algerian War.[12] Transporting to France those hundreds of boxes of correspondence, photographs, books, and everything else that make up his archive, did Poinssot reflect on the connections between archaeology and colonialism, between the knowledge he had sought, produced, and promoted throughout his career and the power structures that had sustained it? Another of my memorable moments in the Poinssot archive came during a conversation with a fellow researcher, a Tunisian on a brief trip to Paris—all the time he and his employer could afford—to consult archives pertaining to his own patrimony. Without apparent resentment, he observed that these were the true archives of the Bardo Museum, or at least an indispensable complement to them. I could not but be conscious that my good fortune in being able to work on Tunisia in Paris bears its own colonial legacy.

"Glozel For Ever": 1968–1974–2021

Dorothy Garrod's recollections of Glozel, which I have cited at several points in this book, appeared in *Antiquity* in September 1968, a few months before her death. A headnote to the article makes clear that it was a kind of confection, an edited transcript of an interview the journal's editor, the Cambridge archaeologist Glyn Daniel (1914–1986), had conducted with Garrod for a BBC television program about the Glozel affair.[13] A man of wide interests, author of works of detective fiction as well as histories of archaeology, and a frequent television presenter, Daniel had a great curiosity about Glozel. He was well aware that O. G. S. Crawford, the founding editor of *Antiquity*, had played a role in the affair, publishing an article denouncing Glozel as a forgery in June 1927, in what was only the journal's third issue.[14] In 1974 *Antiquity* published an article by an international group of scientists who had subjected Glozel artifacts to thermoluminescence and concluded, tentatively, that they dated to the Gallo-Roman period.[15] "Now we seem to have three parties, not only the Glozelians and anti-Glozelians," Daniel writes in his editorial introducing the article, "but the La Tène/Gallo-Roman modified Glozelians of this paper. It is all fascinating." He goes on to say that for the moment he remains "neutral, or, as the Production Editor reminds us, as neutral as ever a Welshman can be who has already written and lectured about Glozel as a classic case of forgery. Nonplussed is the word, not neutral."[16] Worrying about the possible conflict between "two sets of facts," those of archaeology and of physical science, Daniel does not comment on the odd concordance of the new theory with Camille Jullian's view of Glozel as a Gallo-Roman sorcerer's den.

In December 1927, a French-language newspaper published an article with the English headline "Glozel For Ever," signifying both the international dimensions of the controversy and the sense that it would never go away.[17] Anyone who studies Glozel in the age of Google knows of the regular recurrence of interest in the site, as though the "glozelitis" referred to by one wag at the height of the controversy were an endemic virus subject to periodic surges. Through an odd concatenation of archival moments, the minor—as these things go—kerfuffle unleashed by the 1974 *Antiquity* article can be tracked in an unexpected place: the National Archaeology Museum (MAN) in St. Germain-en-Laye, outside of Paris, the museum once directed by Salomon Reinach, and later by Poinssot's former deputy Raymond Lantier. Since 2018 the MAN archives have held the papers of Dorothy Garrod, gifted to the museum by her literary executor and close friend, the archaeologist Suzanne Cassou de St. Mathurin (1900–1991).[18]

Years after Garrod's death, Daniel, clearly a personal friend of hers and St. Mathurin's, sent the latter a stream of notes, statements, and copies of correspondence about Glozel, to the point where Glozel seems almost literally to haunt Garrod's companion. Declining an invitation to join Daniel and a BBC team planning to film at Glozel in the summer of 1974, St. Mathurin explains that she does not want to infringe on the territory of the prehistorian Henri Delporte (1920–2002), then a curator at the MAN. But she also warns Daniel, "Be careful yourself. Glozel is and has always been a wasps' nest. The pitfalls are far more numerous than they were fifty years ago."[19] Around the same time, she politely but briskly referred Hugh McKerell, one of the lead authors of the thermolumiscence article, who wrote her on Daniel's advice seeking information on the material confiscated during the 1928 raid on the Fradins' farm, to the Paris Police Prefecture, perhaps chuckling to herself at the wild goose chase that might ensue. Yet the following year St. Mathurin wrote the son of the anti-Glozelian lawyer Maurice Garçon, seeking information about the technical appraisals conducted in the 1920s, as well as a copy of his father's argument in the case, "missing from the dossier." Although, she tells the younger Garçon, her views on Glozel as a hoax had not changed, she had visited the site and the "Musée Fradin" to refute any charges of bias.[20]

The documents in St. Mathurin's copious Glozel dossier bear obvious traces of 1970s information gathering: poor-quality photostats, the telltale marks where archivists removed rusting staples, the works. Somehow St. Mathurin obtained a copy of a 1985 letter from the president of a pro-Glozelian association to the publisher Larousse, protesting the treatment of Glozel as a fraud in one of its dictionaries and asking that this be modified in light of recent scientific testing.[21] By this time Daniel had long since tired of Glozel. With the physical evidence coming into question, the correspondence he sent to St. Mathurin includes a sequence of increasingly vehement letters to and from McKerrell and his Danish collaborator Vagn

Mejdahl. Near the end of this exchange, McKerrell and Mejdahl sent, over six single-spaced pages, a detailed rebuttal to Daniel's July 1976 account of the affair in the general-interest magazine *The Listener*.[22] The telling asymmetry here—a private letter responding to an article on the public record—demonstrates once again the imbrication of archaeology and the media and the advantages of the access and credentials that Daniel, both a Cambridge don and a media personality, enjoyed. A few months later, Daniel, hoping to mend fences with McKerrell during a forthcoming trip to Edinburgh, wrote a conciliatory but frank letter in which he told McKerrell, "[You are] a scientist treating your material unscientifically. You are dating a body of material, but you have no proof that any of it came from an undisturbed ancient level." By this time Daniel had come to the widely shared view that the material being subjected to new testing was "stuff pushed into Glozel partly from other sites and partly manufactured in the 20s."[23] In February 1977, with McKerrell and Mejdahl no longer replying to communication from Daniel or other scholars, Daniel wrote St. Mathurin that "the whole second affair of Glozel is over." Two years later, he learned from Mejdahl that McKerrell had given up his scientific career for one in business.[24]

The Garrod/St. Mathurin papers do not offer a clear explanation for St. Mathurin's continued interest in Glozel, but they contain some clues. The original *Antiquity* article on thermoluminescent testing had featured two French nuclear physicists as coauthors, and over a year before the article appeared in *Antiquity*, news of the preliminary results had appeared in the popular weekly *Paris-Match*.[25] The news attracted the attention of government ministers, and in 1983, with the encouragement of the Socialist minister of culture, Jack Lang, a new series of verification digs and testing began at Glozel. Given the recurrent interest in Glozel, the choice of the obscure *Revue archéologique du Centre de la France* as the venue for the summary report of the findings in 1995 counts as a symbolic gesture in itself, as resolute a signal of the professional norms and standards governing the research as the report's own technical language and charts. The report concludes that Glozel had never contained a Neolithic layer, that evidence suggested some manipulation of the site, and that "the hypothesis of a prehistoric civilization has not been verified and should probably be definitively discarded."[26] Although St. Mathurin did not live to read these words, they would surely not have surprised her.

But another reason for St. Mathurin's interest can also be discerned: the recurrent media focus on Glozel. As one of his French informants, the Rennes archaeologist Pierre-Roland Giot, wrote Daniel in 1974, "Not only the local defenders of Glozel, but the general press (including the journalists of the O.R.T.F.) very often take occasion to promote Glozel. This has always caused a difficult situation; not only the books of all those farceurs and curios . . . influence many people who get persuaded that the official prehistory and the official archaeology are rubbish."[27] Any archaeologist with even a remote connection to Glozel, then, might have felt a responsibility to keep up with the news and be able to present an informed opinion.

The reference to the O.R.T.F., the soon to be dissolved government radio and television monopoly, calls attention to a recent addition to the media landscape, television, and its enthusiastic rediscovery of Glozel. Whether in whole documentaries or segments of programs with titles like "Mysteries" or "Strange, You Said?" (Vous avez dit Bizarre?), Glozel offered a reliable, even irresistible attraction.[28] Jean-Paul Demoule, one of the coauthors of the 1995 report and later a leading government archaeologist, has described the way broadcast journalists truncated and distorted his and professional archaeologists' comments on Glozel, to enhance the drama and create an impression of enduring mystery.[29] This is no doubt true, but it perhaps underestimates the appeal of one of the most enduring figures in the controversy, Emile Fradin, who lived to the age of 104, dying only in 2010. Standing in front of the farmhouse-museum where he had lived all his life, or providing voice-over to a dramatic series of period images including headlines announcing his arrest, Fradin embodied the values long associated with the French peasantry: steadfastness, simplicity, shrewdness. But if the camera, microphone, and interviewers tended to portray him as a kind of sage and throwback to a simpler time, they also allowed a seed of doubt to persist, based in equally hoary images of clever farmers, especially those with some property of their own, prepared to cut corners to better their lot in life.[30] Doing their best to conceal their structural bias toward Fradin, these shows endeavored to leave the illusory impression that viewers could decide the truth for themselves.

In a 2021 article on Glozel in the leftist daily *Libération*, part of a series on "Archaeological Disappointments" (Déconvenues archéologiques), Demoule attributes the 1980s resurgence of interest to politics: "The new right," he asserts, "took hold of Glozel to say that something was being hidden from us, that writing had been invented in Europe, in the heart of the Allier, and not in Mesopotamia." On this view, Glozel reenergized a long-standing conspiracy theory, "the peasant little guy against official science."[31] The radio and television image of Glozel in the 1980s and 1990s bears a passing resemblance to that promoted by Morlet and the pro-Glozel media in the late 1920s. Although the 1970s dating controversy prompted a great deal of media coverage, Demoule points to the 1980s, a period that saw both the first long-term Socialist administration in France, the fourteen-year presidency of François Mitterrand (1981–1995), and the emergence of the far right Front National under Jean-Marie Le Pen. Espousing an often racist ideology that targeted immigrants, especially those from France's former colonies, the Front National enjoyed its first local and national electoral successes in the mid-1980s. This is not to suggest that those taking an interest in Glozel or covering it in the media all espoused far-right views. Rather, the politics of the time increased receptivity to the twin picture of a thriving, autochthonous Neolithic culture and, in the twentieth century, an obtuse and dishonest scientific community refusing to acknowledge its existence.

In September 1975, when the thermoluminescence controversy was still in its early days, *Antiquity* published a short communication by a young archaeologist called Colin Renfrew entitled "Glozel and the Two Cultures," a reference to a 1959 lecture by the scientist turned novelist C. P. Snow on the divide between science and the humanities.[32] Renfrew, who would eventually succeed Daniel in the Cambridge chair once held by Garrod, recounts his puzzlement over Glozel in its latest phase. Having attended a conference at Oxford that spring, Renfrew begins from a neutral position and conveys his openness to the findings of McKerrell and Mejdahl, whose presentations at the conference he praises. Confessing (he uses that very word) his own inability "to take Glozel entirely seriously," he acknowledges that this attitude might signal to some the reluctance of archaeologists to come to terms with the data of hard science. "But," Renfrew goes on, "archaeological reasoning is not necessarily entirely subjective," and he enumerates "three coherent principles with which genuine archaeological sites almost invariably conform." These principles can be summarized as parallelism with other sites, contiguity with documented cultures from the region, and consistency with authenticated objects from the same epoch. While, of course, new and unique discoveries do occur in archaeology, the passage of a half century from the discoveries has not helped Glozel's case, since in that time "no persuasive and comparable finds from good contexts have been made elsewhere."[33] For Renfrew the only solution to the mystery would be a serious, methodical new excavation—an argument that would soon enough bear fruit in France.[34]

Renfrew's defense of a distinctive and rigorous archaeological method echoes, in ways no less significant for being predictable, the international commission's original report of 1927, and chimes with Laurent Olivier's assertion that excavation rather than collecting forms the core of professional archaeological practice today. In his remarks on Glozel, Renfrew observes that when Snow's book on the two cultures appeared, many archaeologists saw their own field as a bridge between science and the humanities. Since that time, he writes, what he calls "the discipline of archaeological science"—essentially the application of methods from the hard sciences to archaeological data—had grown considerably, but archaeologists had not yet begun to take sufficient account of it.[35] Nearly forty years later, the distinction between "archaeological science" and archaeology has become much blurrier. Technological advances have transformed every stage of archaeological practice, from discovery via LiDAR (light detection and ranging) to DNA analysis of specimens, to digital visualization and reconstruction. At the same time, archaeologists remain committed to theorizing the interpretive methods they see as equally central to what they do. And the media, from blogs to interactive touch screens to video games, stand ready to relay their news.

Objects, Knowledge, and the Archive: 2022–

Three urns stand before me in a row, in equal measure inviting and inscrutable (plate 5). A quick inspection makes clear that at some point in the past, perhaps when digitizing collection records, the British Museum has simply transposed, without converting, inches and centimeters: the online record gives the height of an object as nine centimeters, when it is obviously nine inches.[36] Further contact, with gloved hands, registers some intriguing contrasts among the urns: differences in texture, from rougher to smoother, subtle gradations of color—none of which really comes through in the online photographs—or differences in the fit of the lids. The question of which way the lid fits puzzled me and the staff member who assisted me a few days earlier at the Cambridge Museum of Archaeology and Anthropology (MAA), where I was examining another Phoenician cinerary urn. In London, I notice another difference, of a museological kind. All of the vessels I am studying come with human remains, now in plastic bags. In Cambridge, the staff member had taken the bags from the urn and placed them at the far end of the table, asking whether I wished to remove the remains from their bags (I did not). At the British Museum, the bags remain inside the urns; no one instructs me on how to take them out, or for that matter enjoins me not to. Except for the largest, the gentle swelling and narrow necks of the urns makes reaching the bags difficult, and again I leave them alone.

My visits to museum storage rooms came about by accident. Both the MAA and the BM have searchable online databases that notionally include paper documentation as well as material objects. In both cases, searches yielded almost nothing archival in the classic sense, though the helpful biographical note about Harden—listed as a kind of secondary donor of the BM objects, on behalf of Kelsey and "the French archaeological authorities"—led me to his papers at the Society of Antiquaries. The Cambridge urn (one of two; the other is missing) intrigued me because the object record lists "Byron Khun de Prorok" as its donor, in 1924, one of his early seasons in Tunisia.[37] The inventory of the MAA's archive does not cover or break down museum correspondence from the 1920s, so the day after my inspection of the urn I went through several years of it, to see if a letter to or from the colorful, elusive "Prorok" proposed or explained this gift, of an object to which he never had proper title. The search came up empty.

Only later did I come to understand these visits as a kind of proof-of-concept test of my understanding of the archaeological archive. Mirjam Brusius's and Kavita Singh's collection of "tales from the crypt" has made a convincing case for the significance of museum storage to histories of collecting, knowledge production, and much else in the "exhibitionary complex."[38] But this book has been only peripherally concerned with museums, and those were not the kinds of insights I was seeking. If the biography of an object includes stages of fabrication, use, exchange, collecting, transfer, and display/storage, I was looking for information about the collecting and transfer stages, for which the objects alone, even in the light of the extensive scholarly literature about them and the cult of child sacrifice they served, would not suffice.[39] Perhaps, to a trained eye, some of the physical features I did notice, for example, a hairline crack on the handle of one of the BM urns, could suggest an accident in handling at some earlier time. My eye, however, did not receive that kind of training. Linked to only the most rudimentary contextual information, these objects struck me, at first, as much for the distinctness of archaeological knowledge from (my) historical practice as for their connection.

And yet. Do these urns, and the countless antiquities preserved in museums and storage rooms around the world, really differ so much from paper archives? Historians also confront, repeatedly, documents of surpassing strangeness, as hard to read as to parse, full of gaps and silences, garrulous about the obvious and mute about what we seek to understand. Carolyn Steedman has described the archive, following Bachelard, as "the counting house of dreams," a place "to do with longing and appropriation . . . with wanting things that are put together, collected, collated, named in lists and indices; a place where a whole world, a social order, may be imagined."[40] Historians share this dream with archaeologists, whose ambitions have grown with the expansion of their technological arsenal, but for whom the reconstruction of societies is usually a substitute for, not a complement to, the kinds of insights into individual lives that historians seek to offer.[41] While both historians and archaeologists acknowledge the ultimate unreachability of that imagined past, we have—perhaps for that very reason—carved out quite different protocols for approaching it. But each field has much to learn from the other: from archaeology's commitment to its unearthed objects, historians can understand the archival documents we read and interpret as the product of multiple cultural practices. From historians, archaeologists can perhaps gain some insight into the weight of history on their practice, notably with regard to the persistent tension in their field between science and spectacle.

Historians and archaeologists alike may take inspiration from contemporary artists like Mark Dion, Michael Rakowitz, and Jean-Luc Moulène, who have incorporated the rewards, challenges, limitations, and ironies of archaeology into their art. Dion's work includes mixed-media installations like *Concerning the Dig* (2013; plate 6), in which print media—books, magazines, postcards—form as important and striking an element as physical tools and ostensible finds; he has also produced collaborative works with members of the public such as *The Dig*, an actual event along the banks of the Thames in London.[42] Rakowitz's work uses commercial packaging material from the Middle East to reconstruct ancient monuments that have been moved to the West, from the lamasu on his celebrated Fourth Plinth project in Trafalgar Square to the Ishtar Gate in Chicago's Museum of Contemporary Art. The Chicago work, *May the Arrogant Not Prevail* (2010; plate 7), replicates an Iraqi replica of the original gate, which is now in the Pergamon Museum in Berlin. This literal, pointedly ironic repackaging of archaeological objects, part of Rakowitz's multisite project *The Invisible Enemy Should Not Exist*, prompts reflection about patterns and gaps in the acquisition, display, and study of such objects.[43]

Finally, consider Jean-Luc Moulène's 2005 installation in the Louvre's own archaeological "crypt," the remnants of the medieval castle unearthed, then preserved, during the creation of the so-called Grand Louvre in the 1980s. Entitled *Le Monde, Le Louvre* (plate 8), Moulène's piece juxtaposes reproductions of statues of divinities from the museum's collection with stacks of free supplements to the daily newspaper *Le Monde*, containing the same reproductions, as well as a video of his own journey into the depths of the Louvre. The "original" photographic images come from the Louvre's searchable database; Moulène did not photograph the actual objects. Thus the whole work, as Sophie Berrebi has persuasively argued, seeks to unsettle assumptions about the truth value of the document.[44] Here archaeological knowledge and its objects have become one with spectacle, inseparable from the media that present them to the world at large. Yet to make the relationship traced in this book into art does not resolve the tension at its heart, which is certain to change as media change, and to endure as long as humans' fascination with our past.

Notes

Abbreviations

ADA	Archives départementales de l'Allier, Moulins
AIBL	Académie des inscriptions et belles-lettres, Paris
AN	Archives nationales de France
BI	Bibliothèque de l'Institut de France, Paris
BMA	Bibliothèque Méjanes, Aix-en-Provence, Fonds Salomon Reinach
CADN	Centre des Archives diplomatiques de Nantes
DAA	Direction des antiquités et des arts (Tunisian Antiquities Service)
DAN Alaoui	*Catalogue du Musée Alaoui* (Tunis; now the Bardo Museum), 1897; supplement 1, 1910; supplement 2, 1922 (Paris: Ernest Leroux)—part of *Description de l'Afrique du Nord*
DAN Lavigerie	*Musée Lavigerie de Saint-Louis de Carthage, collection des Pères-Blancs formée par le R.P. Delattre*, part 1, 1900; part 2, 1899; supplement (two parts bound as one), 1913–1915 (Paris: Ernest Leroux)—part of *Description de l'Afrique du Nord*
FKP	University of Michigan, Ann Arbor, Bentley Historical Library, Francis Kelsey Papers
FP	Institut national d'histoire de l'art, Paris, Archives 106, Fonds Poinssot
MAN	Musée d'archéologie nationale, St. Germain-en-Laye (formerly Musée d'antiquités nationales)
MANA	Musée d'archéologie nationale, Archives
NSN	Antonin Morlet and Emile Fradin, *Nouvelle station néolithique*, 5 fascicules: 2, *L'alphabet de Glozel*; 3, *Le glozelien*; 5, *L'art animalier de Glozel* (Vichy: Octave Belin, 1925–1928; fascicule 3, Vichy: Imprimerie Wallon, 1926)
SA	Society of Antiquaries, London, Archives
SRA	Région Auvergne–Rhône Alpes, Direction régionale des affaires culturelles, Service régional de l'Archéologie, Site de Clermont–Ferrand, Archives

Introduction

1. R. de la Porte, "Encore Carthage!" *La Tunisie française*, 7 June 1924. The articles he refers to are "Aux ruines de Carthage," *Lectures pour tous*, April 1924, 826–32, and J. Jaubert

de Benac, "Delenda Carthago: Les ruines de Carthage au pillage," *L'Illustration*, 11 August 1923, 118–21; the latter article will be discussed in chapter 1.

2. Grosclaude, "Les déclassés de Glozel," *Journal des débats*, 19 February 1928. In French the verbs rhyme: "L'archéologie est une science où l'on cafouille autant que l'on y fouille."

3. The literature on the Tutankhamun excavation and its mediatization is considerable. See especially Christina Riggs, *Treasured: How Tutankhamun Shaped a Century* (London: Atlantic Books, 2021); Elliott Colla, *Conflicted Antiquities: Egyptology, Egyptomania, Egyptian Modernity* (Durham, NC: Duke University Press, 2007), chap. 4; Donald Malcolm Reid, *Contesting Antiquity in Egypt: Archaeologies, Museums and the Struggle for Identities from World War I to Nasser* (Cairo: American University in Cairo Press, 2015), chap. 2.

4. The Académie also benefited from private gifts and bequests that gave it a certain financial independence. See Jean Leclant, "Histoire de l'Académie des inscriptions et belles-lettres," in *Histoire des cinq Académies: Textes . . . rassemblés à l'occasion du bicentenaire de l'Institut de France, octobre 1995*, ed. Institut de France (Paris: Perrin, 1995), 128–29.

5. See Marc-Antoine Kaeser, "On the International Roots of Prehistory," *Antiquity* 76, no. 291 (2002): 170–77, https://doi.org/10.1017/S0003598X0008995X; Kaeser, "Une science universelle, ou 'éminemment nationale'? Les congrès internationaux de préhistoire (1865–1912)," *Revue germanique internationale (Evry)*, no. 12 (2010): 17–31, https://doi.org/10.4000/rgi.248; Nathalie Richard, *Inventer la préhistoire: Les débuts de l'archéologie préhistorique en France* (Paris: Vuibert, 2008), 97–102.

6. Bjørnar Olsen, Michael Shanks, Timothy Webmoor, and Christopher Witmore, *Archaeology: The Discipline of Things* (Berkeley: University of California Press, 2012), 1.

7. Magness continues, "Not publishing would be like conducting an experiment and then not publishing your results," thus making clear the relation between media and scientific rigor. "UNC Archaeologist Finishes Dig at Ancient Jewish Synagogue Adorned with Biblical Mosaics," interview with Jodi Magness by Liz Schlemmer, WUNC Arts and Culture, 11 August 2023, https://www.wunc.org/arts-culture/2023-08-11/unc-archaeologist-dig-ancient-jewish-synagogue-biblical.

8. Guy Debord, *The Society of the Spectacle*, trans. Donald Nicholson-Smith (1967; New York: Zone Books, 1994), 24; Debord, *Comments on the Society of the Spectacle*, trans. Malcolm Imrie (1988; London: Verso, 1998), 3.

9. Nathan Schlanger, "Ancestral Archives: Explorations in the History of Archaeology," introduction to a special section of *Antiquity* 76 (2002): 127–31, 128.

10. See Wiktor Stoczkowski, "L'histoire de l'archéologie peut-elle être utile aux archéologues?" in "L'archéologie comme discipline?," ed. Philippe Boissinot, special issue of *Le Genre humain*, no. 50 (2011), 221–34.

11. Marc-Antoine Kaeser, "Biography as Microhistory: The Relevance of Private Archives for Writing the History of Archaeology," in *Archives, Ancestors, Practices: Archaeology in the Light of Its History*, ed. Nathan Schlanger and Jarl Nordbladh (New York: Berghahn Books, 2008), 9–20, 10–11.

12. See Margarita Díaz-Andreu and Marie Louise Stig Sørensen, eds., *Excavating Women: A History of Women in European Archaeology* (London: Routledge, 1998); Díaz-Andreu et al., *The Archaeology of Identity: Approaches to Gender, Age, Status, Ethnicity and Religion* (London: Routledge, 2005); Debbie Challis, *The Archaeology of Race: The Eugenic Ideas of Francis Galton and Flinders Petrie* (London: Bloomsbury Academic, 2013).

13. Philip L. Kohl, Mara Kozelsky, and Nachman Ben-Yehuda, "Introduction," in *Selective Remembrances: Archaeology in the Construction, Commemoration, and Consecration of National Pasts* (Chicago: University of Chicago Press, 2008), 2, https://doi.org/10.7208/chicago

/9780226450643.001.0001; see also Philip Kohl and Clare Fawcett, eds., *Nationalism, Politics, and the Practice of Archaeology* (Cambridge: Cambridge University Press, 1995); Margarita Díaz-Andreu and Timothy Champion, eds., *Nationalism and Archaeology in Europe* (Boulder, CO: Westview Press, 1996); Lynn Meskell, ed., *Archaeology under Fire: Nationalism, Politics and Heritage in the Eastern Mediterranean and Middle East* (London: Routledge, 1998).

14. Margarita Díaz-Andreu, *A World History of Nineteenth-Century Archaeology: Nationalism, Colonialism, and the Past* (Oxford: Oxford University Press, 2007), 210. Among many related titles, see Nadia Abu El-Haj, *Facts on the Ground: Archaeological Practice and Territorial Self-Fashioning in Israeli Society* (Chicago: University of Chicago Press, 2001); Zainab Bahrani, Edhem Eldem, and Zeynep Çelik, eds., *Scramble for the Past: A Story of Archaeology in the Ottoman Empire, 1753–1914* (Istanbul: SALT, 2011).

15. Recent work includes Aaron Tugendhaft, *The Idols of ISIS: From Assyria to the Internet* (Chicago: University of Chicago Press, 2020); Raphael Greenberg and Yannis Hamilakis, *Archaeology, Nation and Race: Confronting the Past, Decolonizing the Future in Greece and Israel* (Cambridge: Cambridge University Press, 2022); Justin Jacobs, *The Compensations of Plunder: How China Lost Its Treasures* (Chicago: University of Chicago Press, 2020). On the Parthenon sculptures, see William St. Clair, *Lord Elgin and the Marbles* (Oxford: Oxford University Press, 1998).

16. See, for example, Sam Holley-Kline, "Archaeology, Land Tenure, and Indigenous Dispossession in Mexico," *Journal of Social Archaeology* 22 (2022): 255–76, https://doi.org/10.1177/14696053221112608; William Carruthers, *Flooded Pasts: UNESCO, Nubia, and the Recolonization of Archaeology* (Ithaca, NY: Cornell University Press, 2022).

17. Colla, *Conflicted Antiquities*, 180. See also Amara Thornton, *Archaeologists in Print* (London: UCL Press, 2018); Joan Gero and Dolores Root, "Public Presentations and Private Concerns: Archaeology in the Pages of *National Geographic*," in *The Politics of the Past*, ed. Peter Gathercole and David Lowenthal (London: Routledge, 1994), 19–37, https://doi.org/10.4324/9780203167892.

18. On the press and other media, see Richard, *Inventer la préhistoire*, 148–62; Bonnie Effros, *Uncovering the Germanic Past: Merovingian Archaeology in France, 1830–1914* (Oxford: Oxford University Press, 2012), 288–349, covering worlds' fairs, the press, and the arts. On the 1867 exposition, Donald Reid, *Whose Pharaohs? Archaeology, Museums, and Egyptian National Identity from Napoleon to World War I* (Berkeley: University of California Press, 2002), 128–30; Ève Gran-Aymerich, *Les chercheurs de passé, 1789–1945: Aux sources de l'archéologie* (Paris: CNRS, 2007), 153, 199. The latter volume incorporates two books by Gran-Aymerich previously published by CNRS: a historical study, *Naissance de l'archéologie moderne, 1798–1945* (1998), and *Dictionnaire biographique d'archéologie: 1798–1945* (2001).

19. See, for example, Brian Molyneaux, ed., *The Cultural Life of Images: Visual Representation in Archaeology* (London: Routledge, 1997); Stephanie Moser and Sam Smiles, eds., *Envisioning the Past: Archaeology and the Image* (Malden, MA: Blackwell, 2005). On imaging as a problem in nineteenth-century science, see Jennifer Tucker, "The Historian, the Picture, and the Archive," *Isis* 97 (March 2006): 111–20.

20. Exceptions include Frederick N. Bohrer, *Photography and Archaeology* (London: Reaktion, 2011); Christina Riggs, *Photographing Tutankhamun: Archaeology, Ancient Egypt, and the Archive* (London: Bloomsbury Visual Arts, 2018). I cite both works extensively, but their focus differs from mine.

21. Steven Shapin and Simon Schaffer, *Leviathan and the Air-Pump: Hobbes, Boyle, and the Experimental Life*, 2nd ed. (1985; Princeton, NJ: Princeton University Press, 2011), xliii.

22. Sheila Jasanoff, "Genealogies of STS," *Social Studies of Science* 42, no. 3 (2012): 435–41,

439, https://doi.org/10.1177/0306312712440174. My thanks to Charlotte Bigg for this reference.

23. Shapin and Schaffer, *Leviathan*, 332; they reaffirm the centrality of this insight in the introduction to the 2011 edition, xl–xli.

24. See Nicole Chevalier, *La recherche archéologique française au Moyen-Orient: 1842–1947* (Paris: Éd. Recherche sur les Civilisations, 2002); Nabila Oulebsir, *Les usages du patrimoine: Monuments, musées et politique coloniale en Algérie, 1830–1930* (Paris: Maison des sciences de l'homme, 2004); Bonnie Effros, *Incidental Archaeologists: French Officers and the Rediscovery of Roman North Africa* (Ithaca, NY: Cornell University Press, 2018).

25. See Alain Schnapp, *The Discovery of the Past*, trans. Ian Kinnes and Gillian Varndell (New York: Harry N. Abrams, 1997), 222–56, 310–24; Richard, *Inventer la préhistoire*, 1–90; Marc Groenen, *Pour une histoire de la préhistoire: Le paléolithique* (Grenoble: J. Millon, 1994), 37–94.

26. On France's lag, see Alain Schnapp, "French Archaeology: Between National Identity and Cultural Identity," in Díaz-Andreu and Champion, *Nationalism and Archaeology in Europe*, 48–67.

27. See Frederick N. Bohrer, "Inventing Assyria: Exoticism and Reception in Nineteenth-Century England and France," *Art Bulletin* 80 (1998): 336–56, https://doi.org/10.2307/3051236; Gran-Aymerich, *Les chercheurs de passé*, 186–90.

28. See Gran-Aymerich, *Les chercheurs de passé*, 149–53; Effros, *Uncovering the Germanic Past*, 264–88; Arnaud Bertinet, *Les musées de Napoléon III: Une institution pour les arts (1849–1872)* (Paris: Mare & Martin, 2015), 311–52.

29. Gran-Aymerich, *Les chercheurs de passé*, 203–21.

30. Gran-Aymerich, *Les chercheurs de passé*, 285–354; Richard, *Inventer la préhistoire*, 109–34; Effros, *Uncovering the Germanic Past*, chaps. 3 and 6.

31. Effros, *Uncovering the Germanic Past*, 59–87, 210–15. Effros gives as examples Alexandre Bertrand and Salomon Reinach, successive directors of the Musée d'antiquités nationales, who were both trained classicists. On local learned societies in general, see Stéphane Gerson, *The Pride of Place: Local Memories and Political Culture in Nineteenth-Century France* (Ithaca, NY: Cornell University Press, 2003).

32. The term "heroic age" is that of William H. Stiebing Jr., *Uncovering the Past: A History of Archaeology* (New York: Oxford University Press, 1993), esp. 23–25. In his chronology, the heroic age consists of two "phases," a long early stage of travel and discovery, and a shorter period in which "archaeology comes of age"—roughly 1860–1925, with some variation depending on the subfield.

33. Gran-Aymerich, *Les chercheurs de passé*, 388–89; Schnapp, "French Archaeology," 60. On heritage protection legislation and the failure to include archaeology, see Arnaud Hurel, *La France préhistorienne de 1789 à 1941* (Paris: CNRS, 2007), 180–203; Astrid Swenson, *The Rise of Heritage: Preserving the Past in France, Germany and England, 1789–1914* (Cambridge: Cambridge University Press, 2013), 289–92.

34. Gran-Aymerich, *Les chercheurs de passé*, 329.

35. Schnapp, "French Archaeology," 48–49.

36. Salomon Reinach, "La méthode en archéologie," *La revue du mois* 11 (1911): 279–92; Effros, *Uncovering the Germanic Past*, 216–17.

37. Christophe Charle, *Le siècle de la presse, 1830–1939* (Paris: Seuil, 2004), 247–67, 289–307.

38. Jean-Noël Jeanneney, *L'argent caché: Milieux d'affaires et pouvoirs politiques dans la France du XXe siècle*, 2nd rev. ed (Paris: Seuil, 1984), 205–30, Blum quoted on 230.

39. Myriam Bacha, *Patrimoine et monuments en Tunisie: Sauvegarde et mise en valeur pendant le Protectorat, 1881–1920* (Rennes: Presses Universitaires de Rennes, 2013); Clémentine Gutron, *L'archéologie en Tunisie (XIXe–XXe Siècles): Jeux généalogiques sur l'antiquité* (Paris and Tunis: Karthala/IRMC, 2010).

40. Concise treatments of Glozel as a hoax include Richard, *Inventer la préhistoire*, 91–93; Jean-Paul Demoule, *On a retrouvé l'histoire de France: Comment l'archéologie raconte notre passé* (Paris: Robert Laffont, 2012), 201–5; Arnaud Hurel, *L'abbé Breuil: Un préhistorien dans le siècle* (Paris: CNRS, 2011), 312–17; Paul Bahn and Colin Renfrew, "Garrod and Glozel: The End of a Fiasco," in *Dorothy Garrod and the Progress of the Palaeolithic: Studies in the Prehistoric Archaeology of the Near East and Europe*, ed. William Davies and Ruth Charles (Oxford, England: Oxbow Books, 1999), 76–83. The defender of Glozel's authenticity is Joseph Grivel, notably in *La préhistoire chahutée: Glozel (1924–1941)* (Paris: L'Harmattan, 2003), as well as in volumes on later discussions of Glozel.

41. Kenneth E. Foote, "To Remember and Forget: Archives, Memory, and Culture," *American Archivist* 53, no. 3 (1990): 378–92, https://doi.org/10.17723/aarc.53.3.d87u013444j3g6r2.

42. Patrick Wright, "How Privatization Turned Britain's Red Telephone Kiosk into an Archive of the Welfare State," 207–14, and Judith E. Endelman, "'Just a Car': The Kennedy Car, the Lincoln Chair, and the Study of Objects," 245–52, both in *Archives, Documentation, and Institutions of Social Memory: Essays from the Sawyer Seminar*, ed. Francis Xavier Blouin and William G. Rosenberg (Ann Arbor: University of Michigan Press, 2007).

43. Examples include Bonnie Effros, *Merovingian Mortuary Archaeology and the Making of the Early Middle Ages* (Berkeley: University of California Press, 2003); Morgan Pitelka, *Handmade Culture: Raku Potters, Patrons, and Tea Practitioners in Japan* (Honolulu: University of Hawai'i Press, 2005); and David Spafford, *A Sense of Place: The Political Landscape in Late Medieval Japan* (Cambridge, MA: Harvard University Press, 2013).

44. Francis X. Blouin and William G. Rosenberg, *Processing the Past: Contesting Authority in History and the Archives* (Oxford: Oxford University Press, 2013), 24.

45. Florence Bernault, "Suitcases and the Poetics of Oddities: Writing History from Disorderly Archives," *History in Africa* 42 (2015): 272, https://doi.org/10.1017/hia.2015.5.

46. Marlene Manoff, "Theories of the Archive from Across the Disciplines," *Portal: Libraries and the Academy* 4, no. 1 (January 2004): 9–25, https://doi.org/10.1353/pla.2004.0015.

47. This is one of the starting points of Jacques Derrida, *Archive Fever: A Freudian Impression*, trans. Eric Prenowitz (Chicago: University of Chicago Press, 1996). For an illuminating if skeptical gloss on this work, see Carolyn Steedman, *Dust: The Archive and Cultural History* (New Brunswick, NJ: Rutgers University Press, 2002), 1–16.

48. Allan Sekula, "Reading the Archive," in *Blasted Allegories: An Anthology of Writings by Contemporary Artists*, ed. Brian Wallis (Cambridge, MA: MIT Press, 1987), 118.

49. For a probing discussion of the French case, see Jennifer S. Milligan, "'What Is an Archive?' in the History of Modern France," in *Archive Stories*, ed. Antoinette Burton (Durham, NC: Duke University Press, 2005), 159–83, https://doi.org/10.1215/9780822387046-008.

50. Schlanger, "Ancestral Archives," 130. On the materials stored in archaeological archives, see Schlanger and Nordbladh, "General Introduction: Archaeology in the Light of Its Histories," in *Archives, Ancestors, Practices*, 3.

51. Christopher Tilley, "Excavation as Theatre," *Antiquity* 63 (1989): 275–80, https://doi.org/10.1017/S0003598X00075992.

52. Olsen et al., *Archaeology*, 12–16, 196–209.

53. Joseph Déchelette, *Manuel d'archéologie préhistorique, celtique et gallo-romaine*, 4 vols. (1908–1914; Paris: Picard, 1924), 1:viii.

54. Ann Laura Stoler notes that the commitment to source materials distinguishes archival from ethnographic approaches: "Colonial Archives and the Arts of Governance: On the Content in the Form," in Blouin and Rosenberg, *Archives, Documentation, and Institutions*, 268.

55. On history's scientificity, which may be more of a preoccupation in France than in the Anglophone world, see Étienne Anheim, *Le Travail de l'histoire* (Paris: Éditions de la Sorbonne, 2018), 12–13.

56. Michel Foucault, "Nietzsche, Genealogy, History," in *Language, Counter-Memory, Practice: Selected Essays and Interviews*, trans. Donald F. Bouchard and Sherry Simon (Ithaca, NY: Cornell University. Press, 1977), 139–67, 139.

57. Michel Foucault, *The Archaeology of Knowledge and the Discourse on Language*, trans. A. M. Sheridan Smith (1969; New York: Pantheon, 1972), 130.

58. Shapin and Schaffer, *Leviathan*, 4–6. The terms they use are "member's accounts" and "stranger's accounts"; they advocate for a position combining insider knowledge with the distance of an outsider.

59. For inventories of these records, see Marie-Elisabeth Antoine and Suzanne Olivier, *Inventaire des papiers de la Division des sciences et lettres du Ministère de l'instruction publique et des services qui en sont issus: sous-série F17*, 2 vols. (Paris: Archives nationales, 1975, 1981). The printed volumes have been incorporated into the Archives Nationales' online catalog.

60. Notably the Tunisian Protectorate archives at the Nantes Branch of the Archives Diplomatiques, as well as the Archives Nationales, because the Ministry of Education continued to fund excavations in Tunisia.

61. The inventory of the Fonds Poinssot is available at http://www.calames.abes.fr/Pub/inha.aspx?fullText=fonds+poinssot%20-%20details?id=FileId-3397#, consulted 30 April 2022.

62. For information on the Poinssot family and their contributions to the archaeology of North Africa, see María Fernández Portaencasa, "Julien Poinssot and His Descendants: Three Generations of Discoveries Which Unravelled the Ancient Religions of North Africa," *Revista de Historiografía*, no. 36 (2021): 177–217, https://doi.org/10.20318/revhisto.2021.6555.

63. See Aurélien Caillaud and Sébastien Chauffour, "Un fonds peut en cacher un autre … et un autre … et un autre … : Traitement et inventaire des archives Poinssot," in *Autour du fonds Poinssot*, ed. Monique Dondin-Payre, H. Jaïdi, S. Saint-Amans, and M. Sebaï (Paris: Publications de l'Institut national d'histoire de l'art, 2017), https://doi.org/10.4000/books.inha.7151.The title alludes to a warning sign at French train crossings that has been the source of many metaphors: "Un train peut en cacher un autre."

64. John Randolph, "On the Biography of the Bakunin Family Archive," in Burton, *Archive Stories*, 209–31; Kaeser, "Biography as Microhistory," 12–13.

65. This decision is recorded in AN, F17 17263, and discussed in detail in chapter 2.

66. One key repository, the archives of the Fradin family, on whose property the "discovery" was made and the excavations carried out, is open only to researchers vetted by an association whose members espouse the authenticity of the site. I chose not to submit to this process and thus did not have access to these archives.

67. See Anke te Heesen, *The Newspaper Clipping: A Modern Paper Object*, trans. Lori Lantz (Manchester: Manchester University Press, 2014).

68. On archaeological epistemology, see Eugenio Donato, "The Museum's Furnace: Notes toward a Contextual Reading of *Bouvard and Pécuchet*," in *Textual Strategies: Perspectives in Post-Structuralist Criticism*, ed. Josué V. Harari (Ithaca, NY: Cornell University Press, 1979), 213–38; see also Susan Stewart, *On Longing: Narratives of the Miniature, the Gigantic, the Souvenir, the Collection* (Durham, NC: Duke University Press, 1993 [1984]).

69. For a concise definition of the history of photographs, an approach that emphasizes the materiality of photographs and their framing devices, and its relation to the history of photography, see Catherine E. Clark, "Capturing the Moment, Picturing History: Photographs of the Liberation of Paris," *American Historical Review* 121 (2016): 826–29, https://doi.org/10.1093/ahr/121.3.824.

70. Olsen et al, *Archaeology*, 6.

71. See, for example, Muriam Haleh Davis, *Markets of Civilization: Islam and Racial Capitalism in Algeria* (Durham, NC: Duke University Press, 2022), 9.

72. Owen White, *The Blood of the Colony: Wine and the Rise and Fall of French Algeria* (Cambridge, MA: Harvard University Press, 2021), 8–9. I do not need to follow White's coinage of the term "Euro-Algerian" to describe settlers, both because Tunisians, unlike Algerians, were never juridically French subjects and because the European settler community in Tunisia does not play a major role in the book.

Chapter 1

1. AN, F17 2943^D, Dossier Cagnat, Cagnat to Charmes, 21 May 1881. Earlier letters explaining his movements, 19 April and 6 May 1881, are in the same dossier.

2. FKP, box 80.2, Stoever to Kelsey, 25 July 1925.

3. Frederick Cooper, *Colonialism in Question: Theory, Knowledge, History* (Berkeley: University of California Press, 2005), 200. The literature on colonialism and its cultural and disciplinary relays is vast; Cooper's work, as well as Ann Laura Stoler, *Carnal Knowledge and Imperial Power: Race and the Intimate in Colonial Rule* (Berkeley: University of California Press, 2002), and Nicholas B. Dirks, ed., *Colonialism and Culture* (Ann Arbor: University of Michigan Press, 1992), are among those that have shaped my understanding of modern colonialism.

4. George R. Trumbull IV, *An Empire of Facts: Colonial Power, Cultural Knowledge, and Islam in Algeria, 1870–1914* (Cambridge: Cambridge University Press, 2009), 17.

5. Houcine Jaïdi, "La création du Service des antiquités de Tunisie: Contexte et particularités," in *Autour du fonds Poinssot*, ed. M. Dondin-Payre, H. Jaïdi, S. Saint-Amans, and M. Sebaï (Paris: Pubications de l'Insitut national d'histoire de l'art, 2017), 31; DOI: 10.4000/books.inha.7157.

6. Bruno Latour, *Science in Action: How to Follow Scientists and Engineers through Society* (Cambridge, MA: Harvard University Press, 1987), 180.

7. Latour, *Science in Action*, 219–23, 222.

8. See, for example, Warwick Anderson, "From Subjugated Knowledge to Conjugated Subjects: Science and Globalisation, or Postcolonial Studies of Science?," *Postcolonial Studies* 12, no. 4 (December 2009): 389–400, https://doi.org/10.1080/13688790903350641; Projit Bihari Mukharji, "Cultures of Fear: Technonationalism and the Postcolonial Responsibilities of STS," *East Asian Science, Technology and Society* 6, no. 2 (June 1, 2012): 267–74, https://doi.org/10.1215/18752160-1626341.

9. Latour's chapter on centers of calculation begins with the explorer Jean-François Lapérouse's appropriation of local knowledge to create a map of Sakhalin Island when he arrived there in 1787 (Latour, *Science in Action*, 215–18). Michael Bravo takes on this encounter explicitly in critiquing Latour; Michael T. Bravo, "Ethnographic Navigation and the Geographical Gift," in *Geography and Enlightenment*, ed. David N. Livingstone and Charles W. J. Withers (Chicago: University of Chicago Press, 1999), 199–235.

10. Simon Schaffer, "Exact Sciences and Colonialism: Southern India in 1900," in *Science*

as Cultural Practice, vol. 1, *Cultures and Politics of Research from the Early Modern Period to the Age of Extremes*, ed. Mortiz Epple and Claus Zittel (Boston: DeGruyter, 2010), 121–40, 121.

11. Joan Gero and Dolores Root, "Public Presentations and Private Concerns: Archaeology in the Pages of *National Geographic*," in *The Politics of the Past*, ed. Peter Gathercole and David Lowenthal (London: Routledge, 1994), 19–37.

12. Dougga was, however, the site of a modern settlement that archaeological excavation would uproot; see Clémentine Gutron, *L'archéologie en Tunisie (XIXe–XXe siècles): Jeux généalogiques sur l'Antiquité* (Paris and Tunis: Karthala/IRMC, 2010), 138–45.

13. For a recent account of the White Fathers' role in Tunisia, see Joseph W. Peterson, *Sacred Rivals: Catholic Missions and the Making of Islam in Nineteenth-Century France and Algeria* (New York: Oxford University Press, 2022), 159–74.

14. On early archaeological excavation in Tunisia, see Joann Freed, *Bringing Carthage Home: The Excavations of Nathan Davis, 1856–1859*, University of British Columbia Studies in the Ancient World 2 (Oxford: Oxbow, 2011); Myriam Bacha, *Patrimoine et monuments en Tunisie: Sauvegarde et mise en valeur pendant le Protectorat, 1881–1920* (Rennes: Presses Universitaires de Rennes, 2013), 32–39, 69–76; Bacha, "La constitution d'une notion patrimoniale en Tunisie, XIXe–XXe siècles: Emergence et apport des disciplines de l'archéologie et de l'architecture," in *Chantiers et défis de la recherche sur le Maghreb contemporain*, ed. Pierre-Robert Baduel (Paris and Tunis: Karthala/IRMC, 2009), 162–64.

15. Joann Freed, "Le Père Alfred-Louis Delattre (1850–1932) et les fouilles archéologiques de Carthage," *Histoire et missions chrétiennes* 4, no. 8 (2008), 82–84; Myriam Bacha, "Paul Gauckler, le père Delattre et l'archevêché de Carthage: Collaboration scientifique et affrontements institutionnels," in Dondin-Payre et al., *Autour du fonds Poinssot*, 2–7, DOI: 10.4000/books.inha.7158.

16. For a thorough discussion of the complicated relationship between Gauckler and Delattre, see Joann Freed, "Father Alfred-Louis Delattre (1850–1932) versus Paul Gauckler (1866–1911): The Struggle to Control Archaeology at Carthage at the Turn of the Twentieth Century," in *Life-Writing in the History of Archaeology: Critical Perspectives*, ed. Clare Lewis and Gabriel Moshenska (London: UCL Press, 2023), 233–63, https://www.jstor.org/stable/j.ctv37mk2fp. See also Bacha, "Paul Gauckler," 9–17; Jaïdi, "La création du Service des Antiquités," 31.

17. On the importance of Flaubert, see Martin Bernal, *Black Athena: The Afroasiatic Roots of Classical Civilization*, vol. 1, *The Fabrication of Ancient Greece, 1785–1985* (New Brunswick, NJ: Rutgers University Press, 1987), 338, 358–59; Peter van Dommelen, "Punic Identities and Modern Perceptions in the Western Mediterranean," in *The Punic Mediterranean: Identities and Identification from Phoenician Settlement to Roman Rule*, ed. Josephine Crawley Quinn and Nicholas C. Vella (Cambridge: Cambridge University Press, 2014), 42–46.

18. The initial discovery at Carthage, in 1921, came two years after a similar discovery at another Punic site at Motya (Sicily); see Brien K. Garnand, "The Use of Phoenician Human Sacrifice in the Formation of Ethnic Identities" (PhD diss., University of Chicago, 2006), especially the introduction and chapter 6, which discusses revisionist work beginning in the late 1980s that challenges the consensus that Carthage practiced infant sacrifice. Poinssot and his deputy Raymond Lantier were the first to publish the Carthage finds; Louis Poinssot and Raymond Lantier, "Un sanctuaire de Tanit à Carthage," *Revue de l'histoire des religions* 87 (1923): 32–68.

19. The figure comes from the French social science portal Persée: https://www-persee-fr.libproxy.lib.unc.edu/authority/217396, consulted 18 July 2023.

20. Examples include A. L. Delattre, *La nécropole des rabs, prêtres et prêtresses de Carthage: Troisième année des Fouilles*, Publications des Pères Blancs 145 (Paris: Feron-Vrau, 1906),

originally published in *Cosmos*; Delattre, *Dix nouvelles années de trouvailles (1915–1925): Carthage terre mariale* (Paris: Feron-Vrau, 1925), http://gallica.bnf.fr/ark:/12148/bpt6k62006082, a stand-alone pamphlet from a period when *Cosmos* was no longer appearing weekly.

21. See, for example, Mabel Moore, *Carthage of the Phoenicians in the Light of Modern Excavations* (London: W. Heinemann, 1905), which recounts Delattre's most recent discoveries; Byron Khun de Prorok, *Digging for Lost African Gods: The Record of Five Years Archaeological Excavation in North Africa* (New York: G. P. Putnam's Sons, 1926), 79–83.

22. AN, F17 2943^D, Marius Vachon, "Une mission artistique à Kerouan," *La France*, 10 October 1881.

23. AN, F17 2943^D, correspondence between Cagnat and Charmes, December 1881–May 1882.

24. Matthew M. McCarty, "French Archaeology and History in the Colonial Maghreb: Inheritance, Presence, and Absence," in *Unmasking Ideology: The Vocabulary, Symbols, and Legacy of Imperial and Colonial Archaeology*, ed. Bonnie Effros and Guolong Lai (Los Angeles: Cotsen Institute of Archaeology Press, 2018), 364–66.

25. McCarty, "French Archaeology and History," 367–72; Jacques Alexandropoulos, "Tourisme et patrimoine à Carthage en 1930: Souvenirs d'un pèlerin du congrès eucharistique," in *Fabrique du tourisme et expériences patrimoniales au Maghreb, XIXe–XXIe siècles*, ed. Cyril Isnart, Charlotte Mus-Jelidi, and Colette Zytnicki (Rabat: Centre Jacques-Berque, 2018), https://doi.org/10.4000/books.cjb.1407, 6–7.

26. See, for example, AN, F17 2943^D, note by G. Perrot (a member of the Institut), 7 March 1882. On the importance of Rome for the military officers excavating in Algeria between 1830 and 1880, see Bonnie Effros, *Incidental Archaeologists: French Officers and the Rediscovery of Roman North Africa* (Ithaca, NY: Cornell University Press, 2018), 86–87, 91–118.

27. Alex Csiszar, *The Scientific Journal: Authorship and the Politics of Knowledge in the Nineteenth Century* (Chicago: University of Chicago Press, 2018), 159–70.

28. AN, F17 2943^D, Cagnat to Minister of Education, 25 June 1882.

29. AN, F17 2943^D, Note from Father Delattre on costs of excavations in Tunisia, n.d. but 1881 from context (dossier 1, doc. 51); Cagnat, report from Djilma, 9 January 1883 (mosque permissions); Cagnat to Charmes, 26 April 1883 (serious men). On military officers as archaeologists in North Africa, see Effros, *Incidental Archaeologists*.

30. *Le nouveau Petit Robert* (Paris: Dictionnaires Robert, 1993), s.v. "réclame": "c'est la réclame qui est à la base de la célébrité de cet écrivain." The citation is credited to the writer and poet Valery Larbaud (1881–1957).

31. Effros, *Incidental Archaeologists*, 126.

32. For summary biographical information, see Ève Gran-Aymerich, *Les chercheurs de passé, 1798–1945: Aux sources de l'archéologie* (Paris: CNRS, 2007), 658–59.

33. René Cagnat, *Carthage, Timgad, Tébessa et les villes antiques de l'Afrique du Nord*, Les villes d'art célèbres (Paris: H. Laurens, 1909).

34. Pierre Bourdieu, *Science of Science and Reflexivity*, trans. Richard Nice (Chicago: University of Chicago Press, 2004), 32–33.

35. The most thorough account of the institutional foundations is Bacha, *Patrimoine et monuments*, 93–103, 151–57. Although the French resident general and, through him, the Protectorate government reported to the Ministry of Foreign Affairs, the Antiquities Service also received funding for specific projects from the French Ministry of Education (Ministère de l'instruction publique), which gave it some supervisory authority over the service.

36. Biographical information from Gran-Aymerich, *Les chercheurs de passé*, and from the website of the Académie des inscriptions et belles-lettres, http://www.aibl.fr/membres

/academiciens-depuis-1663/?lang=fr, consulted 2 January 2018. Cagnat, elected in 1895 at age forty-three, was perpetual secretary from 1916 to his death in 1937. Merlin, elected in 1928 at age fifty-two, was perpetual secretary from 1948 to 1964, when he stepped down a few months before his death.

37. Unfortunately, the Poinssot papers contain few drafts or copies of letters *from* Poinssot. Members of the Poinssot family gathered up papers of other people in their circle, including Gauckler and Merlin. Merlin's correspondence includes seventy-six letters from Poinssot, all but four prior to 1914 (FP, 040, folder 2); Poinssot's correspondence includes nearly 650 letters from Merlin over the period 1901–1963 (FP, 033). The complete inventory of the Fonds Poinssot is available online at http://www.calames.abes.fr/Pub/inha.aspx?fullText=fonds+poinssot%20-%20details?id=FileId-3397#details?id=FileId-3397.

38. See, for example, FP 033, Merlin to Poinssot (then at Mahdia), 11, 14, and 15 May 1909; Merlin (at Mahdia) to Poinssot, 17 June 1910.

39. James A. Secord, "Halifax Keynote Address: Science in Transit," *Isis* 95 (2004): 661.

40. Kapil Raj, "Go-Betweens, Travelers, and Cultural Translators," in *A Companion to the History of Science*, ed. Bernard Lightman (London: John Wiley, 2016), 39–57.

41. AN, F17 2944, dossier Carton, Carton, "Rapport sur un projet de déblaiement d'édifices à Dougga," 1 April 1892; Note to Charmes from "Missions d'Afrique," 12 April 1892.

42. AN, F17 2944, dossier Carton, Carton to Charmes, 7 September 1892; Foreign minister to Minister of education, 5 October 1892, enclosing letter from the French resident general. See also Carton's letters to Gauckler, FP, 191.

43. See, for example, Louis Carton, *La restauration de l'Afrique du Nord* (Brussels: Imprimerie des Travaux Publics, 1898), a twenty-eight-page pamphlet based on a lecture given at the Congrès international colonial de Bruxelles in 1897. His views are a variant of those discussed in Diana K. Davis, *Resurrecting the Granary of Rome: Environmental History and French Colonial Expansion in North Africa* (Athens: Ohio University Press, 2007).

44. FP 191, Carton to Paul Gauckler, 9 February 1896.

45. Louis Carton, "Chronique d'achéologie nord-africaine," *Revue tunisienne* 11 (1904): 142. Carton began his annual chronicles in 1903.

46. AN, F17 2944, dossier Carton, exchange between Carton and Ministry of Education, 5–25 June 1899. In his response to the news, Carton protested that the decision was harmful to scholarship, as it kept him from completing a mission he had begun at the ministry's behest.

47. See, for example, FP, 050, dossier 7, Gauckler to Bernard Roy, 8 September 1904, on his enthusiasm for bringing Poinssot back to Tunisia and the means to accomplish it. This box also includes letters from Poinssot to Gauckler.

48. FP 032, Merlin to Poinssot, 20 August 1911.

49. Bacha, *Patrimoine et monuments*, 218–20.

50. See Hédi Dridi and Antonella Mezzolani Andreose, "'Ranimer les ruines': L'archéologie dans l'Afrique latine de Louis Bertrand," *Nouvelles de l'archéologie*, no. 128 (June 2012): 10–16, and more generally on Bertrand, Patricia M. E. Lorcin, *Imperial Identities: Stereotyping, Prejudice and Race in Colonial Algeria* (London: I. B. Tauris, 1995), chap. 9.

51. Louis Bertrand, "Lettre ouverte à M. le Ministre de l'Instruction publique: Pour les ruines antiques de l'Afrique du Nord," *Journal des débats*, 2 June 1920. Gran-Aymerich is at pains to contrast this attitude to those of professional archaeologists: *Les chercheurs de passé*, 389.

52. Bertrand, "Lettre ouverte."

53. The clearest exposition of the term and its use in colonial discourse can be found in Alice L. Conklin, *A Mission to Civilize: The Republican Idea of Empire in France and West Africa, 1895–1930* (Stanford, CA: Stanford University Press, 1997), esp. 6–7, 41–42.

54. See Clémentine Gutron, "Voyager dans le temps avec un archéologue à travers la Tunisie coloniale: Louis Carton (1861–1924) et sa Tunisie en l'an 2000," in *Explorations et voyages scientifiques de l'Antiquité à nos jours*, ed. C. Demeulenaere-Douyère (Paris: CTHS, 2008), 553; Gutron, *L'archéologie en Tunisie*, 121–22.

55. Bertrand, "Lettre ouverte."

56. Bertrand, "Lettre ouverte."

57. "Trésors honorés: Les ruines de l'Afrique du Nord," *La Dépêche tunisienne*, 14 June 1920; all the passages quoted above were reprinted. The same paper had, a few days earlier, reprinted another article by Bertrand published in the *Écho de Paris*; "Pour les ruines antiques de l'Afrique du Nord," *La Dépêche tunisienne*, 10 June 1920, addresses similar themes but with more emphasis on Bertrand's recent travels in the area. Finally, it should be noted that the *Journal des débats* published a reply to Bertrand's open letter, with a further response by Bertrand, but the reply is concerned almost exclusively with Algeria; Martial Douel, "Pour les ruines antiques de l'Afrique du Nord," *Journal des débats*, 10 July 1920. Douel opines that Bertrand's article was excessively harsh.

58. For example, Louis Carton, *Carthage et le tourisme en Tunisie: Ouvrage publié sous le patronage du Touring-Club de France* (Boulogne: Imprimeries réunies, 1919).

59. See, for example, Louis Carton, *Guide express de Carthage* (Tunis: J. Danguin, 1909), 2.

60. A. Mezzolani Andreose, "Regards de femmes sur la ville de Didon: Le Comité des dames amies de Carthage (1920–1924)," in *Fabrique du tourisme et expériences patrimoniales au Maghreb*, DOI: 10.4000/books.cjb.1407.

61. CADN, 1 TU1 2079, dossier Fouilles de Carthage, ticket and "avis."

62. Dr. Carton, *Pour Carthage!* (Tunis: Imprimerie Rapide, 1922?), 6–7.

63. Carton, *Pour Carthage!*, 3; emphasis in original.

64. BMA, Carton to Reinach, 14 April 1910.

65. FP, 033, Merlin to Poinssot, 8 November 1912.

66. AN, F17 17238, "Annexe à la lettre adressée le 1er novembre, à la Commission de l'Afrique du Nord," nine-page typescript. Handwritten insertions and corrections suggest that this typescript was produced within the ministry as a transcript of a handwritten original.

67. AN, F17 17238, Cagnat to minister of education, 28 November 1921, saying that though the situation "que signale le docteur Carton est certainement intéressante . . . l'Académie ne se croit pas en droit d'intervenir directement auprès de M. le Résident Général"; FP, 033, Merlin to Poinssot, 19 November 1921.

68. FP, 033, Merlin to Poinssot, 17 December 1921.

69. For Saumagne's background and attempts to use precedents from Roman law to protect indigenous property under the colonial regime, see Gutron, *L'archéologie en Tunisie*, 125–32.

70. AN, F17 17238, Charles Saumagne, "Le problème de Carthage," *La Dépêche tunisienne*, 30 July 1923 (clipping).

71. See, for example, R. de la Porte, "Pourquoi il y a une question de Carthage?" *La Tunisie française*, 12 February 1924; "Pour Carthage et les villes d'or," *La Tunisie française*, 2 June 1924, refuting a recent article by Bertrand and using Saumagne's trope of Carthage as a living city for the living; Jacques Denis, "Les grandes inventions: Un défenseur de Carthage," *Le Tunis*

socialiste, 9 March 1924, in response to a letter, published simultaneously, from the secretary-general of a Carthage booster group founded by Carton. Excerpts from Denis's article were published the next day in *La Tunisie française* under the title "La scie cartonginoise" [*sic*], a significant rapprochement between a leftist and a quite conservative newspaper.

72. The *fêtes de Carthage* of 1906 and 1907 will be discussed in detail in chapter 4.

73. FP, 045, Louis Carton, "A propos d'un article sur Carthage," handwritten fair copy, 14 pages, n.d. but referring to Saumagne's article in the *Dépêche tunisienne* by date. Within the Poinssot papers, FP 045, devoted to Carthage excavations, is part of a thematic series on excavations in various places; it is not primarily a correspondence file. The handwriting is not Carton's but is similar to that used in some of Poinssot's official correspondence, suggesting that one of his associates had recopied it before returning the original to Paris as requested.

74. Gutron, "Voyager dans le temps," 553–71. On preservation campaigns before World War I, see Astrid Swenson, *The Rise of Heritage: Preserving the Past in England, France, and Germany, 1789–1914* (Cambridge: Cambridge University Press, 2013), 239–72. On the 1931 Athens Conference on monument preservation, the most significant interwar moment for preservation campaigns, see Lucia Allais, *Designs of Destruction: The Making of Monuments in the Twentieth Century* (Chicago: University of Chicago Press, 2018), 34–70.

75. AN, F17 17238, Directeur de l'enseignement supérieur to Poinssot (draft), 11 September 1923.

76. FP, 045, Carton, "A propos d'un article"; AN, F17 17238, Poinssot to Directeur de l'enseignement supérieur, 4 October 1923.

77. AN, F17 17238, Poinssot to Directeur de l'enseignement supérieur, 4 October 1923. The last sentence is cited in Jaïdi, "La création du Service des Antiquités," 8.

78. AN, F17 17238, Poinssot to Directeur de l'enseignement supérieur, 4 October 1923.

79. CADN, 1 TUI 2079 dossier "Fouilles de Carthage": "Dévastation Sacrilège: Les ruines de Carthage sont mises au pillage," clipping from *Le Matin*, 19 June 1923; letter to senators de Jouvenel and Chenebenoit (draft).

80. J. Jaubert de Benac, "Delenda Carthago: Les ruines de Carthage au pillage," *L'Illustration*, 11 August 1923, 118–21. The article in *L'Illustration* is close enough to that in *Le Matin* to permit the conclusion that Jaubert de Benac, who is mentioned in the latter, sent his magazine article to the daily before its publication.

81. AN, F17 17238, Poinssot to Directeur de l'enseignement supérieur, 1 October 1923, with "Note demandée par M. le Ministre de l'Instruction publique et des Beaux-Arts, au suject de vues parues dans l'Illustration du 11 août 1923."

82. AN, F17 17238, Directeur de l'enseignement supérieur to Merlin (draft), n.d. but enclosing Poinssot's two reports; AN, F17 17256, Merlin to Ministre, 24 October 1923.

83. It is conceivable that Prorok did not on this occasion actually meet Poinssot, whose wife gave birth to a child that same month (see FP, 033, Merlin to Poinssot, 13 May and 5 June 1922), but that would make the deposit of a photograph no less peculiar.

84. FP, 033, Merlin to Poinssot, 30 January 1922.

85. FP, 033, Merlin to Poinssot, 30 January 1922.

86. FP, 033, Prorok to Poinssot, 30 July 1922 (date from context) and 23 March 1923. In the latter, Prorok says he gave seventy-eight lectures in the US on a tour of 30,000 kilometers (18,750 miles).

87. CADN, 1 TUI 2079, dossier "Fouilles de Carthage–Mission Américaine," Ministre des affaires étrangères to Saint, 2 January 1923; Projet de réponse, with cover note from Poinssot, 13 January 1923.

88. CADN, 1 TUI 2079, dossier "Fouilles de Carthage–Mission Américaine," Saint to Ministre des affaires étrangères, 16 March 1923. Saint enclosed a clipping from the *Dépêche tunisienne* of the same date to illustrate "cet état d'esprit."

89. CADN, 1 TUI 2079, dossier "Fouilles de Carthage–Mission Américaine," Jusserand to Ministre des affaires étrangères, 10 February 1923 (copy), enclosing typescript copy of note from Arthur Fairbanks, director, Museum of Fine Arts, Boston, 23 January 1923.

90. CADN, 1 TUI 2079, dossier "Fouilles de Carthage–Mission Américaine," Jusserand to president of Yale, 9 May 1923, transmitted to Saint for his information, 23 May 1923. In a later account of his archaeological (ad)ventures, Prorok notes that he was awarded an endowed lectureship at a professional meeting held at Yale, which could account for this particular element of confusion; Prorok, *Digging for Lost African Gods*, 51.

91. CADN, 1 TUI 2079, dossier "Fouilles de Carthage–Mission Américaine," Jusserand to Ministre des Affaires étrangères, 19 April 1923.

92. CADN, 1 TUI 2079, dossier "Fouilles de Carthage–Mission Américaine," Ministre de l'instruction publique et des beaux-arts to Ministre des affaires étrangères, 13 March 1923 (copy).

93. FKP, 79.2, "Memorandum for French Counsel," 10 February 1924, sets out the legal status of the Archaeological Society.

94. FKP, 79.3, document headed "Appendix: Memorandum regarding proposals for cooperative reseach at Carthage, read as a part of the report of the Research Committee before the Trustees of the Archaeological Society of Washington on February 1st, and approved by vote of the Board of Trustees." The University of Michigan and the Society were each to provide a quarter of the budget; of the $5,000 required of the Society, $3,000 would come from the earlier gift of Robert Woods Bliss.

95. FP, 033, Prorok to Poinssot, 30 July 1922 (date from context), 23 March 1923, and summer 1923 (date conjectural), in which he announces eleven forthcoming talks in England.

96. Kenny is mentioned in FKP, 79.2, "Outline (Confidential)," a two-page typescript from late 1924, and FKP, 79.3, "Minutes of Carthage Exploration Committee," 12 January 1925, where he is credited with the gift of $18,000 for purchase of land in Carthage.

97. FKP, 79.2, "Excavation of Carthage, Proposed Plan of Operations, 1925, Memorandum no. I with Exhibits."

98. FKP, 79.2, Proposed Plan of Operations, 1925, item 4c; FKP, 80.2, Kelsey to Edward Stoever, 28 October 1924, typescript carbon. For an excellent recent study of the photography of the Tutankhamun expedition and its use in publicity, see Christina Riggs, *Photographing Tutankhamun: Archaeology, Ancient Egypt, and the Archive* (London: Bloomsbury Visual Arts, 2018).

99. CADN, 1 TUI 2079, dossier "Fouilles de Carthage–Mission Américaine," Jusserand to Ministre des affaires étrangères, 19 April 1923.

100. FKP, 79.4 "'U.M.' Explorers Find Many Rare Old Relics," *Detroit Free Press*, 23 June 1925; Francis Kelsey, "Carthage: Ancient and Modern," *Art and Archaeology*, 21, no. 2 (February 1926): 54–59; FKP, 79.5, "Americans Ready to Dig at Carthage," *New York Times*, 6 March 1925; "Bones of Babes Sacrificed to Goddess found by 'M' Party," *Detroit News*, 5 July 1925.

101. FP, 033, Merlin to Poinssot, 29 May 1923.

102. FP, 033, Prorok to Poinssot, n.d. but marked in pencil "Eté 1923?." Prorok writes in error-filled but comprehensible French.

103. FP, 033, Prorok to Poinssot, 19 September (1923? added in pencil). There is a similar reference to Jaubert "dit de Bénac" in a letter of 17 October 1923.

104. P. Gielly, "Conférence archéologique de M. le Comte Byron Kuhn de Prorok," *Le petit matin* (Tunis), 11 April 1923. Gielly, a landowner and amateur archaeologist in Carthage, was a supporter of Carton's, and his reference to Poinssot—by title alone, not by name—suggests that the director was shedding crocodile tears on seeing the film's depictions of neglected archaeological sites.

105. Prorok, *Digging for Lost African Gods*, 67–68, 109.

106. Prorok, *Digging for Lost African Gods*, 68–69, 130–35, 130.

107. Prorok, *Digging for Lost African Gods*, 44–55.

108. See Bacha, *Patrimoine et monuments*, 173–74, 283; Stefan Altekamp and Mona Khechen, "Third Carthage: Struggles and Contestations over Archaeological Space," *Archaeologies: Journal of the World Archaeological Congress* 9 (2013): 472–75.

109. FP, 033, Merlin to Poinssot, 30 January 1922.

110. FKP, 79.2, "Memorandum for French Counsel," 10 February 1924; "Excavation of Carthage, Proposed Plan of Operations, 1925, Memorandum no. I with Exhibits."

111. Prorok, *Digging for Lost African Gods*, 58, 125.

112. CADN, 1 TUI 2079, dossier "Fouilles de Carthage–Mission Américaine," "Film américain à Carthage," *Dépêche tunisienne*, 16 or 17 March 1923, clipping.

113. FKP, 79.2, "Outline (Confidential)," late 1924 or early 1925.

114. FKP, 79.1, "Report on Land Deals in Carthage," ca. March 1925 but reporting on contacts in September 1924.

115. FP, 033, Merlin to Poinssot, 26 September 1924; FKP, 80.1, Edward Stoever to d'Erlanger, 16 September 1924.

116. Mezzolani Andreose considers the commissioning of this report one of the great achievements of the CDAC; "Regards de femmes," paragraph 27 and n. 41.

117. FP, 033, Merlin to Poinssot, 19 November and 11 December 1921, 19 December 1922.

118. CADN, 1 TUI 2079, dossier "Fouilles de Carthage–Mission Américaine," M. Carton, "Remerciements du C.D.A.C. à la Mission américaine," *Dépêche tunisienne*, 16 or 17 March 1923.

119. CADN, 1 TUi 2079, dossier "Fouilles de Carthage 1922–1923," Ministre des affaires étrangères to Résident général Tunis, 26 November 1923; AN, F17 17256, Lucien Saint, resident general, to Ministre des affaires étrangères, 8 December 1923; FP, 033, Merlin to Poinssot, 9 and 22 January 1923.

120. FP, 033, Merlin to Poinssot, 19 November 1921 and 6 June and 19 November 1922. On Gsell's friendships with Carton and Bertrand, see Dridi and Mezzolani Andreose, "Ranimer les ruines," 10, 12.

121. FP, 032, Louis Poinssot to Stéphane Gsell (draft), 4 February 1924.

122. CADN, 1 TUi 2079, dossier "Mission Gsell, 1924,"typescript report, p. 2.

123. AN, F17 17156, Note from Coville, Directeur de l'enseignement supérieur, to Chef du cabinet, 4 August 1924, in response to an inquiry from the wife of the minister (Édouard Herriot).

124. On the complexity of land regulation in pre-1881 and protectorate Tunisia, see Geneviève Goussaud-Falgas, *Les français de Tunisie de 1881 à 1931* (Paris: L'Harmattan, 2013), 65–74; on conflicts generated in part by land disputes, see Mary Dewhurst Lewis, *Divided Rule: Sovereignty and Empire in French Tunisia, 1881–1938* (Berkeley: University of California Press, 2014), 50–52, 140–41.

125. On the beginnings of Destour and the 1922 abdication crisis, see Lewis, *Divided Rule*, 119–22; Kenneth J. Perkins, *A History of Modern Tunisia* (Cambridge: Cambridge University Press, 2004), 79–83.

126. The personnel and responsibilities are spelled out in FKP, 79.3, "Carthage Exploration Committee, January 12, 1925" (minutes). In FKP, 79.2, "Excavation of Carthage, Proposed Plan of Operations, 1925, Memorandum no. I with Exhibits," Chabot is mentioned under "Suggested Staff"; that list also includes Merlin, who did not come to Tunisia at this time, and Father Delattre, who was never officially part of the delegation.

127. FKP, 80.1, Poinssot to Prorok, 2 April 1925, typescript copy.

128. FKP, 80.1, Poinssot to Prorok, 2 April 1925. As this typescript was most likely produced by a native English speaker from the original French, I have interpreted *accaper* (a word that does not exist in French) as a typo for *accaparer*, "to monopolize."

129. FKP, 80.1, Kelsey to Poinssot, 17 March 1925, typescript with handwritten note signaling Kelsey's approval; Poinssot to Kelsey, 26 March 1925. The note (to a secretary) suggests that the letter was drafted by a native French speaker, perhaps Chabot.

130. FKP, 80.1, Kelsey to Poinssot, 15 April 1925, typescript copy (original in French). According to this letter, the break occurred in a face-to-face meeting between Poinssot, Kelsey, and Stoever on 14 April.

131. FKP, 80.1, Poinssot to Kelsey, 21 April 1925.

132. "Saving Carthage from Tunis Babbitts: Suburban Real Estate Boom on Site of Ancient Punic Capital Hampers Archaeologists—the New City Threatens to Seal the Old from Research, as in Rome," *New York Times*, 10 May 1925.

133. Prorok, *Digging for Lost African Gods*, 127–29.

134. FKP, 80.1, Kelsey, "Appendix to the Report on the Franco-American Excavations at Carthage, 1925," typescript. Kelsey acknowledges some conflict with the Tunisian laborers, who, despite receiving seven francs per day (one more than when working for the Antiquities Service), staged a strike to protest the surveillance to which the Americans subjected them.

135. FKP, 80.1, Poinssot to Kelsey, 16 April 1925.

136. FKP, 80.1, Kelsey, "Appendix to the Report." Although the report is undated beyond the year, it includes a rumor that Poinssot will be replaced as director, and thus probably dates from before Stoever's July letter, cited at the beginning of the chapter, on Poinssot's protector.

137. FKP, 80.2, Kelsey to Prorok, 12 September 1925 (typescript copy sent to Stoever).

138. "Les américains à Carthage," *Le petit matin*, 11 June 1925, reprinted in the *Courrier de Tunisie* of the same date. On Chabot's character, see Prorok, *Digging for Lost African Gods*, 61.

139. "Une lettre de M. Poinssot," *Courrier de Tunisie*, 19 June 1925. Like Chabot's letter, it was published on the front page.

140. FP, 033, Merlin to Poinssot, 21 June 1925.

141. FP, 033, Merlin to Poinssot, 29 June 1925.

142. FP, 032, typescript of charges relayed to Lantier by Reinach, director of the National Antiquities Museum, St. Germain-en-Laye, with Lantier's response and a cover letter to Poinssot dated 17 August 1923.

143. FP, 032, Lantier to Poinssot, 24 August 1923.

144. FP, 032, Lantier to Poinssot, 22 June 1925.

145. There is extensive documentation of this incident in FP, 004, folder 9, including clippings from *Le petit matin* critical of Poinssot and copies of his own accounts of both incidents. In "Regards de femmes sur la ville de Didon," an article generally sympathetic to Mme. Carton, Mezzolani Andreose describes this incident as one in which the widow struck Poinssot with her umbrella (paragraph 4).

146. CADN, 1 TU1 2079, dossier "M. Poinssot, Directeur des Antiquités et des Arts,"

Carton to the Résident général, 16 July 1922, copy submitted par A. Gounot, président du Syndicat des communes de la Banlieue-Nord de Tunis, 19 July 1922.

147. FP, 033, Merlin to Poinssot, 24 December 1919.

148. A printed copy of the twenty-four-page decree is in AN, F17 17236.

149. Paul Gielly, "Carthage et la science à l'envers," *La dépêche tunisienne*, 11 May 1923. A clipping is in AN, F17 17238,"Fouilles de Carthage."

150. Gielly, "Carthage et la science à l'envers." Gielly confuses the compensation provision in article 30, putting it at one-third of the value of the objects rather than one-quarter, and ignores the indemnification provision in article 7, which is subject to litigation if the parties cannot agree. The piece also does not mention his personal stakes in the matter, though his ownership of the land was hardly a secret.

151. Gutron, *L'archéologie en Tunisie*, 101.

152. Prorok, *Digging for Lost African Gods*, 75–77.

153. See CADN, 1 TU1 2079, dossier "Fouilles de Carthage/Fouilles Diverses," letter from Poinssot to Léonce Bénédite, forwarded by Louis Bertrand to the resident general, 30 July 1924 ("personne répugnante"); FP, 032, Lantier to Poinssot, 17 August 1923 ("immonde").

154. CADN, 1 TU1 2079A, Folder "Fouilles de Carthage: incidents entre M. Poinsot et Mme Carton," postcard with handwritten note from Marie Carton, 29 May 1925.

155. CADN, 1 TU1 2079, dossier "Mission Gsell, 1924,"report.

156. FP, 004, Paul Groseille to unknown correspondent, July 1925 (date from context).

157. FP, 031, Poinssot to "M. le Sécrétaire perpetuel" (draft), 15 December 1934.

158. Latour, *Science in Action*, 254–55.

159. Latour, *Science in Action*, 232.

160. Latour, *Science in Action*, 243. On the beginnings of the *Musées et collections archéologiques* project, see Effros, *Incidental Archaeologists*, 257.

161. See Bjørnar Olsen, Michael Shanks, Timothy Webmoor, and Christopher Witmore, *Archaeology: The Discipline of Things* (Berkeley: University of California Press, 2012), 1–2.

162. Latour, *Science in Action*, 243.

163. Latour, *Science in Action*, 232.

Chapter 2

1. "Examining a Discovery at Glozel," *Paris Times*, undated clipping in AIBL, K51, but early January 1928 from context. The Agence Meurisse photograph, preserved in the Bibliothèque Nationale de France, is cropped in both the *Paris Times* and *L'Humanité*, with the result that the hillside is not visible and the landscape more generic.

2. "Fradin Père et Fils, Conservateurs du Musée de Glozel," *L'Humanité*, 6 January 1928. The headline and caption erroneously refer to the older man as Emile Fradin's father.

3. For a succinct summary, see Jean-Paul Demoule, *On a retrouvé l'histoire de France: Comment l'archéologie raconte notre passé* (Paris: Robert Laffont, 2012), 202.

4. Alex Csiszar, *The Scientific Journal: Authorship and the Politics of Knowledge in the Nineteenth Century* (Chicago: University of Chicago Press, 2018), 170–84.

5. SRA, 4, Morlet to Audollent, 22 October 1927.

6. Reinach held the presidency of the Academy, a rotating and largely ceremonial position, for 1927, a fact that obliged him to a certain public reserve: see SRA, 4, Reinach to Audollent, 30 October 1927. On the last-minute change of plans that led to his selection as president, see BMA, René Cagnat to Reinach, 31 December 1926 and undated but clearly early January 1927.

7. On the importance of the telegraph and of news agencies, see James W. Carey, "Technology and Ideology: The Case of the Telegraph," in *Communication as Culture: Essays on Media and Society*, rev. ed. (New York: Routledge, 2008), 155–77; Terhi Rantanen, *When News Was New* (Malden, MA: Wiley-Blackwell, 2009), 29–56.

8. On the halftone process, see Neil Harris, "Iconography and Intellectual History: The Halftone Effect," in *Cultural Excursions: Marketing Appetites and Cultural Tastes in Modern America* (Chicago: University of Chicago Press, 1990), 304–17; Thierry Gervais, "La Similigravure: Le récit d'une invention (1878–1893)," *Nouvelles de l'Estampe* 229 (March 2010): 8–25. On rotogravure, see Andrés Mario Zervigón, "Rotogravure and the Modern Aesthetic of News Reporting," in *Getting the Picture: The Visual Culture of the News*, ed. Jason E. Hill and Vanessa R. Schwartz (London: Bloomsbury Academic, 2015), 197–205.

9. Arthur Asseraf, *Electric News in Colonial Algeria* (Oxford: Oxford University Press, 2019), DOI 10.1093/oso/9780198844044.003.000, 7.

10. "Les fouilles continuent/Dans la boue des tranchées de Glozel . . . /Où l'on voit M. Salomon Reinach arriver dans un char antique," *L'Intransigeant*, 14 April 1928, clipping in SRA, 4.

11. Bernard Roshco, "Newsmaking," in *News: A Reader*, ed. Howard Tumber (New York: Oxford University Press, 1999), 34; Hill and Schwartz, "General Introduction," in *Getting the Picture*, 7.

12. Christophe Charle, *Le siècle de la presse, 1830–1939* (Paris: Seuil, 2004), 247–67.

13. For a general survey, see Arnaud Hurel, *La France préhistorienne de 1789 à 1941* (Paris: CNRS, 2007); see also Hurel and Noël Coye, eds., *Dans l'épaisseur du temps: Archéologues et géologues inventent la préhistoire* (Paris: Muséum national d'histoire naturelle, 2011).

14. On Boucher de Perthes and his precurseurs, see Alain Schnapp, *The Discovery of the Past*, trans. Ian Kinnes and Gillian Varndell (New York: Harry N. Abrams, 1997), 277–89, 310–14; Nathalie Richard, *Inventer la préhistoire: Les débuts de l'archéologie préhistorique en France* (Paris: Vuibert, 2008), 52–78; Jean-Yves Pautrat, "L'homme antédiluvien: Les vestiges de l'homme et l'avenir des commencements," in Hurel and Coye, *Dans l'épaisseur du temps*, 97–149; Rémi Labrusse, *Préhistoire: L'envers du temps* ([Vanves]: Hazan, 2019), 50–55.

15. On conflicts within the field of prehistory, see Labrusse, *Préhistoire*, 58–63; Richard, *Inventer la préhistoire*, 79–93, 165–200; and, more generally, Claudine Cohen, *L'Homme des origines: Savoirs et fictions en préhistoire* (Paris: Seuil, 1999), chap. 2.

16. A vast literature has grown up around the Altamira controversy. See Benito Madariaga de la Campa, *Sanz de Sautuola and the Discovery of the Caves of Altamira*, trans. Shirley Clarke (Santander: Fundación Marcelino Botín, 2001), 31–81; Oscar Moro Abadía and Francisco Pelayo, "Reflections on the Concept of 'Precursor': Juan de Vilanova and the Discovery of Altamira," *History of the Human Sciences* 23, no. 4 (2010): 1–20; Gregory Curtis, *The Cave Painters: Probing the Mysteries of the World's First Artists* (New York: Knopf, 2006), 48–57.

17. On periodicals incorporating scholarship on prehistory, see Ève Gran-Aymerich, *Les chercheurs du passé, 1798–1945: Aux sources de l'archéologie* (Paris: CNRS, 2007), 152–53; Labrusse, *Préhistoire*, 63. More generally on specialized versus popularizing scientific journalism, see Csiszar, *Scientific Journal*, 206–10.

18. See, for example, Richard, *Inventer la préhistoire*, 148–64.

19. Joshua Nall, *News from Mars: Mass Media and the Forging of a New Astronomy, 1860–1910* (Pittsburgh, PA: University of Pittsburgh Press, 2019), 15–17.

20. Nall, *News from Mars*, chap. 3, esp. 131–33.

21. Both documents can be found in ADA, 36 J 8, the circular in *Département de l'Allier: Bulletin de l'Instruction Publique*, February 1924, 36–37; Picandet to Inspecteur d'Académie,

20 March 1924, typescript copy. The copy of Picandet's letter was sent to the Société d'émulation du Bourbonnais by the Directeur des Beaux-Arts (head of the division of the Ministry of Education and Fine Arts responsible for the circular) in response to a request from the society's vice-president dated 15 October 1926.

22. AN, F17 17263, Clippings dossier, note ("M. de Bar/ Glozel/dossier à classer/garder que les coupures").

23. "La commission des Monuments préhistoriques se prononce contre le 'classement' de Glozel," *Petit Journal*, 29 January 1928; "Glozel ne mérite pas le classement," *L'Oeuvre*, 29 January 1928.

24. AN, F17 17263, Cabinet du directeur, Direction de l'enseignement supérieur, Note pour Monsieur le Ministre, 29 September 1927.

25. Morlet's declaration to the *Petit niçois*, as transcribed in René Dussaud to Camille Jullian, 11 October 1927, BI, Ms 5764; BMA, Morlet to Reinach, 29 January 1928.

26. The file is in Médiathèque du patrimoine et de la photographie, Charenton-le-Pont, H/80/26/2-18.

27. Lisa Gitelman in "*AHR* Conversation: Historical Perspectives on the Circulation of Information," *American Historical Review* 116, no. 5 (December 2011): 1418–19.

28. On the Argus de la Presse, see Boris Dänzer-Kantof and Sophie Nanot, *Le roman vrai de l'Argus de la Presse: De Mata Hari à Internet* (Paris: Hervas, 2000), an anecdotal corporate history that seems to have been commissioned by the company, which is still in existence today.

29. BI, Ms 5764, Dussaud to Jullian, 17 October 1927.

30. The Société d'Emulation materials concerning Glozel in the Allier departmental archives (ADA) are catalogued in the range 36 J 2 to 36 J 9.

31. Lisa Gitelman, *Paper Knowledge: Toward a Media History of Documents* (Durham, NC: Duke University Press, 2014), 2. As Gitelman puts it for her own field, "Media studies must continue to aim at media, in short, not just 'the Media' as such" (19).

32. BMA, Espérandieu to Reinach, 1 January 1927.

33. BMA, Peyrony to Reinach, 20 January 1928.

34. BI, Ms5765, Viple to Jullian, 4 May 1926.

35. *NSN*, 1:46 and figs. 13, 52.

36. See John F. Healey, "The Early Alphabet," in *Reading the Past: Ancient Writing from Cuneiform to the Alphabet*, ed. J. T. Hooker (Berkeley: University of California Press, 1990), 197–257.

37. Biographical details from Gran-Aymerich, *Les chercheurs du passé*, 899–901.

38. SRA, 1, Morlet to Audollent, 3 October 1925. The first letter from Morlet to Jullian in BI, Ms 5765, is dated 25 September 1925.

39. ADA, 36 J 8, Jullian to de Brinon, 1 May 1925.

40. BI, Ms 5764, Grenier to Jullian, 15 and 20 September 1926. On Grenier's (1878–1961) ties to Jullian, see Gran-Aymerich, *Les chercheurs de passé*, 842–43.

41. BI, Ms 5764, Espérandieu to Jullian, 25 and 29 September 1926; quote from 25 September.

42. BI, Ms 5765, Morlet to Jullian, 25 and 30 November, 5 December 1925. The sample was sent to Paris, which Jullian occasionally visited for professional reasons.

43. BI, Ms 5764, Espérandieu to Jullian, 25 and 29 September 1926.

44. SRA 1, Audollent to Jullian (draft), 2 November 1926.

45. BI, Ms 5765, Jullian to Morlet, 15 July 1926.

46. On the meetings of the Academy, in which Reinach publicly disagreed with Jullian, see *Journal des débats*, 7 and 14 November 1926. The newspaper referred to an earlier communication by Jullian but without a specific date.

47. Camille Jullian, "Glozel," *Les Nouvelles littéraires, artistiques et scientifiques*, 20 November 1926.

48. ADA, 36 J 7, Lacarelle to Brinon, 8 January 1927.

49. BI, Ms 5765, Viple to Jullian, 4 May 1926. Strictly speaking, Emile Fradin, to whom he is referring, was the "petit-fils"; he was plowing with his grandfather when they made the initial discovery in 1924.

50. Morlet's resignation letter, addressed to Marcel Génermont, the editor of the society's bulletin, and dated 1 March 1926, is in ADA, 36 J 8. This letter and other documents related to the dispute were reprinted in *Mercure de France*, 1 August 1926.

51. ADA, 36 J 8, Espérandieu to Joseph Loth, 5 May 1925; de Brinon to Espérandieu, 14 August 1925.

52. BI, Ms 5765, Arnold Van Gennep to Jullian, 4 July 1926, quoting Reinach.

53. BI, Ms 5765, Reinach to Jullian, 22 July 1926 and n.d. (early August 1926 from context).

54. BI, Ms 5765, Reinach to Jullian, 26 and 27 August 1926.

55. On de Ricci's statements, which Morlet reported in *Mercure de France* in December, see René Dussaud, *Autour des inscriptions de Glozel* (Paris: Armand Collin, 1927), 10–11.

56. Lucien Chassaigne, "Les tablettes découvertes à Glozel sont-elles gallo-romaines ou préhistoriques?" *Le Journal*, 6 September 1926; the article takes into account de Ricci's view that the discoveries were fakes, but frames Jullian's and Reinach's positions as the main ones in the controversy. On Reinach's report to the Academy on 10 September, see "Académie des Inscriptions: Les fouilles de Glozel," *Le Journal*, 11 September 1926.

57. Comte Bégouën, "Les fouilles de Glozel," *Journal des débats*, 9 September 1926. Bégouën was a newspaper owner and journalist, but he had studied with one of the founders of French prehistory, Emile Cartailhac, and lectured on prehistory at the University of Toulouse.

58. ADA, 36 J 8, Clément to de Brinon, 25 April, 19 May, and 3 and 11 December 1925. On Capitan (1854–1929), see Gran-Aymerich, *Les chercheurs de passé*, 668–69. Capitan was an early visitor to the site, but Morlet and others suspected him of wanting to take it over as scientific director; see BI, Ms 5765, Morlet to Jullian, 12 November 1925.

59. BMA, Pottier to Reinach, 14 September 1926; he is responding to a letter from Reinach: "on jette sa langue au chat." For Pottier's (1855–1934) biography, see Gran-Aymerich, *Les chercheurs de passé*, 1073–74.

60. On civility as underpinning modern scientific exchange, see Steven Shapin and Simon Schaffer, *Leviathan and the Air-Pump: Hobbes, Boyle, and the Experimental Life*, 2nd ed. (1985; Princeton, NJ: Princeton University Press, 2011), 72–76. On the significance of Pottier's work, see Gran-Aymerich, *Les chercheurs de passé*, 293, 299–300.

61. SRA, 1, Bégouën to Audollent, 21 May 1927.

62. SRA, 1, Audollent to Bégouën (draft with corrections), 4 September 1927, and reply, 9 September 1927.

63. SRA, 1, Bégouën to Audollent, 21 May 1927. Leite de Vasconcelos had in fact already commented publicly on the similarities between the Glozel tablets and those of Alvão; see Chassaigne, "Les tablettes découvertes à Glozel," *Le Journal*, 6 September 1926. On Leite de Vasconcelos's private endorsement, BMA, Pottier to Reinach, 14 September 1926.

64. BI, Ms 5765, Mendes Corrêa to Jullian, 15 October 1926.

65. BI, Ms 5765, Mendes Corrêa to Jullian, 4 June and 10 September 1927.

66. On the Scandinavian tests, see BMA, Morlet to Reinach, 26 November 1926; Joseph Loth to Reinach, 1 and 12 December 1927, 15 and 20 February 1928; for the anti-Glozelian view, BI, Ms 4780, Claude Schaeffer to Dussaud, 25 November 1927.

67. BI, Ms 5765, Mendes Corrêa to Jullian, 10 September 1927.

68. BI, Ms 5765, Mendes Corrêa to Jullian, 15 October and 20 December 1926, and 4 June 1927; the greeting in the 1927 letter is "Monsieur et très éminent Collègue."

69. SRA, 1, Mendes Corrêa to Audollent, 21 January 1928.

70. SRA, 1, Bégouën to Audollent, 9 September 1927; ADA, 36 J 8, Breuil to André Vayson de Pradenne (copied by the latter), 2 August 1927. Breuil's letter was published by Vayson de Pradenne in the *Bulletin de la Société préhistorique de France*; see Arnaud Hurel, *L'abbé Breuil: Un préhistorien dans le siècle* (Paris: CNRS, 2011), 314–15.

71. BMA, Pottier to Reinach, 23 July 1927.

72. See Paul Bringuier, "Les découvertes néolithiques de Glozel ne seraient-elles qu'une mystification?" *Le Journal*, 19 September 1927. Dussaud gives his own account of the *comité secret* meeting in a long letter to Jullian, who was not present, dated 17 September 1927, in BI, Ms 5764. For reactions to Dussaud's intervention, see BI, Ms 4850, Pottier to Dussaud, 24 September 1927, and BI, Ms 5765, Reinach to Jullian, 18 September 1927.

73. BI, Ms 4850, Reinach to Dussaud, "Jeudi" (22 September 1927 from context).

74. BI, Ms 4850, Reinach to Dussaud, "Mercredi" (late October 1927 from context). Loth's letter, of 25 October 1927, is in BMA, and Reinach quotes it accurately.

75. BI, Ms 5764, Dussaud to Jullian, 10, 11, and 13 October 1927; quotation from 10 October. The declaration that he was breaking off relations comes in a postscript to the 13 October letter.

76. BI, Ms 5764, Dussaud to Jullian, 10 October 1927.

77. BMA, Cagnat to Reinach, 20 September 1927.

78. BI, Ms 5764, Dussaud to Jullian, 28 October 1927. Dussaud singled out Stéphane Gsell for mention as especially critical of Glozel; he noted that Pottier and the medievalist Emile Mâle were still undecided.

79. BI, Ms 5764, Dussaud to Jullian, 13 October 1927 (on ministries and museums), and 1 January 1928.

80. BI, Ms 4850, Jullian to Dussaud, 8 October 1927. Jullian began publishing his views on Glozel in one of his regular rubrics, "Notes gallo-romaines," in *Revue des études anciennes* 29, no. 2 (April 1927), under the title "Au champ magique de Glozel," and did so in each issue through the end of 1928, seven issues in all.

81. Jullian, "Chronique gallo-romaine," *Revue des études anciennes* 28, no. 1 (January 1926). The "Chronique gallo-romaine" was Jullian's other regular column, offering short news items rather than the longer, single-topic essays in the "Notes." Jullian makes disparaging comments about the Glozelian side in letters to Dussaud in BI, Ms 4850, dated 12, 27, and 29 October 1927, among others.

82. BI, Ms 4850, Jullian to Dussaud, 8 October 1927.

83. BMA, Pottier to Reinach, 25 December 1927. See also BI, Ms 5765, Peyrony to Jullian, 20 November 1927.

84. BMA, Cagnat to Reinach, 20 September 1927.

85. BI, Ms 4850, Reinach to Dussaud, "vendredi soir" (likely early to mid-1927).

86. BMA, Loth to Reinach, 25 October 1927; BI, Ms 5764, Dussaud to Jullian, 30 December 1927.

87. SRA, 4, Morlet to Audollent, 14 and 21 December 1927.

88. BMA, Peyrony to Reinach, 4 January 1928; SRA, 1, Bernard Faÿ (?) to Audollent, 14 January 1928. Faÿ was writing from Copenhagen, where he claimed Danish scholars were indignant about Dussaud's anti-Glozel campaign.

89. BI, Ms 5764, Dussaud to Jullian, 14 October 1927.

90. BMA, Peyrony to Reinach, 20 December 1927. The "Institute" to which Peyrony refers is the Institut de France, comprising the AIBL and four other academies including the Académie française.

91. *Le nouveau Petit Robert* (Paris: Dictionnaires Robert, 1993); "bande" and "cabale" are given as synonyms.

92. Serge Joannidès, "L'affaire de Glozel," *L'avenir*, 24 December 1927.

93. "Le docteur Morlet va répondre à la Commission internationale," *Le Quotidien*, 29 December 1927.

94. G. Archambault, "L'enseignement de Glozel: Quand le renne vivait en Limagne," *Paris-Soir*, 13 November 1927. On the *fait divers*, see Alain Monestier, ed., *Le fait divers*, catalog of an exhibition at the Musée national des arts et traditions populaires, Paris, 1982–1983.

95. BI, Ms 4850, Alfred Boissier to Dussaud, 27 October 1927, second part.

96. A. Vayson de Pradenne, "Chronologie de Glozel," *Bulletin de la Société préhistorique française* 24, no. 9 (September 1927): 300.

97. "Autour d'une controverse scientifique: Le gisement de Glozel," *Le Temps*, 30 November and 4 December 1927.

98. See Henri Breuil, "Henri Bégouën (1863–1956)," *Bulletin de la Société préhistorique française* 54 (1957): 78–81.

99. "Une lettre de M. le comte Bégouën," *Mercure de France*, 1 August 1927, 708.

100. BI, Ms 5764, Dussaud to Jullian, 13 October 1927 ("Le Figaro où les Reinach ont des intérêts"; BI, Ms 4850, Paul Bringuier to Dussaud, 22 September 1927 (date conjectural; several stamps on the document make it difficult to read).

101. BMA, Morlet to Reinach, 9 and 13 February 1928.

102. SRA, 4, Reinach to Audollent, 12 November 1927.

103. BI, Ms 5764, Dussaud to Jullian, 1 January 1928 .

104. J.L., "Autour d'une controverse scientifique: Le gisement de Glozel," *Le Temps*, 21 November 1927.

105. SRA, 4, Lefranc to Audollent, 17 November 1927.

106. The extensive correspondence is in SRA, 4, Drogan (spelling uncertain), editor to Audollent, October–November 1927; the article is Auguste Audollent, "L'énigme de Glozel," *Le Correspondant* 291 (October–December 1927): 440–62, 440.

107. "Une lettre du docteur Morlet," *Journal des débats*, 30 January 1928. See Matthew 7:3–5 and Luke 6:41. "Mote" and "beam" are from the King James Version; revised versions substitute "speck" and "plank," but the basic point is, why obsess over microscopic flaws in your neighbor while ignoring huge ones of your own.

108. Emile Dermenghem, "Le mystère de Glozel: M. Marcelin Boule donne pour la première fois son opinion," *L'Information*, 27 October 1927.

109. V. Méric, "Mon point de vue: La leçon de Glozel," *Le Soir*, 8 January 1928. Méric goes on to wonder what would happen if such minor disagreements were to arise in medicine, preventing doctors from treating a patient.

110. Raoul Bouillerot, "Causerie archéologique: Une opinion sur Glozel," *Progrès de la Cote d'Or*, 25 November 1927; [title obscured], *Manchester Guardian*, 23 November 1927.

111. "L'affaire Glozel: Le monde archéologique en emoi," *La Meuse* (Liège), 3 November 1927.

112. *Le Matin*, 9 November 1927.

113. Marcel Sauvage, "Les problèmes que soulève l'affaire de Glozel," *Réforme*, 5 November 1927. The Argus identifies the place of publication for this paper as Alexandria, but Sauvage's main affiliation was with the Paris daily *L'Intransigeant*, and it is likely this piece was syndicated.

114. "Une interview du C^t Espérandieu: L'authenticité des objets découverts à Glozel est affirmée," *Petit méridional* (Montpellier), 14 November 1927; Eugène Marsan, "De Tauride en Glozel," *Le Figaro*, 30 December 1927.

115. S. Reinach, "Le mirage oriental," *Anthropologie* 4 (1893): 539–78, 699–732.

116. Martin Bernal, *Black Athena: The Afroasiatic Roots of Classical Civilization*, vol. 1, *The Fabrication of Ancient Greece, 1785–1985* (New Brunswick, NJ: Rutgers University Press, 1987), 370–73.

117. Dussaud, *Autour des inscriptions de Glozel*, 25–33.

118. BI, Ms 5764, Dussaud to Jullian, 17 October 1927.

119. BI, Ms 4850, Reinach to Dussaud, 22 September 1927.

120. BI, Ms 5765, Reinach to Jullian, 9 September 1926: "Fichus, les Phéniciens! Coulé, Bérard!" Dussaud dedicated *Autour des inscriptions* to Victor Bérard.

121. On the Saitaphernes affair, see James McAuley, *The House of Fragile Things: Jewish Art Collectors and the Fall of France* (New Haven, CT: Yale University Press, 2021), 73–77.

122. "La bataille pour Glozel: Les antiglozéliens chahutent le cours du professeur Loth," *Petit Journal*, 22 January 1928.

123. BI, Ms 5764, Dussaud to Jullian, 13 October 1927.

124. A. Vayson de Pradenne, *Les fraudes en archéologie préhistorique, avec quelques exemples de comparaison en archéologie générale et sciences naturelles* (Paris: É. Nourry, 1932), 519–73, 571. This substantial volume (675 pages) by one of the leading anti-Glozelians famously contains no mention of Glozel, perhaps for legal reasons; Reinach is, however, skewered in the chapter on the tiara. See Pierre-Paul Bonenfant's introduction to the 1992 edition of *Les fraudes en archéologie* (Grenoble: J. Millon, 1992), "Pour une critique archéologique," 7–10. Bonenfant sees Glozel as the impulsion for Vayson's interest in the larger topic of archaeological fraud.

125. BMA, Pottier to Reinach, 5 October 1927 and 20 October 1928.

126. On omniprésence, see René Aigrain, "Pourquoi y a-t-il une 'énigme' dans l'affaire de Glozel," *Journal de l'Ouest* (Poitiers), 2 December 1927; for veiled anti-Semitism, "Professeur Dusssaud: Antiglozélien," *Carnet de la Semaine*, 15 January 1928; L. Daudet, "Glozel or not Glozel?" *Action française*, 11 and 31 December 1927.

127. Marcel Sauvage, "Glozel: M. Loth annonce des révélations," *L'Intransigeant*, 5 January 1928.

128. Marc Bloch, "Réflexions d'un historien sur les fausses nouvelles de la guerre," *Revue de synthèse historique* 97–99 (1921): 13–35. See also Hélène Guillot, "Fausses nouvelles et Première Guerre mondiale: L'usage de la photographie," in *Les fausses nouvelles: Un millénaire de bruits et de rumeurs dans l'espace public français*, ed. Philippe Bourdin and Stéphane Le Bras (Clermont-Ferrand: Presses universitaires Blaise Pascal, 2018), 143–57.

129. Bloch, "Les fausses nouvelles," 17, 31.

130. See Leonard V. Smith, "Le récit du témoin: Formes et pratiques d'écriture dans les témoignages sur la Grande Guerre," trans. Christophe Prochasson, in *Vrai et faux dans la Grande Guerre*, ed. Prochasson and Anne Rasmussen (Paris: La Découverte, 2004), 295–300; Smith, *The Embattled Self: French Soldiers' Testimony of the Great War* (Ithaca, NY: Cornell University Press, 2007), chap. 1.

131. Gaëtan Sanvoisin, "La commission internationale se prononce contre l'authenticité du gisement," *Le Gaulois*, 24 December 1927; Paul Voivenel, "La glozélite," *La rumeur*, 1 January 1928.

132. Bloch, "Les fausses nouvelles," 15, 29–30. The *s* in "Braisne" would be silent in French.

133. "Vendredi," *Carnet de la Semaine*, 13 November 1927.

134. Stéphane Le Bras, "L'histoire et les historiens face aux fausses informations: identifier, décrypter, et exploiter," in Bourdin and Le Bras, *Les fausses nouvelles*, 199.

135. Vincent Michelot, "De 'fausses nouvelles' à '*fake news*': Itinéraire sémantique américain," in Bourdin and Le Bras, *Les fausses nouvelles*, 179–86, esp. 181–82.

136. Dorothy Garrod, "Recollections of Glozel," *Antiquity* 42 (1968): 174; Henry de Varigny, "L'enquête de Glozel, quatrième journée," *Journal des débats*, 10 November 1927, reprinting the text of the statement; it is also included in the report.

137. "Des pièces rares au tableau," *Loire républicain*, 8 November 1927.

138. "L'authenticité du gisement de Glozel est reconnue à l'unanimité par la commission internationale," *Le Matin*, 8 November 1927; "Scientists' Rich Finds at Glozel: Ancient History Rewritten, Stone-Age Man's Alphabet," *Daily Mail*, 7 November 1927.

139. *Le Quotidien*, 7 November 1927: "Une tête de renne gravée sur un galet atteste l'authenticité du gisement de Glozel."

140. BI, Ms 5764, Reinach to Jullian, 26 August 1927. Reinach's handwriting is extremely difficult to decipher, so the translation represents the sense, with a few words omitted.

141. BI, Ms 5764, Reinach to Jullian, 27 August 1927.

142. "Le gisement préhistorique de Glozel," *Le Matin*, 9 November 1927.

143. "Autour des fouilles de Glozel: Oui, mais que signifie l'expression: 'Nous sommes fixés'?" *L'Oeuvre*, 11 November 1927.

144. "Autour de Glozel: L'opinion de M. le Professeur Bégouën," *L'express du Midi* (Toulouse), 11 November 1927.

145. "Fouilles de Glozel: Rapport de la Commission Internationale," *Revue anthropologique*, nos. 10–12 (1927), supplement, 389–416.

146. "Les fouilles du 'Matin' à Glozel: De vieilles galeries obturées sont ouvertes à la 'Goutte' Barnier; Les Fradin ne veulent pas être traités de faussaires," *Le Matin*, 6 January 1928.

147. "Pour ou contre Glozel," *La Rumeur*, 9 January 1928; Maurice-Verne, "Le petit Glozel de Mistral," *Paris-Soir*, 16 February 1928.

148. "Toujours Glozel," *L'Humanité*, 7 January 1928.

149. See Laurent Martin, *Le canard enchaîné: Histoire d'un journal satirique (1915–2005)* (Paris: Nouveau Monde, 2005); on its political stance between the wars, see esp. 104. For an excellent concise summary, see Pierre Taminiaux, "*Le canard enchaîné*," in *The Columbia History of Twentieth-Century French Thought*, ed. Lawrence D. Kritzman (New York: Columbia University Press, 2006), 690–92.

150. Pierre Bénard, "Les membres de la Commission internationale ont fait, à Glozel, une sérieuse enquête," *Canard enchaîné*, 15 November 1927.

151. "D'une mode néolithique," *Canard enchaîné*, 14 December 1927.

152. Jules Rivet, "Le lotissement de Glozel: Le *Canard* a, lui aussi, entrepris des fouilles préhistoriques," *Canard enchaîné*, 11 January 1928.

153. BI, Ms 5765, Viple to Jullian, 4 May and 31 July 1926.

154. See, for example, the photo caption to "Fradin Père et Fils, Conservateurs du Musée de Glozel," *L'Humanite*, 6 January 1928.

155. For examples of these rumors, see BI, Ms 5765, Pottier to Jullian, 25 and 30 July 1927; Dussaud to Jullien, 17 October 1927.

156. BMA, Espérandieu to Reinach, 28 February 1928.

157. On the Hauser affair, see Hurel, *La France préhistorienne*, 149–77, and on its insertion into the Glozel controversy, 164–65; for Morlet's role in spreading these rumors, see BMA, Morlet to Reinach, 1 and 10 April 1929. The charges about Peyrony were for the most part accurate, but a 1910 investigation had cleared him of wrongdoing.

158. BI, Ms 4850, St. Just Péquart to Dussaud, 20 November 1927.

159. *Le Quotidien* was close to the Cartel des Gauches; see Claude Bellanger, Jacques Godechot, Pierre Guiral and Fernand Terrou, eds., *Histoire générale de la presse française*, 5 vols. (Presses universitaires de France, 1969–1973), 3:569–71.

160. Charles Dauzats, "Le rapport de la commission internationale," *Le Figaro*, 24 December 1927.

161. The most comprehensive article on connections between the church and the anti-Glozelian position is "Le glas de Glozel," *Cité*, 8 January 1928. See also Albert Bayet, "L'Eglise et la science ou chacun chez soi," *Ere Nouvelle*, 9 January 1928 (the clipping date is unclear), which attacked an article in *La Croix* (Diégo, "Les civilisés préhistoriques," 30 December 1927) offering an interpretation of the "civilized" nature of prehistoric man compatible with the Bible. Bayet's piece provoked a reply from "Diégo": "Les dogmes de M. Bayet," *La Croix*, 27 January 1928.

162. "La bataille continue autour de Glozel," *La Liberté*, 25 December 1927; "La comédie de Glozel," *Le Temps*, 25 December 1928. On *La Liberté*'s political orientation, see Bellanger et al., *Histoire générale de la presse*, 3:521.

163. "A propos de Glozel," *La Nation*, 31 December 1928. Bellanger et al., *Histoire générale de la presse*, 3:296, lists *La Nation* as a "journal-fantôme," without any significant circulation, in 1910–1912, but does not mention any political orientation; at the very least, the article indicates it was to the right of the center-left Radicals.

164. "Dans les coulisses de l'archéologie," *La Rumeur*, 10 February 1928; Jean Cabrerets, "A Glozel, chez les Fradin," *Le Quotidien*, 4 January 1928.

165. BMA, Morlet to Reinach, 12 March 1929 and 5 June 1929. Morlet's religious views are discussed further in chapter 5.

166. See Aron Rodrigue, "Totems, Taboos, and Jews: Salomon Reinach and the Politics of Scholarship in Fin-de-Siècle France," *Jewish Social Studies* 10 (2004): 7.

167. "Autour des procès," *Comoedia*, 11 January 1928. The placement of the clipping on the sheet makes it difficult to identify the date with any certainty.

168. Without specifically referring to Dreyfus, an article in *Le Figaro* helpfully observed, "The big difference between a *question* and an *affair* is that one is a discussion and the other is a fight," and Glozel had clearly become the latter; Artigny, "Billet du Matin," *Le Figaro*, 9 January 1928.

169. "Glozel et la musique," *Comoedia*, 9 January 1928; untitled article, *Le National*, 15 January 1928.

170. "J'Accuse Emile Fradin, déclare M. Peyrony, conservateur du musée des Eyzies," *Le Journal*, 7 January 1928.

171. Dussaud claimed that his epigraphic skills exposed Fradin's forgery; BI, Ms 5764, Dussaud to Jullian, 1 January 1928.

172. BMA, Peyrony to Reinach, 20 December 1927 and 4 January 1928. On the allegory of truth as a naked woman emerging from a well, see Norman L. Kleeblatt, ed., *The Dreyfus Affair: Art, Truth and Justice* (Berkeley: University of California Press, 1987), figs. 53 and 54.

173. BI, Ms 5764, Dussaud to Jullian, 10 October 1927. Esterhazy was the army officer

whose clumsy forgery, designed to prove Dreyfus's guilt, made evident the conspiracy against Dreyfus.

174. Henri Simoni, "Ainsi que *l'Oeuvre* avait prévu, la Commission internationale a conclu à la 'non-ancienneté' des documents de Glozel," *L'Oeuvre*, 24 December 1927. On *L'Oeuvre*'s political positioning and reputation, see Charle, *Le siècle de la presse*, 252; Bellanger et al., *Histoire générale de la presse*, 3:566–67.

175. D., "La belle affaire," *L'Oeuvre*, 14 January 1928.

176. "Rapport de la Commission," 413.

177. Garrod, "Recollections of Glozel," 172–77.

178. Garrod, "Recollections of Glozel," 172.

179. "Rapport de la Commission," 390, 400–401.

180. "Rapport de la Commission," 413.

181. "Rapport de la Commission," 413.

182. BI, Ms 5764, Grenier to Jullian, 7 November 1926; Dussaud to Jullian, 14 October 1927.

183. BI, Ms 4850. See, for example, André Vayson de Pradenne to Dussaud, 26 November 1927.

184. On the carnival float in Aix, see *Le petit provençal*, 20 and 22 February 1928; for a photograph, *Excelsior*, 22 February 1927. On the "bal glozélien" in Nice, *Gazette de Nice*, 15 February 1928; *Petit niçois*, 26 February 1928. For the student ritual, a "monôme," see "Les étudiants s'amusent," *Le Quotidien*, 15 January 1928. On the variety show, called *Kif-kif Bouricot*, see reviews in *La liberté*, 4 February 1928; *Comoedia*, 4 February 1928; and *L'Oeuvre*, 22 Feburary 1928. On the Salon des Humoristes, *L'Oeuvre*, 3 March 1928. Some of these performances are discussed in greater detail in chapter 4.

185. There was considerable press coverage of the search, known in French by the judicial term *perquisition*; see, for example, "Les perquisitions de Glozel," *Journal des débats*, 27 February 1928; "Le gisement de Glozel," *Le Temps*, 28 February 1928; Pierre Guitet-Vauquelin, "Un coup de théâtre à Glozel," *Le Matin*, 26 February 1928. For reactions of the Glozelians, see BMA, Espérandieu to Reinach, 28 February 1928 (two letters); BMA, Morlet to Reinach, 2 March 1928; SRA, 4, Reinach to Audollent, 27 February 1928. Morlet was also widely quoted in the press; see, for example, his open letter to the minister of Justice, *L'Intransigeant*, 8 March 1928.

186. Paul Bringuier, "M. Beyle [sic], chef de l'identité judiciaire conclut que Glozel est faux," *Le Journal*, 5 October 1928, clipping in ADA, 36 J 5.

187. The most thorough account of Bayle's tests, which subjected sample Glozel tablets to heating, immersion in water, and microscopic examination, is "Un glozélien savant géologue entend M. Bayle et constate la non authenticité des tablettes saisies chez les Fradin," *Le Journal*, 29 May 1929; see also "Edmond Bayle 1879–1929," *Bulletin de la Société préhistorique française* 26 (1929): 433–34. I have found no evidence that Bayle published the report either on its own or in a periodical, and he was murdered in September 1929 by a disgruntled litigant.

188. The report's submission was covered in all the major dailies; see, for example, Robert Gauthier, "L'affaire de Glozel: Le rapport de M. Bayle est remis au juge d'instruction," *Le Temps*, 11 May 1929; Henri Simoni, "Et voici qu'on reparle de Glozel," *L'Oeuvre*, 11 May 1929. On the Glozelians' response and that of the Société préhistorique, "La querelle de Glozel," *Le Progrès de Lyon*, 12 May 1929; "La querelle glozélienne: Le rapport de la Société préhistorique française," *L'ère nouvelle*, 13 May 1929; Augustin Suisse, "Depuis l'autre semaine: L'hydre fabuleuse de Glozel entre dans une de ces crises périodiques de fureur," *Le populaire*,

15 May 1929; "L'affaire de Glozel: Déclarations de M. Morlet," *Le Temps*, 22 May 1929 (all citations except *Le Temps* and *L'Oeuvre* from clippings in SRA, 7).

189. BMA, Morlet to Reinach, 23 March and 11 October 1928, 13 January 1929; Loth to Reinach, 20 March 1928.

190. BMA, Morlet to Reinach, 15 February and 7 May 1929.

191. Carcopino was elected to the AIBL a year later. For his candidacy, see BI, Ms 7158, Alfred Merlin to Carcopino, 6 December 1929. It was not uncommon for scholars to pose their candidacies several times; this had been Merlin's own case.

192. BMA, Carcopino to Reinach, 29 January and 3 February 1929.

193. Jérôme Carcopino, *Souvenirs de sept ans, 1937–1944* (Paris: Flammarion, 1953).

194. BMA, Carcopino to Reinach, 3 February 1929.

195. Laurent Olivier, "Du musée des antiquités nationales au musée d'Archéologie nationale," in *La fabrique de l'archéologie en France*, ed. Jean-Paul Demoule and Christian Landes (Paris: La Découverte, 2009), 91.

196. "Une conférence de M. Dussaud à Moulins," *Le Figaro*, 4 January 1928.

197. For the text of the legislation, including subsequent modifications, see https://www.legifrance.gouv.fr/loda/id/LEGITEXT000006074257.

198. See Jean-Pierre Reboul, "Genèse et postérité des lois Carcopino," in *La fabrique de l'archéologie en France*, 120–33. On the dispute over authorship, see Elizabeth Campbell Karlsgodt, *Defending National Treasures: French Art and Heritage under Vichy* (Stanford, CA: Stanford University Press, 2011), 135–36. Although subsequently modified, the Carcopino laws remain the basis of French jurisprudence about archaeology to this day.

Chapter 3

1. BMA, Morlet to Reinach, 12 March 1927.

2. BMA, Morlet to Reinach, 19 March 1927, 2 April 1928, 4 and 5 March 1929.

3. BMA, Morlet to Reinach, 24 March 1928, 17 May 1929.

4. BMA, Espérandieu to Reinach, 3 and 14 March 1932. On the sense of the word *fatigué* as "ill," see Laurence Wylie, *Village in the Vaucluse*, 3rd ed. (Cambridge, MA: Harvard University Press, 1974), 187–88.

5. Elizabeth Barry, "The Ageing Body," in *The Cambridge Companion to the Body in Literature*, ed. David Hillman and Ulrike Maude (New York: Cambridge University Press, 2015), 135, https://doi.org/10.1017/CCO9781107256668.

6. Thomas Csordas, "Cultural Phenomenology: Embodiment: Agency, Sexual Difference, and Illness," in *A Companion to the Anthropology of the Body and Embodiment*, ed. Frances E. Mascia-Lees (Oxford: Wiley-Blackwell, 2011), 137, https://doi.org/10.1002/9781444340488.ch8.

7. Tobin Siebers, *Disability Theory* (Ann Arbor: University of Michigan Press, 2008), 68.

8. Chris Mounsey, "Introduction: Variability: Beyond Sameness and Difference," in *The Idea of Disability in the Eighteenth Century*, ed. Mounsey (Lewisburg, PA: Bucknell University Press, 2014), 16–17.

9. Jason S. Farr, *Novel Bodies: Disability and Sexuality in Eighteenth-Century British Literature* (Lewisburg, PA: Bucknell University Press, 2019), 3.

10. Siebers, *Disability Theory*, 23.

11. Steven Shapin and Christopher Lawrence, "Introduction," in *Science Incarnate: Historical Embodiments of Natural Knowledge*, ed. Lawrence and Shapin (Chicago: University of Chicago Press, 1998), 3–4.

12. Christopher Lawrence, "Medical Minds, Surgical Bodies: Corporeality and the Doctors," in Lawrence and Shapin, *Science Incarnate*, 156–201, esp. 193.

13. Bonnie G. Smith, *The Gender of History: Men, Women, and Historical Practice* (Cambridge, MA: Harvard University Press, 1998), 116–29, 127. I am abridging Smith's complex argument about obsession with the archival document and fetishism as historical process.

14. Nathalie Richard, *Inventer la préhistoire: Les débuts de l'archéologie préhistorique en France* (Paris: Vuibert, 2008), 118.

15. See, for example, Bjørnar Olsen, Michael Shanks, Timothy Webmoor, and Christopher Witmore, *Archaeology: The Discipline of Things* (Berkeley: University of California Press, 2012). It is worth noting that the authors are calling for a *return* to a central concern for material things after a period in which the field worked within broader realms of theory and interpretation influenced by poststructuralism.

16. On the community surrounding historians' work, see Joyce Appleby, Lynn Hunt, and Margaret C. Jacob, *Telling the Truth about History* (New York: Norton, 1994), 229.

17. See, for example, Michael Shanks, "Photography and Archaeology," in *The Cultural Life of Images: Visual Representation in Archaeology*, ed. Brian Molyneaux (London: Routledge, 1997), 73–107; Frederick N. Borher, *Photography and Archaeology*, Exposures (London: Reaktion, 2011).

18. Christina Riggs, *Photographing Tutankhamun: Archaeology, Ancient Egypt, and the Archive* (London: Bloomsbury Visual Arts, 2018), 143.

19. On these asymmetries, see Riggs, *Photographing Tutankhamun*, 144; she is here summarizing concepts worked out by Bruno Latour.

20. FP, 033, Merlin to Poinssot, 8 December 1911.

21. Ève Gran-Aymerich, *Les chercheurs de passé, 1789–1945: Aux sources de l'archéologie* (Paris: CNRS, 2007), 821: "Sa situation étant difficile à Tunis, où il se heurte à l'hostilité d'une partie de l'administration locale et de certains particuliers, il donne sa démission en 1905, revient en France et obtient une mission de l'Instruction publique à Rome."

22. Myriam Bacha, *Patrimoine et monuments en Tunisie: Sauvegarde et mise en valeur pendant le Protectorat, 1881–1920* (Rennes: Presses Universitaires de Rennes, 2013), 235–36.

23. FP, 190, dossier 2, folder 4, Witness statements (here anonymized) of C, 4 May 1905 (streetcar ride, books), and T, 7 May 1905.

24. FP, 190, dossier 2, folder 4, Statements of M, 16 May 1905 (approached by intermediary); S, 8 May 1905 (saw other men entering Gauckler's house but claims never to have entered himself); and T, 7 May 1905 (saw other men entering house).

25. FP, 190, dossier 2, folder 4, Statement of T, 7 May 1905.

26. FP, 190, dossier 2, folder 4, Léal report, 10 July 1905.

27. FP, 190, dossier 2, folder 4, Statement of C, 4 May 1905.

28. FP, 190, dossier 2, folder 4, Bayet to Résident Général, Tunis, 11 July 1905 (fair copy).

29. FP, 190, dossier 2, folder 4, Pichon to Bayet, 16 July 1905 (fair copy). See Joann Freed, "Father Alfred-Louis Delattre (1850–1932) versus Paul Gauckler (1866–1911): The Struggle to Control Archaeology at Carthage at the Turn of the Twentieth Century," in *Life-Writing in the History of Archaeology: Critical Perspectives*, ed. Clare Lewis and Gabriel Moshenska (London: UCL Press, 2023), 250–53, https://www.jstor.org/stable/j.ctv37mk2fp.

30. Farr, *Novel Bodies*, 2.

31. See, for example, Eve Kosofsky Sedgwick, *Epistemology of the Closet* (Berkeley: University of California Press, 1990), 2; Heike Bauer, "Literary Sexualities," in Hillman and Maude, *Cambridge Companion to the Body in Literature*, 102–6; Antony Copley, *Sexual*

Moralities in France, 1780–1980: New Ideas on the Family, Divorce, and Homosexuality: An Essay on Moral Change (London: Routledge, 1989), 135–47.

32. See Copley, *Sexual Moralities*, 24.

33. Vernon Rosario, "Pointy Penises, Fashion Crimes, and Hysterical Mollies: The Pederast's Inversions," in *Homosexuality in Modern France*, ed. Jeffrey Merrick and Bryant T. Ragin (New York: Oxford University Press, 1996), 146–63, 154.

34. Robert Aldrich, *The Seduction of the Mediterranean: Writing, Art and Homosexual Fantasy* (London: Routledge, 1995), 124–25.

35. See Aldrich, *Seduction*, 15–16; Joseph Allen Boone, *The Homoerotics of Orientalism* (New York: Columbia University Press, 2014), 54–77.

36. Edward W. Said, *Orientalism* (New York: Random House, 1978); the most frequently cited passage concerning sexual tourism is on 190. For a Maghrebi sociological perspective on recent North African same-sex sexuality, see Malek Chebel, *L'esprit de sérail: Mythes et pratiques sexuels au Maghreb*, rev. ed. (Paris: Payot, 1995), 13–52.

37. On von Gloeden, see Aldrich, *Seduction*, 143–45; Boone, *Homoerotics*, 271–75. Boone also discusses a Tunis photographic firm, Lehnert and Landrock, which had a line of erotic photographs catering to all tastes, but it was established only in 1905, the year of Gauckler's departure.

38. Jonathan Dollimore, *Sexual Dissidence: Augustine to Wilde, Freud to Foucault* (Oxford: Oxford University Press, 1991), 335–39, 338.

39. Boone, *Homoerotics*, 290–97; Dollimore, *Sexual Dissidence*, 340–44.

40. FP, 190, dossier 2, folder 4, Statements of C and M.

41. Jarrod Hayes, *Queer Nations: Marginal Sexualities in the Maghreb* (Chicago: University of Chicago Press, 2000), 23–24.

42. Dollimore, *Sexual Dissidence*, 337–38.

43. Sedgwick, *Epistemology*, 8–11, 11.

44. FP, 190, dossier 2, folder 4, Gauckler to Bernard Roy, 28 May, 4 and 13 June 1905; quote from 4 June.

45. FP, 190, dossier 2, folder 4, Pichon to Bayet, 16 July 1905; Léal, "Note au sujet de M. Gauckler," 10 July 1905.

46. FP, 190, dossier 2, folder 4, Gauckler to Roy, 13 June (mood), 12 September 1905 (death).

47. FP, 190, dossier 2, folder 4, Gauckler to Roy, 4 June 1905. *Urbi et orbi*, to the city and the world, is of course the description of the audience of papal addresses.

48. FP, 191, Carton to Gauckler, 27 February 1904. This archive contains 109 letters from Carton to Gauckler between 1893 and 1906; on the newspaper articles, 7 March 1904. See Freed, "Delattre versus Gauckler," 251.

49. FP, 190, dossier 2, folder 4, Note in Roy's hand, headed "Répondu le 9 juin.".

50. FP, 031, Gauckler to Poinssot, 31 letters between 1901 and 1907.

51. FP, 190, dossier 2, folder 4, Gauckler to Roy, 4 June 1905, and note of reply, 9 June.

52. FP, 190, dossier 2, folder 4, Gauckler to Roy, 9 July 1905.

53. Bauer, "Literary Sexualities," 103.

54. Virginia Woolf, "On Being Ill," *Criterion* 4 (1926): 36–37.

55. In this regard, René Cagnat, in some ways Roy's successor as Poinssot's protector, was obviously relieved when Louis Poinssot finally married in 1921, at the age of forty-one (an earlier engagement had been broken off); see FP, 031, Cagnat to Poinssot, 23 February 1921, in which he congratulates Poinssot and says, "l'ami Roy aurait été bien heureux" (friend Roy would have been very happy).

56. See, for example, Susan Heuck Allen, *Finding the Walls of Troy: Frank Calvert and Heinrich Schliemann at Hisarlik* (Berkeley: University of California Press, 1999); Roger Atwood, "A Monumental Feud," *Archaeology* 58, no. 4 (July–August 2005): 18–25; Craig Childs, *Finders Keepers: A Tale of Archaeological Plunder and Obsession* (New York: Little, Brown, 2010); Marilyn Johnson, *Lives in Ruins: Archaeologists and the Seductive Lure of Human Rubble* (New York: HarperCollins, 2014).

57. FKP, 79.9, Correspondence with John C. Merriam.

58. Farr, *Novel Bodies*, 14–17, 17.

59. FP, 032, Gauckler to Poinssot, 23 April and 18 May 1903 (laryngitis), 11 February 1904 (flu), 13 May 1905 (epidemic).

60. FP, 032, Gauckler to Poinssot, 5 May 1903.

61. FP, 033, Merlin to Poinssot, 31 December 1912.

62. FP, 033, Merlin to Poinssot, 22 May 1913.

63. FP, 031, Cagnat to Poinssot, 6 November 1926; Albertini to Poinssot, 11 April 1927.

64. Janet Browne, "I Could Have Retched All Night: Charles Darwin and His Body," in Lawrence and Shapin, *Science Incarnate*, 240–87, esp. 242–48.

65. Alain Corbin, "Backstage," in *A History of Private Life*, vol. 4, *From the Fires of Revolution to the Great War*, ed. Michelle Perrot, trans. Arthur Goldhammer (Cambridge, MA: Harvard University Press, 1990), 478–79, 655–57.

66. FP, 032, Marçais to Poinssot, 18 January 1936.

67. FP, 032, Marçais to Poinssot, 20 March 1928.

68. FP, 032, Lantier to Poinssot, 11 September 1923.

69. FP, 032, Lantier to Poinssot, 10 September 1921 and 5 August 1923 (the latter as Lantier was beginning his own vacation). Lantier returned to France permanently in 1926, but he continued to collaborate with Poinssot for years thereafter.

70. FP, 191, Carton to Gauckler, 18 September 1896.

71. FP, 031, Gauckler to Poinssot, 23 April 1903. I take the word *travaux* in the second sentence to refer to restoration work on some previously uncovered structures at Dougga; "en travaux" is the common French term for "under construction."

72. FP, 033, Merlin to Poinssot, 12 August 1908.

73. FP, 032, Lantier to Poinssot, 30 June 1923.

74. FP, 032, Lantier to Poinssot, 30 June 1923. On another example of the "magisterial I," see Zeynep Çelik, *About Antiquities: Politics of Archaeology in the Ottoman Empire* (Austin: University of Texas Press, 2016), 139.

75. Nick Shepherd, "'When the Hand That Holds the Trowel Is Black . . .': Disciplinary Practices of Self-Representation and the Issue of 'Native' Labour in Archaeology," *Journal of Social Archaeology* 3 (2003): 334–52.

76. Riggs, *Photographing Tutankhamun*, 149.

77. FP, 048, dossier 2; the dates on both photographs are 27 November–1 December 1904. I have compared fig. 3.4 with portrait photographs of Gauckler, and although there is a resemblance, I am unsure that the European figure is Gauckler.

78. On the punctum, see Roland Barthes, *Camera Lucida: Reflections on Photography*, trans. Richard Howard (1981; New York: Hill and Wang, 2010), 27; Shepherd, "When the Hand," 337. Shawn Michelle Smith offers an important critique of Barthes's theory of the punctum, noting that it "consistently register[s] a sensation of racial or sexual inquietude" and "asks readers to view a race-based paternalism as natural, or beside the point"; see Smith, *At the Edge of Sight: Photography and the Unseen* (Durham, NC: Duke University Press, 2013), 23–38, 23, 25.

79. Riggs, *Photographing Tutankhamun*, 149.

80. FP, 191, Carton to Gauckler, 11 July 1898.

81. FP, 033, Merlin to Poinssot, 19 August 1908 and 17 June 1910.

82. FP, 033, Merlin to Poinssot, 3 December 1919; FP 032, Lantier to Poinssot, 22 June 1925.

83. FP, 032, Lantier to Poinssot, 24 July 1925.

84. Byron Khun de Prorok, *Digging for Lost African Gods: The Record of Five Years Archaeological Excavation in North Africa* (New York: G. P. Putnam's Sons, 1926), 40–43, 57, 77, 100–102, 109.

85. Stephen Quirke, *Hidden Hands: Egyptian Workforces in Petrie Excavation Archives 1880–1924* (London: Duckworth, 2010), 94–96; Bonnie Effros, *Incidental Archaeologists: French Officers and the Rediscovery of Roman North Africa* (Ithaca, NY: Cornell University Press, 2018), 160–62. For an influential anthropological discussion of local knowledge, see Clifford Geertz, *Local Knowledge: Further Essays in Interpretive Anthropology* (New York: Basic, 1983), esp. chap. 3.

86. Prorok, *Digging for Lost African Gods*, 111–12.

87. FP, 032, G. Marçais to Poinssot, 3 November 1921.

88. FP, 032, G. Marçais to Poinssot, 17 November 1922.

89. FP, 032, G. Marçais to Poinssot, 9 July 1925.

90. FP, 032, Cagnat to Poinssot, 29 December 1920.

91. FP, 033, Merlin to Poinssot, 22 January 1917.

92. The entity's formal title was Commission de publication des documents archéologiques de l'Afrique du Nord; it is discussed further in chapter 5.

93. "Séance de la Commission de l'Afrique du Nord, 14 janvier 1908," *Bulletin archéologique du Comité des travaux historiques et scientifiques*, 1908, clxi–clxvi; for references to the other reports, see the index, 463. The pages cited are those taken up by Merlin's report; Roman numerals distinguish what are technically meeting minutes from reports sent independently and vetted by commission members prior to publication.

94. FP, 032, Marçais to Poinssot, 27 December 1921.

95. FP, 032, Merlin to Poinssot, 18 September 1909 and 1 October 1910.

96. A. Merlin and L. Poinssot, "Bronzes trouvés en mer près de Mahdia (Tunisie)," *Monuments et mémoires publiés par l'Académie des Inscriptions et Belles-Lettres, Fondation Eugène Piot*, 17, no. 1 (1909): 29–59; Merlin, "Statuettes de bronze trouvées en mer, près de Mahdia (Tunisie)," *Monuments et mémoires de la Fondation Piot*, 18, no. 1(1911): 5–18. The *Monuments Piot* was often published well after the cover date.

97. "Nouvelles du jour: Académie des Inscriptions et Belles Lettres," *Journal des débats*, 2 October 1910.

98. On the connection between discovery and publication, see Alex Csiszar, *The Scientific Journal: Authorship and the Politics of Knowledge in the Nineteenth Century* (Chicago: University of Chicago Press, 2018), 159–70.

99. This is notably a theme of Marçais's correspondence with Poinssot, for example, FP, 032, 16 May 1933, but see also FP, 032, Lantier to Poinssot, 13 August 1925.

100. FP, 031, Albertini to Poinssot, 16 January 1928. The publication in question was an article for the *Revue algérienne*, a scholarly journal Albertini had helped to revive.

101. FP, 031, Cagnat to Poinssot, 14 January (clearly between 1922 and 1925). The article in question had been announced but not yet published by a periodical of the Accademia nazionale dei Lincei.

102. FP, 032, Lantier to Poinssot, 12 February 1928, recounting a meeting with Rostovtzeff in Paris, and enclosing a copy of a note to Rostovtzeff dated 6 February 1928.

103. FP, 034, Poinssot to Rostovtzeff (typescript copy), 16 January 1928: "les découvertes

faites au cours de fouilles pour lesquelles la France n'a jamais cessé de s'imposer les plus lourds sacrifices, découvertes qui, sans le malheur du temps, eussent déjà été mises à la disposition du monde savant."

104. FP, 191, Carton to Gauckler, 27 February 1904.

105. See BMA, Carton to Reinach, 27 April 1919 and 15 April 1922.

106. FP, 031, Albertini to Poinssot, 16 January 1928; FP, 032, Lantier to Poinssot, 8 September 1922.

107. FP, 032, Marçais to Poinssot, 7 January 1935.

108. Ian Hodder, "Writing Archaeology: Site Reports in Context," *Antiquity* 63 (1989): 268–74, 272, https://doi.org/10.1017/S0003598X00075980.

109. Merlin and Poinssot, "Bronzes trouvés en mer," 34–35.

110. FP, 032, Marçais to Poinssot, 2 November 1923.

111. FP, 032, Marçais to Poinssot, 12 February 1925.

112. FP, 032, Marçais to Poinssot, 19 May 1937 and 10 February 1938.

113. On this series, see Renato Miracco and Claudio Bisogniero, eds., *Giorgio De Chirico: Myth and Archaeology*, exh. cat., Phillips Collection, Washington, DC (Milano: Silvana Editoriale, 2013); for de Chirico's description of the beginnings of the series, see *Giorgio de Chirico: La fabrique des rêves*, exh. cat., Musée d'art moderne de la ville de Paris (Paris: Paris Musées, 2009), 305.

114. On de Chirico's use of mannequins, see Briony Fer, "Surrealism, Myth and Psychoanalysis," in Fer, David Batchelor, and Paul Wood, *Realism, Rationalism, Surrealism: Art between the Wars* (New Haven, CT: Yale University Press/Open University, 1993), 188–91; Elizabeth Cowling and Jennifer Mundy, *On Classic Ground: Picasso, Léger, de Chirico, and the New Classicism, 1910–1930* (London: Tate Gallery, 1990), 81.

115. Lawrence, "Medical Minds."

116. FP, 032, Gauckler to Poinssot, 25 March and 18 February 1903, 8 February 1904, 6 and 13 May 1905.

117. Riggs, *Photographing Tutankhamun*, 146–48, 152–54.

118. Riggs, *Photographing Tutankhamun*, 146.

119. For a clear contemporary explanation of the fraud, see "Falsification de titres hongrois," *Journal des débats*, 10 November 1927. New developments in the affair occupied headlines throughout the following month.

120. On the umbrella as symbol of bourgeois respectability, see Daniel J. Sherman, *Worthy Monuments: Art Museums and the Politics of Culture in Nineteenth-Century France* (Cambridge, MA: Harvard University Press, 1989), 228–29.

121. Browne, "I Could Have Retched All Night," 253.

122. "Une maison de paysans transformée en musée," *Le Peuple*, 10 November 1927.

123. "On fouille à Glozel," *Cri de Paris*, 13 November 1927.

124. Dorothy Garrod, "Recollections of Glozel," *Antiquity* 42 (1968): 174.

125. Fradin appears in the identical outfit in a Meurisse photograph with his grandfather, the two of them studying an object presumably unearthed at Glozel; this photograph appeared in *Le Peuple* and *Paris Times*, 6 January 1928.

126. For an example, see the head shot of Peyrony in the *Petit parisien*, 5 January 1928.

127. "The Woman of the Day," *Paris Times*, 11 November 1927. For more on Garrod's career, see Pamela Jane Smith, "Dorothy Garrod, First Woman Professor at Cambridge," *Antiquity* 74 (2000): 131–36.

128. "L'enquête à Glozel de la commission internationale," *Petit parisien*, 7 November 1927; "Dans le 'champ des morts' à Glozel," *Moniteur du Puy-de-Dome*, 6 November 1927.

129. *Carnet de la Semaine*, 13 November 1927.

130. "Dans trois semaines, décision des experts de Glozel," *Le Matin*, 10 November 1927.

131. "Le gisement préhistorique de Glozel," *Le Matin*, 10 November 1927.

132. See, for example, "La dispute continue à Glozel," *La volonté*, 27 December 1927; "Les fouilles de Glozel: Ça se complique," *Victoire*, 27 December 1927, and Morlet's ex post facto account in A. Morlet, *Petit historique de l'affaire de Glozel* (1932; Marsat: Editions de la Source, 1970), 169–83.

133. "Véhemente intervention du docteur Morlet," *Le Figaro*, 27 December 1927.

134. Jean Cabrerets, "Retour de Glozel: Quelles seront les conclusions de la commission d'enquête?" *Progrès civique*, 19 November 1927. This article is noteworthy for its line drawings of commission members, including two of Garrod, and its passing allusion to the incident between Morlet and Garrod.

135. Ronda, "Glozel, film à épisodes, ou Le traître apparaît," *Homme Libre*, 29? December 1927.

136. Jules Amar, "Autour de Glozel," *La volonté*, 5 January 1928. The French and English words *garrot* (French verb *garroter*) have the same meaning.

137. See Hugues Monod, "AMAR, Jules (1879–1935). Directeur du Laboratoire de Recherches sur le travail musculaire professionnel (1913–1920)," in *Les professeurs du Conservatoire national des arts et métiers: Dictionnaire biographique 1794–1955, A–K*, ed. Claudine Fontanon and André Grelon, Histoire Biographique de l'enseignement 19 (Paris: Institut national de recherche pédagogique, 1994), 97–107, https://www.persee.fr/doc/inrp_0298 -5632_1994_ant_19_1_8396.

138. "Une lettre de M. Bosch-Gimpera," *Journal des débats*, 15 January 1928; in his reply in *Débats*, 18 January 1928, Morlet said he had simply relayed the commission's desire that this incident not be reported, but had not engaged his own word.

139. *Echo de Paris*, 28 January 1928.

140. *Echo de Paris*, 28 January 1928.

141. Garrod, "Recollections of Glozel," 174–75.

142. Garrod's application for the professorship and a recommendation from the abbé Breuil are reprinted in William Davies and Ruth Charles, eds., *Dorothy Garrod and the Progress of the Palaeolithic: Studies in the Prehistoric Archaeology of the Near East and Europe* (Oxford: Oxbow Books, 1999), 15–18. Neither mentions Glozel.

143. See Pamela Jane Smith, *A "Splendid Idiosyncrasy": Prehistory at Cambridge, 1915–50* (Oxford: Archaeopress, 2009), 69–100; William Davies, "Dorothy Annie Elizabeth Garrod (5th May, 1892–18th December, 1968): A Short Biography," in Davies and Charles, *Dorothy Garrod*, 9–10.

144. *Echo de Paris*, 28 January 1928.

145. De Varigny (1855–1934) was a respected naturalist as well as journalist, and he may well have been regarded as the dean of the reporters on the Glozel beat.

146. Garrod, "Recollections of Glozel," 174–75.

147. Elizabeth Edwards, *Raw Histories: Photographs, Anthropology and Museums* (Oxford: Berg, 2001), 16–20.

Chapter 4

1. FP, 004, dossier 2, folder 9, Louis Poinssot, directeur des antiquités et des arts de Tunisie: Contentieux.

2. FP, 004, dossier 2, folder 9, typescript memorandum by Poinssot headed "Note sur un

incident survenu le 10 Mars 1926 entre Madame CARTON et M. L. POINSSOT," 20 March 1926. Although in contemporary sources she is universally referred to as "Madame Carton," I refer to her as Thélu Carton to avoid confusion with her husband.

3. FP, 004, dossier 2, folder 9, undated notes in Poinssot's hand, headed "L'incident du poignet." On the verdict, the same file contains clippings from *Petit matin*, 19 and 24 April 1926; Poinssot was fined 3 francs for "violences légères," the charge of "tapage injurieux" (public disorder) having been dismissed.

4. See, for example, FP, 004, dossier 2, folder 9, Poinssot to Doliveux (carbon), 20 March 1926: "Je dois à la vérité de déclarer que si le 10 Mars il y a eu selon votre expression 'scandale', ce n'est pas moi qui l'ai provoqué ni par mes paroles, ni par mon attitude."

5. FP, 004, dossier 2, folder 9, Poinssot to resident general (typescript copy), 10 June 1926.

6. Hédi Dridi and Antonella Mezzolani Andreose, "'Ranimer les ruines': L'archéologie dans l'Afrique latine de Louis Bertrand," *Nouvelles de l'archéologie*, no. 128 (June 2012): 13; Clémentine Gutron, *L'archéologie en Tunisie (XIXe–XXe siècles): Jeux généalogiques sur l'Antiquité* (Paris and Tunis: Karthala/IRMC, 2010), 175–76. The Bibliothèque Nationale's "Catalogue collectif de France" has no listing for *Le croisé*; my description is based on the cited scholarship and on reviews. The daily *Tunis socialiste* serialized a parodic version of the play by someone called "Baubard de Jénec, de l'Académic-Arton," between early April and mid-May 1926.

7. Mieke Bal, *Travelling Concepts in the Humanities: A Rough Guide*, Green College Lectures (Toronto: University of Toronto Press, 2002), 175. Bal's discussion of the concept obviously is much more extensive and will be developed further below.

8. FP, 004, dossier 2, folder 9, Poinssot handwritten note, undated: "Le poignet–Il a été saisi parce que je craignais le geste feuilletonesque".

9. See Edward Berenson, *The Trial of Madame Caillaux* (Berkeley: University of California Press, 1992).

10. "Supercherie ou découverte: Pour en finir avec Glozel," *Comoedia*, 29 December 1927.

11. FP, 004, dossier 2, folder 9, Poinssot, undated note ("Le poignet–Il a été saisi").

12. FP, 004, dossier 2, folder 9, Poinssot, handwritten note, undated, headed "L'incident du poignet—Pourquoi je pouvais me considérer comme menacé."

13. FP, 004, dossier 2, folder 9, typescript memorandum of 20 March 1926.

14. FP, 004, dossier 2, folder 9, Poinssot, note ("Pourquoi je pouvais").

15. James R. Lehning, *The Melodramatic Thread: Spectacle and Political Culture in Modern France* (Bloomington: Indiana University Press, 2007).

16. FP, 004, dossier 2, folder 9, Poinssot, handwritten note headed "Les incidents du 10 mars 1926." In this document Poinssot singles out Gounot, a prominent member of the French settler community and longtime supporter of the Cartons; Gounot is the main source for the report in *Le Petit matin*, 13 March 1926, "L'incident provoqué par M. Poinssot à la Commission des Théâtres Antiques."

17. Jonathan D. Culler, "Philosophy and Literature: The Fortunes of the Performative," *Poetics Today* 21, no. 3 (2000): 503–19, 513–14; Culler's article is a shrewd and clear analysis of the scholarly literature on the performative beginning with J. L. Austin's *How to Do Things with Words*. See also Judith Butler, *Gender Trouble: Feminism and the Subversion of Identity* (New York: Routledge, 1990), 140.

18. FP, 004, dossier 2, folder 9, typescript memo, 20 March 1926: "Je suis Madame Carton, la présidente des Dames Françaises." The echo of Caillaux's "Je suis une dame" at the moment of her arrest is uncanny; Berenson, *Trial*, 2.

19. Butler, *Gender Trouble*, 136.

20. Bal, *Travelling Concepts*, 97–98.

21. Freddie Rokem, *Performing History: Theatrical Representations of the Past in Contemporary Theatre* (Iowa City: University of Iowa Press, 2000), 5–6.

22. Florence Fix, *L'histoire au théâtre, 1870–1914* (Rennes: Presses Universitaires de Rennes, 2010), 10.

23. Rokem, *Performing History*, 6.

24. Fix, *L'histoire au théâtre*, 20.

25. Roland Barthes, "L'effet de réel," *Communications* 11 (1968) 84–89; trans. Richard Howard as "The Reality Effect," in Barthes, *The Rustle of Language* (Berkeley: University of California Press, 1986), 141–48.

26. Barthes, "Reality Effect," 144–45.

27. Judith Butler, "The Force of Fantasy: Feminism, Mapplethorpe, and Discursive Excess," *Differences* 2, no. 2 (Summer 1990): 105–20, 106.

28. Barthes, "Reality Effect," 145–46.

29. Louis Carton, *De l'utilité des études archéologiques au point de vue de la colonisation dans l'Afrique du Nord* (Paris: Bibliothèque des Annales Economiques, 1890), 6.

30. Stefan Altekamp and Mona Khechen, "Third Carthage: Struggles and Contestations over Archaeological Space," *Archaeologies: Journal of the World Archaeological Congress* 9 (2013): 476.

31. Sarah Bracke, "Nostalgia's Violence," *Forum*, no. 15 (Autumn 2012), http://www.forum journal.org/article/ view/527/815. On nostalgia's varied meanings in the colonial setting, see Thomas Dodman, *What Nostalgia Was: War, Empire, and the Time of a Deadly Emotion* (Chicago: University of Chicago Press, 2018); Renato Rosaldo, "Imperial Nostalgia," *Representations*, no. 26 (Spring 1989): 107–22. Bracke both relies and expands on Rosaldo.

32. See Charles-André Julien, "Colons français et Jeunes Tunisiens (1882–1912)," *Revue d'histoire d'outre-mer* 54 (1967): 87–150, and Taoufik Ayadi, "Insurrection et religion en Tunisie: L'exemple de Thala-Kasserine (1906) et du Jellaz (1911)," in *Révolte et société: Actes du IVe Colloque d'Histoire au Présent, mai 1988* (Paris: Publications de la Sorbonne, 1989), 166–75; the latter is particularly valuable for its critical review of earlier accounts based on Protectorate officials' dismissal of the revolts as religiously motivated.

33. See Ali Mahjoubi, *Les origines du mouvement national en Tunisie (1904–1934)* (Tunis: Université de Tunis, 1982), 122–29; Kenneth J. Perkins, *A History of Modern Tunisia* (Cambridge: Cambridge University Press, 2004), 61–71.

34. For example, Nicola A. Ziadeh, *Origins of Nationalism in Tunisia* (Beirut: American University of Beirut, 1962), 73; Béchir Tlili, *Crises et mutations dans le monde islamo-méditerranéen contemporain (1907–1918)*, 2 vols. (Tunis: Université de Tunis, 1978), 1:47, 2:355.

35. Ann Laura Stoler, "Introduction: 'The Rot Remains': From Ruins to Ruination," in *Imperial Debris: On Ruins and Ruination*, ed. Stoler (Durham, NC: Duke University Press, 2013), 14.

36. Myriam Harry, "Impressions tunisiennes: Autour de l'affaire Thala-Kasserine," *Le Temps*, 23 February 1907. On reactions to this article, see Julien, "Colons français," 96.

37. "Souvenirs de Carthage," *Dépêche tunisienne*, 12 March 1906, as reprinted in *Revue tunisienne* 13 (1906): 266–67.

38. *Par les champs et par les grèves*, the account of Flaubert's 1847 walking tour of the French provinces with his friend Maxime du Camp, was not published until 1881, following Flaubert's death the previous year. See Gustave Flaubert, *Oeuvres complètes de Gustave Flaubert*, vol. 10, *Par les champs et par les grèves; Voyages et carnets de voyages 1* (Paris: Editions

de l'honnête homme, 1973), 104-05, http://ark.bnf.fr/ark:/12148/cb351133616. Carnac is well known for its multiple, extensive alignments of megaliths or standing stones; Flaubert relates with skepticism some of the speculative explanations of their placement.

39. "Souvenirs de Carthage," *Dépêche tunisienne*, 12 March 1906.

40. See Michel Autrand, *Le théâtre en France de 1870 à 1914* (Paris: Honoré Champion, 2006); Mireille Losco-Lena, *La scène symboliste (1890–1906): Pour un théâtre spectral* (Grenoble: ELLUG Université Stendhal, 2010); and, for the comparative context, Michael Hays, *The Public and Performance: Essays in the History of French and German Theater, 1871–1900* (Ann Arbor: UMI Research Press, 1981).

41. Sylvie Humbert-Mougin, "Rêveries sur le théâtre en plein air," in *Théâtres virtuels*, ed. Sylvie Triaire and Pierre Citti (Montpellier: Publications Montpellier 3, Université Paul-Valéry, 2001), 329–39.

42. Sylvie Humbert-Mougin, "La mise en scène du répertoire grec en France du Romantisme à la Belle Époque," in *Le spectaculaire dans les arts de la scène, du Romantisme à la Belle Époque*, ed. Isabelle Moindrot (Paris: CNRS, 2006), 196–204.

43. "Institut de Carthage: Excursion à Carthage (6 mars 1906)," *Revue tunisienne* 13 (1906), 265.

44. All the details on the festival are taken from the lengthy account "La fête" in *Revue tunisienne* 13 (1906). The program is reprinted between pages 430 and 431.

45. The complexities of identifying the Phoenicians will be discussed in detail in chapter 5.

46. Claude Liauzu and Pierre Soumille, "La gauche française en Tunisie au printemps 1906: Le Congrès républicain, radical et socialiste de Tunis," *Le Mouvement social*, no. 86 (January–March 1974): 58–62.

47. Carton, "La genèse; l'organisation, les critiques, les enseignements," *Revue tunisienne* 13 (1906): 468.

48. "Evocations," *Revue tunisienne* 13 (1906): 452–53.

49. Gutron, *L'archéologie en Tunisie*, 124: "qui fait office de morale à l'histoire."

50. Lehning, *Melodramatic Thread*; Françoise Vergès, *Monsters and Revolutionaries: Colonial Family Romance and Métissage* (Durham, NC: Duke University Press, 1999). On the blurring of the maternal and sisterly, see Joan Wallach Scott, *Only Paradoxes to Offer: French Feminists and the Rights of Man* (Cambridge, MA: Harvard University Press, 1996), 115.

51. Charlotte Ann Legg, *The New White Race: Settler Colonialism and the Press in French Algeria, 1860–1914* (Lincoln: University of Nebraska Press, 2021), 67–69 and fig. 3; Harry, "Impressions tunisiennes."

52. "Les discours," *Revue tunisienne* 13 (1906), 433.

53. Carton, "La genèse."

54. Commission des Fêtes de Carthage, *Rapport sur les fêtes données au Théâtre Antique, sous le patronage de l'Institut de Carthage, 2 avril 1907* (Tunis: Imprimerie Rapide, 1909), 63, 17.

55. Carton, "La genèse," 467.

56. "Evocations," *Revue tunisienne* 13 (1906): 453.

57. The contributions from government bodies included 10,000 francs from the municipality of Tunis, 2,000 from the Protectorate government, and 1,000 from the agency that promoted seasonal and long-term French settlement in Tunisia. For full listing, see *Rapport sur les fêtes*, 63.

58. *Rapport sur les fêtes*, 7 (dress rehearsal), 9 (ticketing), 44–46 (brochure).

59. *Rapport sur les fêtes*, 13; "Dernière heure," *Le Journal*, 3 April 1907 (10,000); "De Tunis," *Le Figaro*, 3 April 1907 (7,000); "Théâtres," *Le Temps*, 4 April 1907 (7,000); "Informations,"

La Justice, 5 April 1907 (more than 5,000); Maurice Pottecher, "Les fêtes de Carthage," *Le Figaro*, 9 April 1907 (4,000–5,000).

60. Charles Grandmougin, *La mort de Carthage*, reprinted in *Revue tunisienne* 14 (1907), 189–225. Note that Flaubert's *Salammbô* takes place a century earlier, during a revolt by mercenaries that follows the first Punic War.

61. Alexandra David, "Fête antique au Théâtre romain de Carthage," *Mercure de France* 67 (1 May 1907), 149.

62. *Rapport sur les fêtes*, 18–19.

63. Lucie Delarue-Mardrus, *Mes mémoires* (Paris: Gallimard, 1937), 154–55. Delarue-Mardrus states that she cannot remember the exact year of the Carthage pageant but that it was after the premiere of her *Sapho déspespérée* at Orange. The playwright is mentioned in André Segond, *Les Chorégies d'Orange: De 1869 à nos jours* (Gémenos: Autres temps, 2012), 25, as having a work performed at Orange between 1902 and 1910, but neither the title nor the specific date is given.

64. Lucie Delarue-Mardrus, *La Prêtresse de Tanit*, preface by Boz Léveillez (Reims: A L'Ecart, 1993), 19.

65. Mabel Moore, *Carthage of the Phoenicians in the Light of Modern Excavations* (London: Heinemann, 1905), 146. The text of the official catalog of the Carthage museum, published by the French commission on North African sites in 1899–1900, describes a similar figure, part of a stele, as one of a hundred excavated there, with other examples in the Louvre; *DAN* Lavigerie, 1:9–10 and plate X, no. 4.

66. Delarue-Mardrus, *La Prêtresse de Tanit*, 26: "Elle n'est plus."

67. Eric Still, "Trois voix de Carthage," *La revue nord-africaine*, 13 April 1907, reprinted in *Rapport sur les fêtes*, 17.

68. *Rapport sur les fêtes*, 19. Delvair had also played the title role in Delarue-Mardrus's *Sappho* at Orange.

69. Myriam Harry, in rubric "Théatres," *Le Temps*, 9 April 1907.

70. *Rapport sur les fêtes*, 19.

71. Carton, "La fête," *Revue tunisienne* (1906), 424.

72. See Sabatino Moscati, *Carthage: Art et civilisation* (Milan: Jaca Book, 1983), 169–71, 176 (page numbering for illustration pages is implicit); these pages include three masks from the Bardo (the first being the likely model for the poster) and, on 176, an example from the island of Mozia, the ancient Motya just off the west coast of Sicily.

73. *Rapport sur les fêtes*, 4.

74. For a description of the two other finalists, see *Rapport sur les fêtes*, 43–44.

75. The entry for this work in the official scholarly catalog of the museum posits that the priestess serves a maternal deity, not Tanit; *DAN* Lavigerie, Suppl. 1:15–18 and plate III.

76. Delarue-Mardrus, *La Prêtresse de Tanit*, 16, 17.

77. On the depiction of auxiliary labor in archaeology, see Nick Shepherd, "'When the Hand That Holds the Trowel Is Black . . .': Disciplinary Practices of Self-Representation and the Issue of 'Native' Labour in Archaeology," *Journal of Social Archaeology* 3 (2003): 334–52, and the discussion in chapter 3.

78. Delarue-Mardrus, *La Prêtresse de Tanit*, 41.

79. *Rapport sur les fêtes*, 44.

80. "La fête," *Revue tunisienne* (1906), 427.

81. *Rapport sur les fêtes*, 59–61.

82. *Rapport sur les fêtes*, 24.

83. "La fête," *Revue tunisienne* (1906), 426.

84. Polybius, *The Histories*, vol. 6, *Books 28–39. Fragments*, trans. W. R. Paton, revised by F. W. Walbank and Christian Habicht, ed. S. Douglas Olson (Cambridge, MA: Harvard University Press, 2012), 489 (book 38, part 21); the original Loeb Classical Library edition (1927) is available at http://penelope.uchicago.edu/Thayer/e/roman/texts/polybius/38*.html.

85. David, "Fête antique," 150.

86. David, "Fête antique," 148–49. These excerpts form part of a much more detailed description.

87. David, "Fête antique," 150.

88. Frank Sear, *Roman Theatres: An Architectural Study* (Oxford: Oxford University Press, 2006), 105, 277–78.

89. Myriam Harry, "Carthage (Impressions de voyage),"*Annales politiques et littéraires*, 21 April 1907.

90. Butler, *Gender Trouble*, 136. I am here positing a loose analogy between disciplinary formations and Butler's understanding of gender as performance.

91. Among the newspapers using the term as the title of an article about the raid were *Le Matin* (26 February 1928), *Paris-Midi* (26 February 1928), *Journal des débats* (27 February 1928), and *La Liberté* (28 February 1928). The *Petit Robert*'s definition and usage note for "coup de théâtre" (under "théâtre") includes a citation from Diderot. On Diderot's view of this dramatic device as artificial and to be avoided, see Denis Diderot, *Entretiens sur le fils naturel*, in his *Oeuvres esthétiques*, ed. Paul Vernière (1757; Paris: Garnier frères, 1968), 88; Peter Szondi, "Tableau and Coup de Theatre: On the Social Psychology of Diderot's Bourgeois Tragedy," trans. Harvey Mendelsohn, *New Literary History* 11 (1980): 323–43.

92. "La querelle de Glozel: Ce que dit Mᵉ de Molènes qui défendra les Fradin avec Mᵉ Campinchi," *Paris-Matinal*, 12 January 1928; Eugène Marsan, "Encore Glozel," *Le Figaro*, 3 March 1928.

93. Eugène Labiche and Alphonse Leveaux, *La Grammaire: Comédie-vaudeville en un acte*, 7th ed. (Paris: E. Dentu, 1871). Clément Vautel, "Mon film," *Le Journal*, 22 September 1927, and "Les évènements," *La semaine Vermot*, 27 November 1927, both refer to this play; Henri Duvernois, "Voeux," *Informations*, 3 January 1928, refers more generally to "un acte gai à la façon de Labiche."

94. Duvernois, "Voeux"; Jacques Coutant, "Les nouveaux arguments du docteur Morlet," *La Liberté*, 19 January 1928.

95. G. Stuart, "A la Lune Rousse: Suivez la Fouille! Revue de MM. Bonnaud, Léon-Michel et Eug. Wyl," *Le Soir*, 28 December 1927.

96. "Oeil-de-Paris: 'Gloz . . . ons!', revue en un acte de MM. Jean Rieux et Henri Dumont," *Excelsior*, 1 February 1928; Marcel de Bare, "Encore un Glozel?" *La Liberté*, 6 February 1928, with specific reference to the Dreyfus affair.

97. "Aux Deux-Anes: *Kif-Kif Bourricot!* Revue de René Dorin et Georges Merry," *La Liberté*, 4 February 1928.

98. The critic for *L'Oeuvre* thought Dalio overacted in the part of a "membre de l'Institut gâteux," whereas his counterpart at *Comoedia* considered the actor's performance crucial to the success of the scene: "Théâtre des Deux-Anes," *L'Oeuvre*, 22 February 1928; "'Kif-Kif Bourricot': Revue de MM. R. Dorin et G. Merry," *Comoedia*, 4 February 1928.

99. "Revue de fin d'année 1927," *Le Journal*, 1 January 1928. Reinach follows and briefly converses with the Paris Police Prefect Chiappe, prompted by Chiappe's use of the verb *fouiller*, which can refer to body searches as well as to excavations.

100. "Aux Deux-Anes," *La Liberté*, 4 February 1928.

101. "Glozel au cirque," *Petit Journal*, 21 January 1928. The clown goes by the name Walter.

102. Mark Antliff, *Inventing Bergson: Cultural Politics and the Parisian Avant-Garde* (Princeton, NJ: Princeton University Press, 1993), 4–5.

103. R. D., "'Incroyable légèreté,' 'Coupable ignorance,'" *La Rumeur*, 11 January 1928; this account treats the coughing as probably confected but does not mention chemical agents. Two other accounts, however, mention stinkballs and powders: J. C., "Une offensive de boules puantes interrompt le cours du professeur Loth au Collège de France," *Le Quotidien*, 11 January 1928; Henri Simoni, "Les Fradin vont faire un procès," *L'Oeuvre*, 11 January 1928.

104. Paul Bringuier, "Le vacarme à la conférence de M. Loth sur Glozel s'achève en petite émeute," *Le Journal*, 22 January 1928.

105. Jean-Paul Ariste, "La guerre de Glozel," *La Rumeur*, 23 January 1928.

106. Emile Condroyer, "Le professeur Loth prend la défense de Glozel," *Le Journal*, 4 January 1928.

107. *Paris matinal*, 11 January 1928.

108. "La bataille de Glozel," *Action française*, 9 January 1928. On Action française youth interrupting and sometimes attacking leftist professors, see Eugen Weber, *Action Française: Royalism and Reaction in Twentieth Century France* (Stanford, CA: Stanford University Press, 1962), 157–62, 180–81; Jean-François Sirinelli, *Génération intellectuelle: Khâgneux et normaliens dans l'entre-deux-guerres* (Paris: Fayard, 1988), 220–26, 229–41; Guillaume Gros, "Les jeunes et l'Action française (1914–1939)," in *L'Action française: Culture, société, politique*, ed. Michel Leymarie and Jacques Prévotat (Villeneuve d'Ascq: Presses universitaires du Septentrion, 2019), 217–28, https://doi.org/10.4000/books.septentrion.39150.

109. "'Incroyable légèreté,'" *La Rumeur*; "Les antiglozéliens chahutent le cours du professeur Loth," *Petit journal*, 22 January 1928.

110. Pierre Audiat, "Les professeurs dans le stade," *Paris-Midi*, 26 January 1928. The sentence specifically naming Loth suggests "les grounds de Colombes"; the next sentence, referring to the Cirque Médrano, mentions "un professeur d'orientalisme," which he was not, but this type of error was quite common.

111. *Petit journal*, 22 January 1928; *Action française*, 9 January 1928.

112. *La Rumeur*, 11 January 1928.

113. BMA, Loth to Reinach, 9 January 1928.

114. Simoni, "Les Fradin vont faire un procès," subsection "Les conférences de M. Loth sont interdites," *L'Oeuvre*, 11 January 1928.

115. "On triomphe à Glozel," *Le Moniteur* (Clermont-Ferrand), 28 June 1931. For a concise summary of the various legal cases and verdicts, see Joseph Grivel, *La préhistoire chahutée: Glozel (1924–1941)* (Paris: L'Harmattan, 2003), 275–76.

116. Georges Claretie, "Gazette des Tribunaux," *Le Figaro*, 9 March 1932.

117. "Glozel Test by Lawsuit Draws Nearer," *Paris Times*, 11 January 1928; on this newspaper, published between 1924 and 1929, see Nancy L. Green, *The Other Americans in Paris: Businessmen, Countesses, Wayward Youth, 1880–1941* (Chicago: University of Chicago Press, 2014), 39.

118. J.-S., "L'affaire de Glozel devant la Cour d'Appel: Un tournoi oratoire entre Glozéliens et Antiglozéliens," *L'Avenir du Plateau Central* (Clermont-Ferrand), 26 February 1930.

119. Benjamin F. Martin, *Crime and Criminal Justice under the Third Republic: The Shame of Marianne* (Baton Rouge: Louisiana State University Press, 1990), 252–53.

120. Dianne Dutton, "Le Plaideur comme comédien: L'art dramatique du procès," in *Représentations du procès: Droit, théâtre, littérature, cinéma*, ed. Christian Biet and Laurence Schifano (Nanterre: Université Paris–X Nanterre, 2003), 127–34.

121. Yasco Horsman, *Theaters of Justice: Judging, Staging, and Working Through in Arendt, Brecht, and Delbo* (Stanford, CA: Stanford University Press, 2010), 8. Horsman is paraphrasing the original French of Alain Finkielkraut, *Remembering in Vain: The Klaus Barbie Trial and Crimes Against Humanity*, trans. Roxanne Lapidus with Sima Godfrey (New York: Columbia University Press, 1992), 70, but this term does not appear as such in the translation.

122. Yann Robert, "Des acteurs au barreau, ou l'invention de l'avocat moderne," *Dix-huitième siècle* 49, no. 1 (2017): 119–31, 130, https://doi.org/10.3917/dhs.049.0119.

123. Dutton, "Le Plaideur comme comédien," 128: the pertinent dual meanings are, on the one hand, a lawyer representing a client (as in English) and, on the other, a specific performance of a play or other live event.

124. M. F., "Glozel en cour d'appel," *La Montagne*, 18 February 1930.

125. "A la 12e chambre correctionnelle: Le jugement du procès de Glozel renvoyé à quinzaine," *Le Matin*, 10 March 1932. Although staunchly Glozelian, *Le Matin* was a party to the suit because it had published the remarks by Dussaud it considered defamatory. Torrès explained that the paper had done this deliberately to prompt the lawsuit.

126. AN, 19860089/787, dossier 9bis; the brief was later published as "Querelles glozéliennes" in *Journal des débats*, 13 May 1929. This Audollent argument, for the defamation case in 1930, is in SRA, 4, a forty-two-page handwritten document that begins "La Préhistoire, par définition est une science du passé."

127. *Le Figaro*, 9 March 1932. The story makes reference to the origin story of Glozel with the Fradins plowing their fields.

128. "L'affaire de Glozel à la 12e chambre correctionnelle," *Le Matin*, 9 March 1932.

129. Pierre Bénard, "A la XIIe Chambre: Glozeliens et antiglozeliens se sont affrontés tumultueusement," *L'Oeuvre*, 9 March 1932. Bénard also describes Vayson de Pradenne as "prétentieux."

130. Bénard, "A la XIIe Chambre."

131. "Nouvel épisode de l'affaire de Glozel," *Le Moniteur*, 19 October 1929.

132. SRA, 4, Audollent, "La Préhistoire": "L'histoire de Glozel s'y retrouve, tracée à grands traits. L'on y voit défiler les principaux personnages qui y jouèrent un rôle."

133. As *Le Temps* put it in "Tribunaux: Un procès à propos de l'affaire de Glozel," 10 March 1932, its sober account of the 1932 hearing, "Le procès en diffamation d'Emile Fradin et de son grand-père à M. Dussaud et au *Matin* va rouvrir, semble-t-il, une discussion qui paraissait close."

134. René Benjamin, *Glozel, vallon des morts et des savants* (Paris: Fayard, 1928). Further citations will be in-text.

135. BMA, Prorok to Reinach, 22 September 1928.

136. G. B., "Vieux Pots: Les fouilles définitives de Glozel," *Journal de l'Est*, 18 July 1928.

137. G. B., "Vieux Pots." On Benjamin's politics and role at Vichy, see Weber, *Action Française*, 444.

Chapter 5

1. *DAN* Lavigerie, 1:252–53.

2. On Leroux as one of the three major scholarly publishers in late nineteenth-century France, see Valérie Tesnière, "L'édition universitaire," in *Le Temps des éditeurs: Du romantisme à la Belle époque*, ed. Henri-Jean Martin, Roger Chartier, and Jean-Pierre Vivet, vol. 3 of *Histoire de l'édition française* (Paris: Promodis, 1985), 220–21.

3. Stephanie Moser, "Archaeological Visualization: Early Artifact Illustration and the Birth of the Archaeological Image," in *Archaeological Theory Today*, 2nd ed., ed. Ian Hodder (Cambridge: Polity, 2012), 292–322, 292; Moser, "Archaeological Representation: The Consumption and Creation of the Past," in *Oxford Handbook of Archaeology*, ed. Barry Cunliffe, Chris Godsen, and Rosemary A. Joyce (Oxford: Oxford University Press, 2009), 1072, https://doi.org/10.1093/oxfordhb/9780199271016.013.0034.

4. Frederick N. Bohrer, *Photography and Archaeology* (London: Reaktion, 2011), 15.

5. See, for example, Claire L. Lyons et al., *Antiquity & Photography: Early Views of Ancient Mediterranean Sites*, exh. cat. (Los Angeles: J. Paul Getty Museum, 2005), https://www.getty.edu/publications/virtuallibrary/0892368055.html, which considers the relations between photography and archaeology from 1840 to 1880.

6. Michael Shanks, "Photography and Archaeology," in *The Cultural Life of Images: Visual Representation in Archaeology*, ed. Brian Molyneaux (London: Routledge, 1997), 73–102, 82.

7. Moser, "Archaeological Representation," 1063–64; Moser, "Archaeological Visualization"; Bohrer, *Photography and Archaeology*, 30–33.

8. Elizabeth Edwards and Janice Hart, "Introduction," in *Photographs Objects Histories: On the Materiality of Images*, ed. Edwards and Hart (London: Routledge, 2004), 2–3. See also Edwards, *Raw Histories: Photographs, Anthropology and Museums* (Oxford: Berg, 2001).

9. See, for example, Sabine T. Kriebel and Andrés Mario Zervigón, eds., *Photography and Doubt* (Abingdon: Routledge, 2017).

10. Stefanie Klamm, "Reconfiguring the Use of Photography in Archaeology," in *Hybrid Photography: Intermedial Practices in Science and Humanities*, ed. Sara Hillnhuetter, Stefanie Klamm, and Friedrich Tietjen (Abingdon: Routledge, 2021), 114–27, https://doi.org/10.4324/9781003157854.

11. Edwards, *Raw Histories*, 62.

12. Bohrer, *Photography and Archaeology*, 50; Moser, "Archaeological Representation," 1064.

13. As Bohrer puts it, "Archaeological photography pairs the technology of picturing absence with the science of deciphering absence and recuperating from it." Bohrer, *Photography and Archaeology*, 8.

14. On archaeological reconstructions, see Cathy Gere, *Knossos and the Prophets of Modernism* (Chicago: University of Chicago Press, 2009), chap. 4; Kostis Kourelis, "Byzantine Houses and Modern Fictions: Domesticating Mystras in 1930s Greece," *Dumbarton Oaks Papers* 65/66 (2011): 297–331; Maria Gabriella Micale, "European Images of the Ancient Near East at the Beginnings of the Twentieth Century," in *Archives, Ancestors, Practices: Archaeology in the Light of Its History*, ed. Nathan Schlanger and Jarl Nordbladh (New York: Berghahn Books, 2008), 191–204. On Flaubert, see above, page 131.

15. See Michael Shanks, *The Archaeological Imagination* (2012; London: Routledge, 2016), esp. 15, 145–49. Although they are certainly related, Shanks discusses the archaeological imagination as a faculty and process employed by archaeologists, whereas I am considering the archaeological imaginary as a storehouse and image bank.

16. See, for example, Debbie Challis, *The Archaeology of Race: The Eugenic Ideas of Francis Galton and Flinders Petrie* (London: Bloomsbury Academic, 2013).

17. The scholarly literature on ethnography is vast. For French ethnography in the late nineteenth and early twentieth centuries, see Nélia Dias, *Le Musée d'ethnographie du Trocadéro, 1878–1908: Anthropologie et muséologie en France* (Paris: CNRS, 1991), 17–31; Alice L. Conklin, *In the Museum of Man: Race, Anthropology, and Empire in France, 1850–1950*

(Ithaca, NY: Cornell University Press, 2013), chap. 1; Julia Kelly, *Art, Ethnography and the Life of Objects: Paris, c. 1925–35* (Manchester: Manchester University Press, 2007), chap. 1.

18. Matthew McCarty, "Pre-Roman Libyan Religion: Colonial Ethnography and the Problem of Religious 'Survivals,'" *Revista de Historiografía*, no. 36 (2021): 127–48, 130, https://doi.org/10.20318/revhisto.2021.6553.

19. Rémi Labrusse, *Préhistoire: L'envers du temps* ([Vanves]: Hazan, 2019), 139. For the persistence of stereotyped images of early man, see Stephanie Moser and Clive Gamble, "Revolutionary Images: The Iconic Vocabulary for Representing Human Antiquity," in Molyneaux, *Cultural Life of Images*, 184–212.

20. On the term "civilization" and its use in and beyond archaeology, see David Wengrow, *What Makes Civilization? The Ancient Near East and the Future of the West* (Oxford: Oxford University Press, 2010), 1–14.

21. On the many problems posed by the category of ancient Phoenicians, see Josephine Crawley Quinn, *In Search of the Phoenicians* (Princeton, NJ: Princeton University Press, 2018). The fall of Carthage, and the several centuries of war with Rome that preceded it, were of course the subject of significant historical writing by Greek and Roman authors, notably Polybius and Appian.

22. Camille Jullian, "Plaidoyer pour la préhistoire," *Revue bleue: Revue politique et littéraire*, 5th ser., 8, no. 24 (1907): 737–44, 737, also cited in Labrusse, *Préhistoire*, 50. This text was Jullian's inaugural lecture as a professor at the Collège de France.

23. Bill Brown, "Thing Theory," *Critical Inquiry* 28, no. 1 (2001): 5, https://doi.org/10.1086/449030.

24. Serge Lewuillon, for example, has argued that "the contribution of archaeological drawing to the progress of science was nothing less than the invention of the object itself." Lewuillon, "Archaeological Illustrations: A New Development in 19th Century Science," *Antiquity* 76, no. 291 (2002): 223–34, 231; https://doi.org/10.1017/S0003598X00090025.

25. David Jenkins, "Object Lessons and Ethnographic Displays: Museum Exhibitions and the Making of American Anthropology," *Comparative Studies in Society and History* 36, no. 2 (1994): 242–70, https://doi.org/10.1017/S0010417500019046.

26. See Carolina López-Ruiz, *Phoenicians and the Making of the Mediterranean* (Cambridge, MA: Harvard University Press, 2021), 15. The Carthaginians or Punic people were themselves colonists from the eastern Mediterranean.

27. The title page is not uniform throughout, but on most volumes it reads, *Description de l'Afrique du Nord entreprise par ordre de M. le Ministre de l'Instruction publique et des beaux arts: Catalogue des Musées et collections archéologiques de l'Algérie et de la Tunisie.*

28. The classic account of the *Description de l'Egypte* in these terms is Edward Said, *Orientalism* (New York: Pantheon, 1978), 84–87. For a concise history of the expedition and publication, with reproductions of some of its illustrations, see Yves Laissus, *Description de l'Égypte: Une aventure humaine et éditoriale* (Paris: Réunion des musées nationaux, 2009).

29. For a list of all the museums covered in the series with dates of publication, see Myriam Bacha, *Patrimoine et monuments en Tunisie: Sauvegarde et mise en valeur pendant le Protectorat, 1881–1920* (Rennes: Presses Universitaires de Rennes, 2013), 343. The most authoritative entry in the catalog of the Bibliothèque nationale de France, https://catalogue.bnf.fr/ark:/12148/cb38901197r (consulted 5 June 2022) lists twenty-three volumes but links to twenty-eight individual catalog entries. The Bardo museum catalogs do not bear the "Museums and Archaeological Collections" series title, but they are clearly part of the *Description* project.

30. On the complicated early history of the committee, in French the Commission de publication des documents archéologiques de l'Afrique du Nord, which was founded in 1884 and went through several name changes before taking that given here in 1893, see Jehan Desanges, "La commission dite 'de l'Afrique du Nord' au sein du CTHS: Origine, évolution, perspectives," in *Numismatique, langues, écritures et arts du livre, spécificité des arts figurés: Actes du VIIe colloque international sur l'histoire et l'archéologie de l'Afrique du Nord, réunis dans le cadre du 121e Congrès des sociétés historiques et scientifiques, Nice, 21 au 31 octobre 1996*, ed. Serge Lancel (Paris: CTHS, 1999), 12–14.

31. R.-H. Bautier stresses the importance of the North Africa Committee's publications within the work of CTHS in "Le Comité des travaux historiques et scientifiques," in Martin et al., *Le Temps des éditeurs*, 224–25.

32. *Le nouveau Petit Robert* (Paris: Dictionnaires Robert, 1993). The identification is from the website https://www.dicocitations.com/reference_citation/94092/Les_Romantiques.php.

33. For convenience, I use "volume" here to refer to the book or books devoted to a given museum and conceived as a whole; a single volume can consist of several physical books. The books were likely shipped unbound, with binding up to the recipient, which accounts for some discrepancies between libraries. My analysis is based on physical examination of the volumes in the Classics Faculty Library at the University of Cambridge; the digitized versions available on Gallica elide some of the differences.

34. The Bardo catalog has no list of abbreviations; the reader is presumed to know, for example, that *C.I.L.* stands for *Corpus Inscriptionum Latinarum*, a comprehensive record of Latin inscriptions begun by German scholars in the mid-nineteenth century. Philippe Berger, author of part 1 of the Carthage catalog, does provide a list of abbreviations: *DAN* Lavigerie, 1:7–8.

35. *DAN* Alaoui, 146, object K-3: "Époque des guerres puniques."

36. *DAN* Lavigerie, 1:200 (plate XXIX).

37. *DAN* Alaoui, 146.

38. *DAN* Lavigerie, 1:xiii.

39. *DAN* Lavigerie, 1:200.

40. Geoffrey Belknap, *From a Photograph: Authenticity, Science and the Periodical Press, 1870–1890* (London: Bloomsbury Academic, 2016), esp. chaps. 1–2. On the technology of print illustration in the nineteenth century, see Michel Melot, "Le texte et l'image," in Martin et al., *Le temps des éditeurs*, 287–311; Laureline Meizel, *Le livre illustré par la photographie: Histoires d'une invention* (Paris: Bibliothèque nationale de France, 2023). On the relation between photographs and other visual media in archaeology, see Bohrer, *Photography and Archaeology*, 22, 40–42; Moser, "Archaeological Representation."

41. *DAN* Alaoui, 61: sculpture, nos. 113–650, 651–655.

42. *DAN* Alaoui, 61.

43. *DAN* Alaoui, 145 (no. I-122); 67 (no. C-828). The next entry also uses the word "grossièrement."

44. *DAN* Lavigerie, 1:96.

45. On connections between archaeology and racial science, see Bruce Trigger, *A History of Archaeological Thought*, 2nd ed. (New York: Cambridge University Press, 2006), chap. 5; Challis, *Archaeology of Race*; Bonnie Effros, "Berber Genealogy and the Politics of Prehistoric Archaeology and Craniology in French Algeria (1860s–1880s)," *British Journal for the History of Science* 50, no. 1 (March 2017): 61–81.

46. *DAN* Alaoui, 62 (nos. C-656–740).

47. *DAN* Alaoui, 240 (no. M-268).

48. Trigger, *Archaeological Thought*, 121–47; Alain Schnapp, *The Discovery of the Past*, trans. Ian Kinnes and Gillian Varndell (New York: Harry N. Abrams, 1997), 298–303.

49. Labrusse, *Préhistoire*, 70–79.

50. Joseph Déchelette, *Manuel d'archéologie préhistorique, celtique et gallo-romaine*, 4 vols. (1908–1914; Paris: Picard, 1924), 1:2.

51. For a transnational survey, see Richard McMahon, "The History of Transdisciplinary Race Classification: Methods, Politics and Institutions, 1840s–1940s," *British Journal for the History of Science* 51, no. 1 (2018): 41–67, https://doi.org/10.1017/S0007087417001054. Reliable discussions of scientific racism and racial science as practiced in France include William B. Cohen, *The French Encounter with Africans: White Response to Blacks, 1530–1880* (1980; Bloomington: Indiana University Press, 2003), 210–62; Martin S. Staum, *Nature and Nurture in French Social Sciences, 1859–1914 and Beyond* (Montréal: McGill–Queen's University Press, 2011), esp. chaps. 2–3; Conklin, *In the Museum of Man*, 22–29. On prehistory and theories of evolution, see Stéphane Tirard, "Les théories de l'évolution et l'épaisseur du temps," in *Dans l'épaisseur du temps: Archéologues et géologues inventent la préhistoire*, ed. Arnaud Hurel and Noël Coye (Paris: Muséum national d'histoire naturelle, 2011), 39–51; Claudine Cohen, *L'homme des origines: Savoirs et fictions en préhistoire* (Paris: Seuil, 1999), 141–75.

52. Trigger, *Archaeological Thought*, 144–47; Marc Groenen, *Pour une histoire de la préhistoire: Le Paléolithique* (Grenoble: J. Millon, 1994), 97–105, 155–225. It should also be noted that geological time scales remained controversial in clerical circles throughout the nineteenth century.

53. *DAN* Alaoui, 227. The Postumia were a patrician family, several of whose members were granted the concession to mint money for the Roman Republic.

54. *DAN* Lavigerie, 1:149 (plate XXIV); 1:98–101 (plate XV), 98. Magna Graecia refers to areas of coastal southern Italy colonized by Greeks from the eighth to the fifth centuries BCE. For more on the problematic concept of "Orientalizing" influences in ancient Mediterranean art and a proposed new usage, see López-Ruiz, *Phoenicians and the Making of the Mediterranean*, chap. 3.

55. *DAN* Alaoui, 62–63.

56. *DAN* Lavigerie, Suppl., part 1, 31.

57. *DAN* Lavigerie, 1:145.

58. *DAN* Lavigerie, 1:vii.

59. *DAN* Lavigerie, Suppl., part 1, vii. Like the original, the supplementary Carthage catalog appeared in two parts, in 1913 and 1915, the first devoted to Punic, the second to Roman antiquities.

60. *DAN* Alaoui, Suppl. 1, vi. The 1910 volume had multiple authors including Gauckler (for the mosaics), Poinssot, and Merlin.

61. *DAN* Lavigerie, Suppl., 15 (plate III). The precise date of its discovery is given under the previous entry.

62. Stéphane Gsell, *La Civilisation carthaginoise*, vol. 4 of *Histoire ancienne de l'Afrique du nord*, 8 vols. (Paris: Hachette, 1913–1928), 471.

63. Biographical details from Ève Gran-Aymerich, *Les chercheurs de passé, 1798–1945: Aux sources de l'archéologie* (Paris: CNRS, 2007), 845–47.

64. Donald Harden, *The Phoenicians* (London: Thames & Hudson, 1962), 242; Pierre Cintas, *Manuel d'archéologie punique* (Paris: Picard, 1970), vii–viii.

65. Khaled Melliti, *Carthage: Histoire d'une métropole méditerranéenne* (Paris: Perrin, 2016).

66. See, for example, Stéphane Gsell, *L'Etat carthaginois*, vol. 2 of *Histoire ancienne de l'Afrique du nord*, 2:6–7; Gsell, *La Civilisation carthaginoise*, 4:131, 167–69 (note the use, on 167, of the term "documents archéologiques").

67. René Dussaud, *Les Civilisations préhelléniques dans le bassin de la Mer Égée*, 2nd ed. (Paris: P. Geuthner, 1914), 324, 363.

68. Dussaud, *Les Civilisations préhelléniques*, 447; he concludes, "Les renseignements archéologiques concordent donc avec les données anthropologiques pour déterminer les mêmes successions ethniques" (449).

69. Déchelette, *Manuel*, 1:202.

70. See, among many other examples, Dussaud, *Les Civilisations préhelléniques*, 218–19, 233 (on di Cesnola); Gsell, *L'Etat carthaginois*, 2:64, 86 (stratigraphy and contextual dating); Déchelette, *Manuel*, 1:213, 301 (on stratigraphy). Di Cesnola (1832–1904) was the American consul in Cyprus before becoming the first director of New York's Metropolitan Museum in 1879. He had sold the museum his massive collection of Cypriot antiquities in 1873. See Katharine Baetjer and Joan R. Mertens, "The Founding Decades," in *Making the Met, 1870–2020*, ed. Andrea Bayer and Laura D. Corey, exh. cat. (New York: Metropolitan Museum of Art, 2020), 38–40.

71. Quinn, *In Search of the Phoenicians*; Josephine Crawley Quinn and Nicholas C. Vella, eds., *The Punic Mediterranean: Identities and Identification from Phoenician Settlement to Roman Rule* (Cambridge: Cambridge University Press, 2014); Asher Kaufman, *Reviving Phoenicia: The Search for Identity in Lebanon* (London: I. B. Tauris, 2004).

72. Jonathan Prag, "*Phoinix* and *Poenus*: Usage in Antiquity," in *The Punic Mediterranean*, 11–23, 22.

73. Nicholas Vella, "The Invention of the Phoenicians: On Object Definition, Decontextualization and Display," in Quinn and Vella, *Punic Mediterranean*, 24–41, 30, 34. By "ambivalent" Vella seems to be referring to the mixture of respect and denigration that twentieth-century scholars have afforded the Phoenicians.

74. See Sabatino Moscati, *The Phoenicians* (New York: Abbeville, 1988); Quinn, *In Search of the Phoenicians*, 22–24; Vella, "Invention of the Phoenicians," 35–39.

75. López-Ruiz, *Phoenicians and the Making of the Mediterranean*, 15–20, 17, 18.

76. The exact date and circumstances of the founding of the colonial settlement remain mysterious; see Melliti, *Carthage*, 25–36. On Carthaginian coinage, which actually dates to around 410 BCE, see Quinn, *In Search of the Phoenicians*, 86–90.

77. Gsell, *La Civilisation carthaginoise*, 4:109, 216–17.

78. Gsell, *La Civilisation carthaginoise*, 4:215.

79. Gsell, *La Civilisation carthaginoise*, 4:220.

80. For summaries of recent research on Carthaginian child sacrifice, see Quinn, *In Search of the Phoenicians*, 91–112; Matthew M. McCarty, "The Tophet and Infant Sacrifice," in *The Oxford Handbook of the Phoenician and Punic Mediterranean*, ed. Brian R. Doak and Carolina López-Ruiz (Oxford: Oxford University Press, 2019), 309–25, https://doi.org/10.1093/oxfordhb/9780190499341.013.21.

81. Gsell, *La Civilisation carthaginoise*, 4:405–9, 409.

82. *DAN* Lavigerie, 1, 5. López-Ruiz refutes this notion in *The Phoenicians and the Making of the Mediterranean*.

83. *DAN* Lavigerie, 1, 6.

84. *DAN* Lavigerie, Suppl., 4–8, *grossier* on 7, citation from 4.

85. *DAN* Lavigerie, Suppl., 9–19, 18; flaws in technique are discussed on pages 10 and

14, describing two of the priest sarcophagi, which Boulanger attributes to a Greek artist working in Carthage.

86. Gsell, *La Civilisation carthaginoise*, 4:486.

87. On the sources of early alphabetic writing in the West, see Florian Coulmas, *The Writing Systems of the World* (Oxford: Basil Blackwell, 1989), 57–90, 137–70; Steven R. Fischer, *A History of Writing* (London: Reaktion, 2001), 68–105; and the following essays in Peter T. Daniels and William Bright, eds., *The World's Writing Systems* (New York: Oxford University Press, 1996): Robert K. Ritner, "Egyptian Writing, " 73–83; M. O'Connor, "Epigraphic Semitic Scripts," 88–107, and Pierre Swiggers, "Transmission of the Phoenician Script to the West," 261–70.

88. For a discussion of the Ahiram sarcophagus, with references to the author's extensive work on it, see Reinhard G. Lehmann, "Calligraphy and Craftsmanship in the Ahirom Inscription: Considerations on Skilled Linear Flat Writing in Early First Millennium Byblos," *MAARAV: A Journal for the Study of the Northwest Semitic Languages and Literatures* 15, no. 2 (2008): 119–64. On the decoration of the sarcophagus and its importance for Phoenician art, see Glenn E. Markoe, "The Emergence of Phoenician Art," *Bulletin of the American Schools of Oriental Research* 279 (August 1990): 13–26, https://doi.org/10.2307/1357205.

89. Philippe Berger, *Histoire de l'écriture dans l'antiquité* (Paris: Imprimerie nationale, 1891), vii.

90. Berger, *Histoire de l'écriture*, 115.

91. Berger, *Histoire de l'écriture*, 115–28, 169–87; Dussaud, *Les Civilisations préhelléniques*, 432–37.

92. Berger, *Histoire de l'écriture*, 116–17.

93. Gsell, *La Civilisation carthaginoise*, 4:181–82, 182.

94. Gsell, *La Civilisation carthaginoise*, 4:485.

95. Gsell, *La Civilisation carthaginoise*, 4:497.

96. Gsell, *La Civilisation carthaginoise*, 4:498.

97. For a sober critique of Gsell's reliance on contemporary ethnography in his treatment of Berber religion, see McCarty, "Pre-Roman Libyan Religion," 135–42.

98. On the *Mercure*, see Élisabeth Parinet, "L'édition littéraire: 1890–1914," in *Le Livre concurrencé: 1900–1950*, ed. Henri-Jean Martin, Roger Chartier, and Jean-Pierre Vivet, vol. 4 of *Histoire de l'édition française* (Paris: Promodis, 1986), 163–66.

99. *NSN*, 3:46. The second sentence starts (and comprises) a new paragraph. The image is figure 52. In all instances, the emphasis is in the original.

100. Pierre Montet, "Nouvelles archéologiques: Les fouilles de Byblos en 1923," *Syria. Archéologie, Art et Histoire*, 1923, 334–44; Montet, "Les fouilles de Byblos en 1923," *L'Illustration*, 3 May 1924, 402–5. Dussaud himself published the *editio princeps*, the first printed transcription of the inscription, in *Syria*: Dussaud, "Les inscriptions phéniciennes du tombeau d'Ahiram, roi de Byblos," *Syria* 5, no. 2 (1924): 135–57.

101. "Autour des décovertes de Glozel," *MdF*, 15 August 1926, 183–88.

102. Déchelette, *Manuel*, 1:240.

103. Frederick N. Bohrer, "Edges of Art: Photographic Albums, Archaeology, and Representation," in *Art and the Early Photographic Album*, ed. Stephen Bann (Washington, DC: National Gallery of Art, 2011), 221–36, 228. See also Klamm, "Reconfiguring the Use of Photography."

104. *NSN*, 3:8, fig. 2; fig. 4 on p. 10 shows three perforated pebbles in a single frame.

105. Morlet, "Glozel: Le premier âge de l'argile," *MdF*, 1 October 1927, 104–11.

106. In fascicule 3, for example, fig. 26, an incised bowl, seems to have been subject to raking light or flash; figs. 34 and 35, both inscribed tablets, have been printed so that the signs appear as black on a light background. Each of these illustrations occupies a full page (pp. 32–33).

107. See "Station néolithique de Glozel: Idoles phalliques et bi-sexuées," *MdF*, 15 September 1926, 562–67.

108. Charles Omessa, "La conclusion," *La Liberté*, 28 December 1927.

109. Morlet sets out these characteristics in "Station néolithique de Glozel: Idoles phalliques et bi-sexuées," *MdF*, 15 September 1926, 562–67, with images of the two types.

110. Morlet, "Au champ des morts de Glozel," *MdF*, 1 August 1927, 599.

111. *NSN*, 3:27.

112. For examples of the face cups, see Alexandra Villing et al., *The BP Exhibition: Troy, Myth and Reality*, exh. cat. (London: Thames & Hudson/British Museum, 2019), 146, fig. 124.

113. Morlet, "La décoration céramique," *MdF*, 15 October 1926, 262, 265, citation 262. On efforts to use ceramics in the dating of prehistoric sites, see Noël Coye, "Humanité et pots cassés: La tentative céramologique des préhistoriens français (1900–1918)," in *Les Politiques de l'anthropologie: Discours et pratiques en France (1860–1940)*, ed. Claude Blanckaert (Paris: L'Harmattan, 2001), 231–67.

114. *NSN*, 3:54.

115. *NSN*, 1:9–10 (neolithic tomb); Morlet, "Connexion du néolithique ancien avec le paléolithique final," *MdF*, 1 May 1927, 581 (lack of stratigraphy, a point made in many places). After a later excavation found what he deemed another tomb, Morlet described it as also without stratigraphy: "Au champ de morts de Glozel," *MdF*, 15 August 1927, 95.

116. *NSN*, 3:47–49 (transitional epoch, ceramics produced by a nonnomadic group); Morlet, "Le travail de l'os à Glozel," *MdF*, 1 July 1927, 68–69 ("ce peuple resté chasseur"); Morlet, "Connexion du néolithique," 578–85 (reindeer, disagreement with Déchelette).

117. *NSN*, 5:37–38; Salomon Reinach, *Description raisonnée du Musée de Saint-Germain-en-Laye: Antiquités nationales*, 2 vols. (Paris: Firmin-Didot, 1889), 1:168. On the wider art historical significance of this phrase, of which the first clause is from Ovid, see Maria Stavrinaki, "Enfant né sans mère, mère morte sans enfant: Les historiens de l'art face à la préhistoire," *Cahiers du Musee national d'art moderne*, no. 126 (2014): 4–13. Morlet quotes the Latin in "Connexion du néolithique,"585.

118. *NSN*, 1:20.

119. On the nomenclature of archaeological periods developed by the French prehistorian Gabriel de Mortillet and based on sites that define them, see Trigger, *Archaeological Thought*, 149–54; Nathalie Richard, *Inventer la préhistoire: Les débuts de l'archéologie préhistorique en France* (Paris: Vuibert, 2008), 168–75. "Magadalenian" is a period label still in use by prehistorians today.

120. *NSN*, 2:20; Morlet, "Le travail de l'os," 67–70.

121. Labrusse, *Préhistoire*, 70–79, 78; Morlet, "Le travail de l'os," 66–78.

122. Morlet, "La décoration céramique," 262, 267; *NSN*, 1:13 (Australian comparison); Morlet, "Connexion du néolithique," 582.

123. Morlet, "Connexion du néolithique," 580.

124. Morlet, "Le travail de l'os," 66 (mistaken terminology); *NSN*, 2:3 (difficulties of dating).

125. *NSN*, 2:20 (indecipherability); Morlet, "De quelques groupements dans les inscriptions de Glozel," 15 September 1927, 590: "On ne peut malheureusement espérer, par suite du trop long recul dans le temps, trouver un jour la pierre de Rosette des inscriptions de Glozel."

126. *NSN*, 2:12. The number of Glozelian letters would eventually grow to around ninety.

127. *NSN*, 2:18–19. Morlet repeats this point in "Invention et diffusion de l'alphabet néolithique," *MdF*, 1 April 1926, 37–38, by which time the number of identified Glozelian letters has reached eighty-six.

128. *NSN*, 2:18. There is a paragraph break between the two quoted sentences.

129. *NSN*, 2:19–20; Morlet, "Invention et diffusion," 43, 46. The quote attributed to Homo is actually from Jullian and concerns agriculture and spoken language but not writing.

130. *NSN*, 2:19–20.

131. *NSN*, 2:8.

132. Morlet, "Connexion du néolithique," 584.

133. *NSN*, 2:18.

134. *NSN*, 3:48. The same sentence, with very minor modifications, appears at the end of the last fascicule (*NSN*, 5:37).

135. *NSN*, 3:49.

136. Dussaud, *Les Civilisations préhelléniques*, esp. chap. 6.

137. Déchelette, *Manuel*, 1:218 (argumentation), 313 (on resemblance not proving filiation).

138. Déchelette, *Manuel*, 1:312–14.

139. *NSN*, 3:49.

140. Gérard Noiriel, *Immigration, antisémitisme et racisme en France, XIXe–XXe siècle: Discours publics, humiliations privées* (Paris: Fayard, 2007), 290–300.

141. Noiriel, *Immigration*, 305; Ethan B. Katz, *The Burdens of Brotherhood: Jews and Muslims from North Africa to France* (Cambridge, MA: Harvard University Press, 2015), 63.

142. Katz, *Burdens of Brotherhood*, 55–56. The Ligue des droits de l'homme was founded at the time of the Dreyfus affair.

143. See Noiriel, *Immigration*, 314–16; Mary Dewhurst Lewis, *The Boundaries of the Republic: Migrant Rights and the Limits of Universalism in France, 1918–1940* (Stanford, CA: Stanford University Press, 2007), 188–89, 198–210; Clifford Rosenberg, *Policing Paris: The Origins of Modern Immigration Control between the Wars* (Ithaca, NY: Cornell University Press, 2006), 77–99, 148–49, 153–67.

144. Rosenberg, *Policing Paris*, 116.

145. Noiriel, *Immigration*, 300, 369.

146. Even in a letter disparaging some clerics, however, Morlet singles out one priest he describes as an "ardent glozélian"; BMA, Morlet to Reinach, 18 October 1928. On Reinach's work on behalf of the Jewish community in France and beyond, see Aron Rodrigue, "Totems, Taboos, and Jews: Salomon Reinach and the Politics of Scholarship in Fin-de-Siècle France," *Jewish Social Studies* 10 (2004): 1–19. A recent collection of Reinach's letters focuses largely on his contacts in Russian archaeology and above all in world Judaism; it does not include any correspondence related to Glozel: Salomon Reinach, *Correspondance, 1888–1932: Un polygraphe sous le signe d'Amalthée*, ed. Boris Czerny (Paris: Honoré Champion, 2020).

147. Victor Basch, "Une protestation de la Ligue des Droits de l'Homme," *MdF*, 15 June 1928, with an editorial note that this is the second such letter; BMA, Morlet to Reinach, 14 March 1928: "Par contre la Ligue des droits de l'homme nous a fait demander de renseignements sur la perquisition." Basch's letter specifically refuses to take a stand on the authenticity of the Glozel finds.

148. See Herman Lebovics, *True France: The Wars over Cultural Identity, 1900–1945* (Ithaca, NY: Cornell University Press, 1992), chap. 1. Marin played a bit part in the Glozel affair: as president of the Institut international d'anthropologie in 1927 he approved the

appointment of the verification commission and was the nominal addressee of its report. A furious reference to him in a 1929 letter from Morlet as a "super-Jesuit" full of clerical bigotry, however, suggests that Marin himself did not embrace the Glozelian cause; BMA, Morlet to Reinach, 14 June 1929. Morlet is reacting to a pamphlet by Marin over which he believes the Fradins could sue. A search in the catalog of the Bibliothèque Nationale de France did not turn up any such publication.

149. René Dussaud, *Glozel à l'Institut*, La Controverse de Glozel 2 (Paris: Paul Catin, 1928), 9.

150. BMA, Morlet to Reinach, 15 March 1928 (Hurons); Morlet to Reinach, 18 August 1928 ("métèques").

151. Sandrine Sanos, *The Aesthetics of Hate: Far-Right Intellectuals, Antisemitism, and Gender in 1930s France* (Stanford, CA: Stanford University Press, 2013), 58–62, 61.

152. Dussaud, *Les Civilisations préhélleniques*, 5. He is referring to uncertainty among text-based historians after the discovery of Mycenean art.

153. Pierre Montet, *Byblos et l'Égypte: Quatre campagnes de fouilles à Gebeil, 1921–1922–1923–1924* . . . , 2 vols., Haut-Commissariat de La République française en Syrie et au Liban, Service Des Antiquités et Des Beaux-Arts. Bibliothèque Archéologique et Historique, t. XI (Paris: P. Geuthner, 1928–1929).

Epilogue

1. For a comprehensive biography, see John G. Hurst, "Donald Benjamin Harden 1901–1994," *Proceedings of the British Academy* 94 (1996): 513–39. Harden was born in Dublin, his father the latest in a long line of Church of Ireland (Protestant) ministers, but he himself was educated and spent most of his career in England.

2. Correspondence in SA, HAR 003; for his low opinion of the PhD, Kelsey to Harden, 9 February 1927.

3. SA, HAR 003, Kelsey to Harden, 9 July 1926. The correspondence includes copies of letters Kelsey sent to abbé Chabot, hoping he could help resolve the impasse over the urns.

4. SA, HAR 013, Harden to Poinssot (copy), 3 October 1933; OFB (illegible signature), Museum of Classical Archaeology, University of Michigan, to Harden, 13 December 1933.

5. SA, HAR 013, Poinssot to Harden, 18 October 1933.

6. D. B. Harden, "The Pottery from the Precinct of Tanit at Salammbo, Carthage," *Iraq* 4, no. 1 (1937): 59, https://doi.org/10.2307/4241606. This was actually Harden's second publication on the pottery, the first having appeared in the *American Journal of Archaeology* in 1927.

7. Donald Harden, *The Phoenicians* (London: Thames & Hudson, 1962), 113.

8. Harden, *Phoenicians*, 218. Nicholas Vella cites this passage as evidence of the continuity of Harden's views with those of earlier scholars; see Vella, "The Invention of the Phoenicians: On Object Definition, Decontextualization and Display," in *The Punic Mediterranean: Identities and Identification from Phoenician Settlement to Roman Rule*, ed. Josephine Crawley Quinn and Nicholas C. Vella (Cambridge: Cambridge University Press, 2014), 34–35.

9. FP, 003, folder 2, Condolence note from Jacques Revault, French Embassy in Tunis, to Claude Poinssot, 4 December 1967 (quoting Prosper Ricard, director of the Service des arts indigènes in Morocco).

10. FP, 003, folder 2, Claude Poinssot to Merlin (draft), 25 June 1962.

11. This information, along with valuable insights into Poinssot's election to the Academy, was communicated to me by his daughter-in-law, Paulette Poinssot, in an interview in

Paris, June 2014. A dossier assembled in FP, 003, makes clear that Poinssot was considering administrative maneuvers to stave off retirement in 1942, though in the end he did retire at the age of sixty-three. Carcopino was not Poinssot's direct superior but had influence over the choice of his successor.

12. See Kenneth J. Perkins, *A History of Modern Tunisia* (New York: Cambridge University Press, 2004), 142–44; Werner Klaus Ruf, "The Bizerta Crisis: A Bourguibist Attempt to Resolve Tunisia's Border Problems," *Middle East Journal* 25 (1971): 201–11.

13. Dorothy Garrod, "Recollections of Glozel," *Antiquity* 42 (1968): 172. The headnote is unsigned.

14. O. G. S. Crawford, "L'Affaire Glozel," *Antiquity* 1, no. 3 (1927): 181–88, https://doi.org/10.1017/S0003598X00000387.

15. Hugh McKerrell et al., "Thermoluminescence and Glozel," *Antiquity* 48, no. 192 (December 1974): 265.

16. Glyn Daniel, "Editorial," *Antiquity* 48, no. 192 (December 1974): 261–64.

17. "Glozel For Ever," *Avenir du Luxembourg*, 27 December 1927, clipping in MANA, 2018/001 028.

18. The inventory of the Fonds Dorothy Garrod is available at https://archives.musee-archeologienationale.fr/index.php/dorothy-garrod-4. Although St. Mathurin's own papers technically comprise a separate collection from Garrod's, still only partially cataloged, archivists made the wise decision to include all Glozel-related material in the Garrod "sous-fonds," that is the discrete Garrod papers within the larger collection bequeathed by St. Mathurin.

19. MANA, 2018/001 031, St. Mathurin to Daniel (carbon), 31 July 1974. St. Mathurin and Daniel always wrote each other in English.

20. MANA, 2018/001 031, St. Mathurin to P.-M. Garçon (carbon), 1 May 1975: "Ces pièces manquent au dossier."

21. MANA, 2018/001 031, President of Association pour la sauvegarde et la protection des collections de Glozel to Editions Larousse (copy), 17 June 1985.

22. MANA, 2018/001 031, McKerrell and Mejdahl to Daniel (copy), 2 September 1976.

23. MANA, 2018/001 031, Daniel to McKerrell, 15 October 1976.

24. MANA, 2018/001 031, Daniel to St. Mathurin, 9 February 1977; Mejdahl to Daniel (copy), 6 June 1979.

25. MANA, 2018/001 031, Pierre-Roland Giot to Daniel (copy), 1 July 1974. Giot was a prehistorian at the Université de Rennes; this letter was a detailed account of the state of knowledge about Glozel among French archaeologists.

26. Jean-Pierre Daugas et al., "Résumé des recherches effectuées à Glozel entre 1983 et 1990, sous l'égide du Ministère de la Culture," *Revue archéologique du Centre de la France* 34 (1995), 251–59, 258.

27. MANA, 2018/001 031, Giot to Daniel (copy, original English), 1 July 1974.

28. These and other program titles can be accessed via the online catalog of the Institut National de l'Audiovisuel, ina.fr, with some available for viewing or listening via the site.

29. Jean-Paul Demoule, *On a retrouvé l'histoire de France: Comment l'archéologie raconte notre passé* (Paris: Robert Laffont, 2012), 203–4.

30. On cunning and cupidity as components of the image of the peasantry in the nineteenth century, see James R. Lehning, *Peasant and French: Cultural Contact in Rural France during the Nineteenth Century* (Cambridge: Cambridge University Press, 1995), 18–20. It is also a theme of twentieth-century peasant narratives, especially in relation to landowners; see, e.g., Emile Guillaumin, *The Life of a Simple Man*, ed. Eugen Weber, trans. Margaret

Crosland (1904; Hanover, NH: University Press of New England, 1983), 36–38, 44–46, 152–54.

31. Florian Bardou, "Déconvenues archéologiques (2/6): La fausse ère de Glozel," *Libération*, 16 August 2021.

32. C. P. Snow, *The Two Cultures and the Scientific Revolution* (Cambridge: Cambridge University Press, 1959). The book has been through a number of editions and produced several responses, notably from the celebrated literary scholar F. R. Leavis.

33. Colin Renfrew, "Glozel and the Two Cultures," *Antiquity* 49 (September 1975): 219–22, 220, https://doi.org/10.1017/S0003598X00103461. A marked-up typescript draft of this article can be found in MANA, 2018/001 031.

34. Renfrew has his own place in the history of archaeology; for a concise summary, see Ian Hodder and Scott Hutson, *Reading the Past: Current Approaches to Interpretation in Archaeology*, 3rd ed. (Cambridge: Cambridge University Press, 2003), 36–39, 245.

35. Renfrew, "Glozel and the Two Cultures," 221. Renfrew revisits the "second affaire Glozel" in a 1999 article with Paul Bahn, "Garrod and Glozel: The End of a Fiasco," in *Dorothy Garrod and the Progress of the Palaeolithic: Studies in the Prehistoric Archaeology of the Near East and Europe*, ed. William Davies and Ruth Charles (Oxford: Oxbow Books, 1999), 76–83. The article takes the fraud as certain and urges "the archaeological scientist" to investigate what went wrong with the thermoluminescent dating in the 1970s.

36. "Burial urn; human skeletal remains," https://www.britishmuseum.org/collection/object/W_1927-0108-1, consulted 10 August 2022. This object has the museum number 118333; the same mistake appears on the records of the other two urns I examined, 118334 (acquisition number 1927-0108-2) and 118336 (1927–0108-4).

37. "Cinerary urn with cover," https://collections.maa.cam.ac.uk/objects/4548736, accessed 10 August 2022.

38. Mirjam Brusius and Kavita Singh, eds., *Museum Storage and Meaning: Tales from the Crypt* (Milton Park, UK: Routledge, 2018). The term "exhibitionary complex" comes from a 1988 article by Tony Bennett, which has been reprinted in several volumes, including his own *The Birth of the Museum: History, Theory, Politics* (London: Routledge, 1995).

39. For some of the pertinent literature on Phoenician child sacrifice, see chap. 1, n18, and chap. 5, n80. On object biography, Igor Kopytoff's foundational article is still useful: Kopytoff, "The Cultural Biography of Things: Commoditization as Process," in *The Social Life of Things: Commodities in Cultural Perspective*, ed. Arjun Appadurai (Cambridge: Cambridge University Press, 1986), 64–91.

40. Carolyn Steedman, *Dust: The Archive and Cultural History* (New Brunswick, NJ: Rutgers University Press, 2002), 80–81.

41. Of course with the new technologies in its arsenal, archaeology increasingly has the ability to shed light on individual lives. A notable example is the discovery of the remains of the English king Richard III in Leicester in 2012, his identity eventually confirmed via DNA testing.

42. *Concerning the Dig* was created for and first shown at an exhibition at the Museum of Contemporary Art Chicago in 2013; the catalog includes a planning drawing for the work. See Dieter Roelstraete, *The Way of the Shovel: On the Archaeological Imaginary in Art* (Chicago: Museum of Contemporary Art/University of Chicago Press, 2013), 100–103. On the *Tate Thames Dig* project, see https://www.tate.org.uk/art/artworks/dion-tate-thames-dig-t07669/digging-thames-mark-dion, accessed 11 August 2022.

43. See Roelstraete, *Way of the Shovel*, 186–91; Claire Armitstead, "Fourth Plinth: How a Winged Bull Made of Date Syrup Cans Is Defying Isis," *Guardian*, 26 March 2018, https://

www.theguardian.com/artanddesign/2018/mar/26/michael-rakowitz-invisible-enemy-should-not-exist-fourth-plinth-winged-bull-date-syrup-cans-defying-isis; Raffi Khatchadourian, "Art of Return," *New Yorker*, 24 August 2020, 46–55, https://www.newyorker.com/magazine/2020/08/24/michael-rakowitzs-art-of-return.

44. Sophie Berrebi, "Not-So-Transparent Things," in Roelstraete, *Way of the Shovel*, 266–76; for the work itself, see 166–73.

Index

Académie des inscriptions et belles-lettres (AIBL): archaeology education, 6; Cagnat and, 66–68, 180; Delattre and, 21; Glozel and, 53; Merlin and, 110, 195; L. Poinssot and, 195
Action française, 75, 155, 188, 189, 242n108
actor-network theory, 17
Adelswärd-Fersen, Jacques d', 97
AIBL (Académie des inscriptions et belles-lettres). *See* Académie des inscriptions et belles-lettres (AIBL)
Albertini, Eugène, 102, 111
Alexandropoulos, Jacques, 22
Altamira (Spain), cave art, 55, 181
Altekamp, Stefan, 130
Amar, Jules, 121–22
Andreose, Antonella Mezzolani, 28
Antiquity (journal), 196, 198
archaeological archives: definition, 8; embodiment and, 92–93; professionalization of history, 9
Archaeological Association of Washington, 36, 39
archaeological photography: Carthage and, 104–8; Dougga and, 113–15; Fradin family and, 117–19; Glozel and, 116–19; photographic binary and, 160. *See also* visual media
archaeology: archaeological habitus, 49–50; archaeological imaginary, 161; archival accumulation, 58–59; astronomy and, 55–56; as colonial science, 48–49; colonialism and, 3, 6–8, 16; definition, 2; as ecology of practices, 2, 9, 49; ethnographic references, 169–70; exploration and, 5; history of, 3–4, 5–7, 55; labor of, 108–9; media and, 69–70; misogyny in, 119–23; national archaeology, 6; photographic records, 94; prehistory and, 55; professionalization, 9; rivalries in, 111–12; scholarly publication and, 110–12; scholarship of, 4; scientific networks and, 16–17, 24, 44, 48, 56, 60–69; scientific status of, 4–5, 48–50; stratigraphy and, 170
archival moments, definition, 13–14
Audollent, Auguste: Bégouen and, 64; Glozel and, 62, 71; media and, 71; newspaper clippings and, 58–59
Audollent, Dominique, 149, 151–53

Bacha, Myriam, 7, 26–27, 95
Bal, Mieke, 126, 128
Bar, de (government official), 57
Bardo Museum (Tunis), 22; L. Carton and, 45–46; catalogs, 164–68, 173; collections, 138, 140, 170, 172; Poinssot papers, 196
Bardo Treaty, 15
Barrès, Maurice, 189
Barry, Elizabeth, 91–92
Barthes, Roland: reality effect, 129, 152; textuality, 92
Basch, Victor, 189
Bauer, Heike, 100
Bayet, Charles, 96, 98–99
Bayle, Edmond, 88

Bégouën, Henri: A. Audollent and, 64; controversy and, 120; fake news and, 78; Glozel and, 63, 64; Loth and, 148; media and, 70
Belknap, Geoffrey, 166
Bénard, Pierre, 152
Benjamin, René: *Glozel: Vallon des morts et des savants*, 153–55; political views, 155
Bérard, Léon, 57
Bérard, Victor, 73
Berger, Philippe: Carthage catalog, 165, 171, 176; Phoenician writing, 177–78
Bernal, Martin, 72–73
Bernault, Florence, 8
Berrebi, Sophie, 203–4
Bertillon, Alphonse, 88
Bertrand, Louis, 27–28
Bizerte crisis (1961), 195–96
Bliss, Robert Woods, 35
Bloch, Marc, 12, 75–76
Blouin, Francis, 8
Blum, Léon, 7
Bohrer, Frederick, 159–60, 181, 207n20
Bosch-Gimpera, Pere: Garrod and, 122; Morlet and, 122; name, 76
Botta, Paul-Émile, 5
Boucher de Perthes, Jacques, 55
Boulanger, André, 176
Boule, Marcelin, 71
Bourdieu, Pierre, 23, 49
Bravo, Michael, 17
Breuil, Henri (abbé): Garrod and, 119, 122; Glozel and, 65, 69
Bringuier, Paul, 70
British Museum, Carthaginian objects in, 201
Brown, Bill, "thing theory," 162, 168
Browne, Janet, 115–16
Brusius, Mirjam, 202
Butler, Judith: gender, 128; performativity, 143; reality effect, 129

Cagnat, René: Académie des inscriptions et belles-lettres (AIBL) and, 66–68; archaeological collections, 48; career, 23; Carthage and, 39–40; L. Carton and, 127; Charmes and, 15; Delattre and, 22; as epigraphist, 17; Glozel and, 66–67; missions, 21–23; museum catalogs and, 48, 164; L. Poinssot and, 16, 50, 232n55; scholarly elite and, 25; scholarly publication and, 110, 163–64, 172–73; in Tunisia, 15; Tunisian Antiquities Service (DAA) and, 30; visual media and, 172–73
Caillaux, Henriette, 126–27
Cambridge Museum of Archaeology and Anthropology, 201–2
Canard enchaîné, Le, 80–82
Carcopino, Jérôme: Glozel and, 88–89; L. Poinssot and, 195
Cartailhac, Émile, 184
Carter, Howard, 115
Carthage: American excavations, 2, 36–39, 41–44; Archaeological Association of Washington and, 36; child sacrifice, 176; colonial era, 178–79; Delattre and, 18, 20–21, 171–72; excavation, 18, 20–21, 36–39; Glozel and, 53–54; Lavigerie and, 38; masks, 157–58; ostensible neglect of, 2, 26, 28, 30, 32; pageants (see *fêtes de Carthage*); photographic interpretation, 104–8; Punic religion, 178; state power, 2, 23, 40–41; Tunisian Antiquities Service (DAA) and, 38; University of Michigan and, 36; White Fathers and, 18
Carthage Ladies' Friends Committee (CDAC), 28, 40
Carthage Museum, catalogs, 164–68
Carton, Louis: archaeological missions, 25–26; Carthage and, 28–29, 30–32, 39–40, 46, 107–8; Carthage Ladies' Friends Committee (CDAC) and, 28–29; collection, 46; colonialism and, 50, 129–30, 134; criticism of, 31, 40; Dougga and, 25; excursion of, 131–32; *fêtes de Carthage* and, 134–38, 140–41; Gauckler and, 26, 100; looting and, 46, 108; outsider status, 29; pageants and, 132, 134–38, 140–41; photographic archive, 135; pictured, 46–47, fig. 1.4; L. Poinssot and, 28, 46, 100; publication and, 103; rivalries, 111; Roman achievements and, 129–30; Société des amis de

Carthage and, 28; Tunisian Antiquities Service (DAA) and, 29–30; Tunisian archaeology and, 27
Carton, Marie Thélu, 44–45, 125–28
Catholicism: Glozel and, 83, 188; in Tunisia, 18, 21, 132–33. *See also* White Fathers
CDAC (Carthage Ladies' Friends Committee). *See* Carthage Ladies' Friends Committee (CDAC)
Cesnola, Luigi Palma di, 174
Chabot, Jean-Baptiste: excavations and, 15; Franco-American Mission and, 41; Merlin and, 43; L. Poinssot and, 43, 193
Charcot, Jean-Baptiste, 97
Charmes, Xavier, 15
Chassaigne, Lucien, 63
Chirico, Giorgio de, *The Archaeologists*, 114, plate 1
Cintas, Pierre, 174
Claretie, Georges, 149
clippings, press, 11–12, 37, 58–60, 66
Colla, Elliott, 4
colonialism: actor-network theory and, 17; archaeology and, 3, 7–8, 45–46; Gauckler and, 94–95, 98; imperial governance and, 27; indigenous dispossession, 25–26, 46–47; institutions and, 16; local labor and, 107; normative framework, 101; pageants and, 134, 142; Protectorate regime (Tunisia), 15, 25, 41; same-sex relationships and, 94; science and, 24, 31, 48, 50; scientific networks and, 16, 24–25, 48; Tunisian resistance to, 41, 130; visual media and, 161, 172
controversy studies, 10
Cooper, Frederick, 16
Corre, F., 182
coup de théâtre, 144, 151, 241n91
Coutant, Jacques, 144
Crawford, O. G. S., 196
Csiszar, Alex, 22, 53
Csordas, Thomas, 92
Culler, Jonathan, 128

DAA (Direction des antiquités et des arts). *See* Tunisian Antiquities Service (DAA)
Dalio, Marcel, 146
Daniel, Glyn, 196–98
Daudet, Léon, 75
David-Néel, Alexandra, 136, 141–42
Debord, Guy, 3
Déchelette, Joseph: archaeological archives and, 9; *Manuel d'archéologie préhistorique*, 162, 174, 181; migration and, 186–87; prehistory and, 170; views, 187
Delarue-Mardrus, Lucie, 136; *La prêtresse de Tanit* (play), 137–38, 140, 142–43
Delattre, Alfred-Louis: Cagnat and, 22; Carthage and, 18, 20, 21; Carthage catalog, 171; Gauckler and, 18; *La Croix* and, 21; Punic art and, 176
Delvair, Jeanne, 137–38
Demoule, Jean-Paul, 199
Dépêche tunisienne, 39, 45–46
Deperet, Charles, 91
Description de l'Afrique du Nord, 48, 163, 173, 180. *See also* Bardo Museum (Tunis); Carthage Museum
Destour (Tunisia), 41
Diaz-Andreu, Margarita, 3
Dion, Mark, 203; *Concerning the Dig*, 203, plate 6
Direction des antiquités et des arts (DAA). *See* Tunisian Antiquities Service (DAA)
Doliveux, Henri, 49, 104, 125
Dollimore, Jonathan, 97–98
Dougga (Tunisia): Bardo and, 171; L. Carton and, 25; excavation of, 18, 23, 103; local labor and, 103–4, 109; L. Poinssot and, 103; scholarly publication and, 111; visitors to, 113–14, fig. 3.6
Dreyfus, Alfred: conviction, 84; retrial, 89
Dreyfus affair, 84–86
Dussaud, René: archival moments and, 58–59; *Les civilisations préhélleniques dans le bassin de la Mer Egée*, 174; clippings, 77–78; on Dreyfus affair, 85; Glozel and, 58–59, 66–67, 89–90; "hermaphrodite idols" and, 80; legal proceedings, 88, 149; media and, 152; on Phoenician writing, 177–78; Phoenicians and, 73; publications, 174, 180; and S. Reinach, 66, 152
Dutton, Dianne, 149, 151

Edwards, Elizabeth: photographic interpretation, 124; visual media and, 160
Effros, Bonnie, 6–7, 23, 109
Elias, Norbert, 92
embodiment, 92–93
Erlanger, Rodolphe d', 39
Espérandieu, Emile: Fradin family and, 82; Glozel and, 61–62, 67, 72, 82–83; medical excuse, 91; newspaper clippings and, 59

"fake news," 76–77
Farr, Jason, 92, 96–97, 101
fêtes de Carthage: analysis of, 138; L. Carton and, 134–38, 140–41; criticism of, 136–37; goals, 129–30; organization, 135–36; performance, 132–34, 136–37; photographic record, 134–35, fig. 4.1, fig. 4.2; posters, 138–40, fig. 4.3, plate 3
Figaro, Le, 84, 120, 141; as establishment daily, 71, 188
Fix, Florence, 128–29
Flaubert, Gustave: Carthage and, 19; on ruins, 131; *Salammbô*, 19, 132; visual media and, 161
Flot, Louis, 138–39
Foucault, Michel: on the body, 92; disciplinary archives and, 10
Fradin, Emile: accusations, 84; death, 197; Dussaud and, 152; fraud case, 149; indictment, 88; photographs of, 116–18, figs. 2.1, 3.8, 3.10; values, 199
Fradin family: archives, 210n66; *coup de théâtre* and, 144; Glozel and, 51–52, 115–16, 152; legal proceedings, 88, 149; media and, 152; monetization, 81–82; photograph of, 116–17, fig. 3.8; photographic interpretation, 117–19; police search, 87; stereotyping, 146
French National Archives: archaeological archives and, 10; Cagnat and, 21; on Glozel, 57–58

Garçon, Maurice, 88, 149–52, 197
Garrod, Dorothy: career, 119, 123–24; criticism of, 119–21; Glozel and, 86, 119–20, 123–24, 196; name, 76; papers, 197; photographic interpretation by, 118–19

Gauckler, Paul: Bardo catalog, 165; Carthage and, 18; colonial authority and, 24; colonial officials and, 100–101; departure, 98–100; excavation workers and, 109; "illness," 101–2; photographs of, 115; relations of, 26; rivalries, 111; sexuality, 12, 95–96, 97–98; suicide, 95; Tunisian Antiquities Service (DAA) and, 23
Gennep, Arnold van, 70
Gero, Joan, 17
Gide, André, 97–98
Gielly, Paul, 45–46
Giot, Pierre-Roland, 198–99
Gitelman, Lisa, 59
Gloeden, Wilhelm von, 97
Glotz, Gustave, 185
Glozel: Académie des inscriptions et belles-lettres and, 66–67; Alvão inscriptions, 65; archival traces, 11–12, 57–60; Carthage and, 53–54; cartoons, 74, 80–81, 84, 115, 121, 144–45, figs. 2.3, 2.5, 2.6, 3.11, 4.4, plate 2; Dreyfus affair and, 83–86; ethnographic comparisons, 184; "fake news" and, 76–77; Glozelian art, 181–83, 184; Glozelian performance, 143, 144–45, 146; inscriptions, 65; international commission, 2, 53, 77–78; judicial phase, 88; legal proceedings, 149–53; media coverage, 67, 69–73, 77–80, 82–83, 124, 199–200; networks and, 56; newspaper clippings, 59–60; performativity and, 153–54; photographic interpretation, 116–19; political codes and, 188–89; pottery, 157, 159; renewed controversy (1970s–1980s), 196–200; scientific testing, 88, 198–99; skepticism of, 62–63; state involvement in, 57; state power and, 2; on television, 199; thermoluminscent testing, 196–98; writing system, 184–86
Gran-Aymerich, Ève, 6–7, 95
Grandmougin, Charles, 136; *La Mort de Carthage* (play), 136–37, 141
Grenier, Albert, 61
Gsell, Stéphane: Carthage and, 40, 175–76, 178; L. Carton and, 40, 46; *Histoire ancienne de l'Afrique du Nord*, 162, 173–74; L. Poinssot and, 40, 46–48
Guitet-Vauquelin, Pierre, 78–79

Gunter, Ann, 178
Gutron, Clémentine, 7, 31, 134

Hamal-Nandrin, Joseph: Garrod and, 122; Glozel and, 120
Hanotaux, Gabriel, 93
Harden, Donald: Gsell and, 173–74; L. Poinssot and, 194–95; at University of Michigan, 193–95
Harry, Myriam: on 1907 *fête de Carthage*, 137–38; Tunisia and, 130–31,134
Hawes, Charles, 174
Hayes, Jarrod, 98
Henriot, Émile, 164
heritage: archaeology as discipline, 48–49; 1920 Tunisian code and, 45–46, 48
Héron de Villefosse, Antoine, 165, 172
Hill, Jason, 54–55
Hodder, Ian, 112
Homo, Léon, 185
Horsman, Yasco, 149, 151
Humbert-Mougin, Sylvie, 132

illness: archaeologists and, 103–4; coenesthesis and, 103; Gauckler and, 96, 101–2; L. Poinssot and, 102
Illustration, L' (magazine), 32, 37, 112, 180
immigration to France, 188–89, 200
INHA (Institut national d'histoire de l'art). See Institut national d'histoire de l'art (INHA)
Institut de Carthage: excursion, 131–32; *fêtes de Carthage* and, 132–36
Institut de France, 11, 80, 225n90. See also Académie des inscriptions et belles-lettres (AIBL)
Institut national d'histoire de l'art (INHA), 24

Jaïdi, Houcine, 16
Jasanoff, Sheila, 4
Jaubert de Benac, J., 126; *L'Illustration* and, 32, 37; Lantier and, 44
Jenkins, David, 163
Jeunes Tunisiens, 130
Journal des débats, 28, 71, 111; as establishment daily, 59, 62, 71; Garrod controversy and, 122, 124

Jullian, Camille: career, 60–61; death, 91; Glozel and, 60–62, 67–68; as model, 184; on prehistory, 161–62; work of, 60–61
Jusserand, Jules, 35–36

Kaeser, Marc-Antoine, 3
Katz, Ethan, 188
Kaufman, Asher, 175
Kelsey, Francis: Franco-American Mission and, 41; Harden and, 193; L. Poinssot and, 41–43; publicity and, 37; Tunisia and, 15; University of Michigan and, 36
Kelsey papers, 37
Kenny, William F., 36, 39
Khair-al-Din al-Tunisi, 16
Khetchen, Mona, 130
Klamm, Stefanie, 160

Labiche, Eugène, *La Grammaire* (play), 144
labor: in archaeology, 105–10; photographic record, 105–7, figs. 3.1–3.4
Labrusse, Rémi, 161, 184
Lacarelle, Count de, 62
Lang, Jack, 198
Lansing, Robert, 36
Lantier, Raymond: colonial authority and, 24; at Dougga, 109; Jaubert de Benac and, 44; L. Poinssot and, 24, 103, 111; work of archaeology and, 49
Lapérouse, Jean-Francois, 211n9
La Porte, R de., 1, 215n71
Lartet, Louis, 55
Latour, Bruno: actor-network theory, 17; networks and, 16; scientists and, 48
Lavigerie, Cardinal Charles, 18, 38
Legg, Charlotte, 134
Leite de Vasconcelos, José, 64–65
Le Pen, Jean-Marie, 200
Lewis, Sinclair, 42
Libération (newspaper), 199
Ligue des droits de l'homme and, 189
looting, 33, 45–46, 108. See also heritage
López-Ruiz, Carolina, 175
Loth, Joseph: Glozel and, 67, 73, 82–83; Glozel articles, 88; lectures by, 146–49; media and, 75

Louis IX, 18
Louvre Museum, 73, 167; curators' views on Glozel, 67

Magness, Jodi, 2
Mahdia excavations, 103, 108; in publications, 110, 112
MAN (National Archaeology Museum). *See* National Archaeology Museum (MAN)
Mangan, Valentin, 97
Marçais, Georges: embodiment and, 112–13; excavations and, 109–10; L. Poinssot and, 103, 110, 112–13
Mardrus, Joseph-Charles, 136
Marin, Louis, 189, 251–52n148
Marsan, Eugène, 72
Martin, Benjamin, 149
Masaryk, Tomas, 104–5
Massis, Henri, 189
Matin, Le (newspaper), 32, 72, 82; excavations near Glozel, 77–80; on Garrod, 120; on Glozel trials, 152
McCarty, Matthew, 22, 161
McKerell, Hugh: Daniel and, 198; St. Mathurin and, 197
Mejdahl, Vagn, 197–98
Mendes Corrêa, António, 65
Mercure de France: *fêtes de Carthage* and, 136; illustrations in, 181; as outlet for Morlet, 70, 179–80, 181
Méric, Victor, 71–72
Merlin, Alfred, 110; Académie des inscriptions et belles-lettres (AIBL) and, 110, 195; archaeological heritage and, 45; and archaeological work, 49; career, 23; L. Carton and, 26; M. Carton and, 127; Chabot and, 43; colonial authority and, 24; colonial bureaucracy and, 100–101; excavation programs, 108–9; Gauckler and, 95; Mahdia excavations, 104, 108, 112; L. Poinssot and, 23–24, 50, 102, 104, 195; scholarly publication and, 110–11; in Tunisia, 12, 104; Tunisian Antiquities Service (DAA) and, 23, 30
Mezzolani, Antonella, 28
Michelot, Vincent, 76
Millet, Jean-Francois, 51

"mirage oriental, Le." *See* Reinach, Salomon
Mitterrand, François, 200
Molière, 144
Mommsen, Theodor, 60
Montet, Pierre, 177, 180
Morlet, Antonin: anti-Catholicism, 83; archaeological discipline and, 184; career, 91; controversy, 120; courtroom appearances, 151; ethnic slurs, 189; findings, 52, 61–62, 181–86; E. Fradin and, 180–81; Garrod, incident with, 122–24; Glozel and, 58, 68, 87–88; Glozelian alphabet, 185; Glozelians, descriptions of, 163, 183–84, 185–86; "hermaphrodite idols" and, 80, 183; legal proceedings and, 88, 149; media and, 71–72, 88; pamphlets by, 157, 180–81; Phoenicians and, 186–87; photograph of, 116–17; practice of, 184; publications, 179–82; publicity and, 189–91; S. Reinach and, 91, 187; views of, 188–89; visual media and, 182–83
Moser, Stephanie, 158–60
Moulène Jean-Luc, *Le Monde, Le Louvre*, 203, plate 8
Mounsey, Chris, 92
Museum of Contemporary Art Chicago, 203, plates 6–7

Nall, Joshua, 55–56
Napoleon III, 5
National Antiquities Museum. *See* National Archaeology Museum (MAN)
National Archaeology Museum (MAN), 49, 53, 197
National Front (France), 200
Neolithic groups, 186–87
New York Times, 42
North Africa Committee (French Ministry of Education), 29, 33, 110; archaeological objects and, 164; scholarly networks and, 43

Oeuvre, L' (newspaper), 82, 85–86
Olivier, Laurent, 89
Olsen, Bjørnar, 13; *Archaeology: The Discipline of Things*, 9, 94, 231n15
Omessa, Charles, 182

Oriental mirage. *See* Reinach, Salomon: "Le mirage oriental"
Otham, Amor Ben, 130–31
outdoor theater, 131–32

Péquart, St. Just, 82
Père Lachaise, cemetery of, 144
performance history, 128–29
performativity, 126, 128–29
Pergamon Museum (Berlin), 203
Perrot, Georges, 28
Peyrony, Denis: controversy and, 120; E. Fradin and, 84; Glozel and, 68–70; images of, 119; S. Reinach and, 59
Phoenicians: alphabetic writing, 177–79, 185–86; cultural advancement, 174; Harden on, 194; identity of, 175; stereotyped, 176–77
Picandet, Adrienne, 57
Pichon, Stephen, 96–97, 99
Poinssot, Claude, 195
Poinssot, Julien, 99
Poinssot, Louis: American excavations and, 41–42, 45–46; archaeology as science, 48–49; career of, 23–24; Carthage and, 38; L. Carton and, 28, 31–32, 46, 100; M. Carton and, 44–45, 125–28; colonial authority and, 24, 50; colonial officials and, 100–101; criticism of, 31, 32–33, 40; Doliveux and, 49; Dougga and, 23–24; Gsell and, 46–48; health concerns, 102–3; on heritage and looting, 33, 45–46; land acquisition and, 38–39, 41–42; Mahdia and, 24, 112; Merlin and, 23–24, 43–44, 102; networks of, 15–16, 193; photographs of, 115; Poinssot archive, 10–11, 99–101, 195–96; on *L'Illustration* article, 32; publicity and, 43–44; scholarly research and, 32; Tunisia and, 12, 16; Tunisian Antiquities Service (DAA) and, 15–16, 23, 26, 110. *See also* Poinssot papers
Poinssot, Paule, 99, 195
Poinssot papers: background, 10–11, 24; colonial science and, 24; Gauckler and, 24, 99; signed photograph and, 33–35; Tunisian archaeological archives and, 10
Polyeucte (Corneille), 132–33

Pottecher, Maurice, 141
Pottier, Edmond, 63–64, 67, 75
Prag, Jonathan, 175
prehistory: early development in France, 6, 55, 161; in popular media, 55–56, 199; as subfield of archaeology, 122; three-ages theory, 170; uncertainty of, 72, 184
Prorok, Byron Khun de: disparages "Arab" workers, 109; Franco-American Mission and, 41; in Glozel novel, 153; identity of, 35; Jusserand and, 35–36; photograph of, fig. 1.3; as publicist for archaeology, 35, 37–38; racism toward, 189; scientific archaeology and, 38; Tunisian Antiquities Service (DAA) and, 37
Prunière, Jean, 150, 153

Quinn, Josephine, 174–75
Quirke, Stephen, 109

Raffalovich, Marc-André, 97
Raj, Kapil, 25
Rakowitz, Michael, *May the Arrogant Not Prevail*, 203, plate 7
Randolph, John, 11
Ranke, Leopold von, 93
reality effect (Roland Barthes), 129
Reinach, Joseph, 84, 188
Reinach, Salomon: anti-Semitism and, 75; A. Audollent and, 151–52; Carcopino and, 88–89; caricature, 73–75, 152, fig. 2.3; death, 91; Dreyfus Affair and, 84; Dussaud and, 66, 152; in fiction, 155; French archaeology and, 7; Glozel and, 63, 66–69, 72–73, 77, 82–83, 88, 191; in the media, 70, 75, 152; "La méthode en archéologie," 7; "Le mirage oriental," 73, 187; newspaper clippings and, 59; on origins of civilization, 72, 186; portrayed on stage, 146; scholarly publication and, 110; tiara of Saitaphernes and, 73–75, fig. 2.3; White Fathers and, 18
Reinach, Théodore, 91
religion. *See* Catholicism
Renan, Ernest, 5–6, 178
Renfrew, Colin, 200–201
Reyer, Ernest, *Salammbô* (opera), 132
Ricci, Seymour de, 63

Richard, Natalie, 6, 94
Riggs, Christina, 94, 207n20; on archaeological photographs, 106, 115
Rivet, Paul, 122
Rostovtzeff, Michael, 111
Roy, Bernard: as colonial official, 99, 100–101; depictions of, 115; L. Poinssot and, 99–100

Said, Edward, 97
Saint, Lucien, 35
Saitapharnes, tiara of, 73–75
Salammbô (Flaubert novel), 19, 131. See also Reyer, Ernest
same-sex relationships, 95–96, 97–98
Sanos, Sandrine, 189
Saumagne, Charles, 30
Sauvage, Marcel, 72
Schaffer, Simon: actor-network theory and, 17; controversy studies and, 10; scientific controversies and, 4–5
Schlanger, Nathan, 3, 9
Schliemann, Heinrich, 183
Schnapp, Alain, 7
science: archaeology as, 48; as "communicative action" (Secord), 24; definition, 2–3; geographical expansion of, 50
Sear, Frank, 142
Secord, James, 24
Sekula, Allan, 9
Shanks, Michael, 160, 244n15
Shapin, Steven: controversy studies, 10; embodiment and, 93; scientific controversies and, 4–5
Shepherd, Nick, 105
Shetelig, Haakon, 184
Siebers, Tobin, 92
Singh, Kavita, 202
Smith, Bonnie, 93
Société des amis de Carthage. See Carthage Ladies' Friends Committee (CDAC)
spectacle, definition of, 3
St. Mathurin, Suzanne Cassou de, 197–98
Steedman, Carolyn, 202
Stiebing, William H. Jr., 208n32
Stoever, Edward: Franco-American Mission and, 41; L. Poinssot and, 41–42; Tunisia and, 15–16

Stoler, Ann Laura, 130
symbolism (literary), 137

Teilhard de Chardin, Pierre, 122
television, and Glozel, 199
Temps, Le (newspaper), 71, 83, 137; as establishment daily, 71
Thierry, Augustin, 93
Thomsen, Christian Jürgensen, 170
tiara of Saitapharnes, 73–75
Tilley, Christopher, 9
Torrès, Henry, 152
Touring Club de France, 28
Trumbull, George, 16
Tunisian Antiquities Service (DAA): 1920 heritage code and, 45; archaeological archives and, 10; Carthage and, 32, 38; L. Carton and, 29–30; founding of, 23; funding of, 213n35; Gauckler and, 18; L. Poinssot and, 40, 43; and work of archaeology, 49

Universal Expositions. See World's fairs
University of Michigan, 36, 193–94

Vachon, Marius, 21
Vayson de Pradenne, André: archaeological frauds and, 75; Garrod and, 122; media and, 70
Vella, Nicholas, 175, 178
Vergès, Françoise, 134
Verne, Maurice, 78
Viple, Joseph, 60; Glozel and, 81–82; Jullian and, 60, 62–63
visual media: background, 158–60; Bardo and, 164–69, 170, 172–73; Cagnat and, 172–73; Carthage museum and, 164–69, 171–73; colonialism and, 161, 172; Punic art and, 176. See also archaeological photography; Glozel: Glozelian art

White Fathers: Carthage and, 18; Delattre and, 21; Lavigerie and, 18; S. Reinach and, 18
Woolf, Virginia, 100
world's fairs, 4

Zola, Emile, 84